Waste Management, Control, Recovery and Reuse

Waste Management, Control, Recovery and Reuse

The International Edition of the 1974 Australian Waste Conference,
Held at the University of New South Wales

Edited by
N. Y. KIROV, D.Sc.
Associate Professor and Head of Department of Fuel Technology,
University of New South Wales, Sydney, Australia

Organized by

THE DEPARTMENT OF FUEL TECHNOLOGY
UNIVERSITY OF NEW SOUTH WALES

In Association with

THE LOCAL GOVERNMENT ASSOCIATION OF NEW SOUTH WALES
THE AUSTRALIAN INSTITUTE OF HEALTH SURVEYORS
THE CLEAN AIR SOCIETY OF AUSTRALIA & NEW ZEALAND
THE INSTITUTE OF FUEL (AUSTRALIA MEMBERSHIP)
THE LOCAL GOVERNMENT ENGINEERS' ASSOCIATION OF NEW SOUTH WALES

ann arbor science PUBLISHERS INC.
POST OFFICE BOX 1425 • ANN ARBOR, MICH. 48106

P.O. Box 1425, Ann Arbor, Michigan 48106 USA

Library of Congress Catalog Card Number 74-33798
ISBN 0-250-40100-2

Printed in the United States of America with typesetting by Kingsway Printers Pty. Limited, Caringbah, N.S.W., Australia

FOREWORD

Encouraged by the success of the earlier two Conferences held in 1967 and 1971, the Waste Disposal Committee decided to hold yet another — The 1974 Australian Waste Management and Control Conference.

The Conference will be held over three days on 17th to 19th July, 1974, in the Science Theatre in the University of New South Wales in Sydney. It is being organised by the Department of Fuel Technology of the University, in association with five other professional organisations: the Local Government Association of N.S.W., the Australian Institute of Health Surveyors, the Clean Air Society of Australia and New Zealand, the Australian Membership of the Institute of Fuel and the Local Government Engineers Association of N.S.W.

The Committee is indebted to the Honorable Sir John B. Fuller, M.L.C., the Minister for Planning and Environment, for so readily agreeing to officiate at the opening of the Conference and address the delegates.

Since the 1971 Conference many developments have taken place both in Australia and overseas and there has been a growing awareness and much public concern about the environment and its pollution. Consequently, the thirty-nine papers presented in this volume cover a wider range of subjects than was the case at the previous two Conferences. These papers are to be briefly presented by the authors at nine technical sessions, allowing ample time for discussion. The subjects to be considered include legislative aspects and waste management developments in the various Australian States, the Capital Territory, the U.K. and the U.N. Centre, surveys and advances in solid waste practices, waste collection and handling services, treatment and disposal of liquid and industrial wastes, and salvage and resource recovery from wastes.

It is gratifying to note that several of the papers present research data and developments based on work carried out in Australia since our last Conference. Yet, in some ways, this could be considered as our first International Conference as several of the contributions discuss plants and processes in the U.K., Bangkok, Singapore, Hong Kong and the Philippines. There is an account of overseas developments in waste control, based on reports resulting from the five travelling scholarships which were awarded by the Conference Committee during 1972.

I am also happy to report that four of the contributions (21, 29, 30 and 31) are from students attending the M.App.Sc. course in Environmental Pollution Control at the University, and were initiated by class assignments in the subject "Unit Operations in Waste Management."

Finally, I wish to express my thanks and appreciation to the authors, the University, the members of the Organising Committee, the representatives of the associated organisations, and all those who, with their interest, advice and assistance in various ways, have made my task as an editor of the published papers and as Chairman of the Organising Committee, a very pleasant one. Their enthusiasm augurs well for the success of our third symposium — The 1974 Waste Management and Control Conference.

N. Y. KIROV,
Conference Chairman

Department of Fuel Technology,
The University of New South Wales,
Kensington, N.S.W., 2033.

CONFERENCE ORGANISING COMMITTEE

Chairman:

A/Prof. N. Y. Kirov, Head, Department of Fuel Technology, University of New South Wales

Vice-Chairmen:

Mr. H. K. Toner, Chief Health Inspector, Waverley Municipal Council
Mr. W. T. Cooper

Hon. Secretary:

Dr. G. D. Sergeant, Senior Lecturer, Dept. of Fuel Technology, University of New South Wales

Hon. Treasurer:

Dr. K. S. Basden, Senior Lecturer, Dept. of Fuel Technology, University of New South Wales

Social Secretary:

Dr. D. Barrett, Senior Lecturer, Dept. of Fuel Technology, University of New South Wales

Dr. T. P. Maher, Senior Tutor, Dept. of Fuel Technology, University of New South Wales

Mr. J. Smits, Professional Officer, Dept. of Fuel Technology, University of New South Wales

CONFERENCE ADVISORY COMMITTEE

The above named and

Representing the Institute of Fuel:

Mr. H. R. Goode, Chief Combustion Engineer, The Shell Co. of Australia

Representing the Clean Air Society of Australia and New Zealand:

Mr. J. Court, Air Pollution Control Branch, N.S.W. State Pollution Control Commission

Representing the Local Government Engineers' Association of N.S.W.:

Mr. K. J. Gunn
Mr. J. R. McClelland

Representing the Local Government Association of N.S.W.:

Ald. J. F. Carr, O.B.E.
Ald. J. C. Merrington

Representing the Australian Institute of Health Surveyors:

Mr. H. K. Toner

LIST OF PAPERS

OFFICIAL OPENING

Wednesday, 17th July, 1974

Conference Chairman: A/Prof. N. Y. Kirov

Address by The Hon. Sir John B. Fuller, M.L.C., N.S.W. Minister for Planning and Environment.

Reply on behalf of the University and the Organising Committee by the Vice-Chancellor and Principal of the University of New South Wales, Prof. R. H. Myers.

11.15 a.m.-12.45 p.m.: FIRST TECHNICAL SESSION — LEGISLATIVE ASPECTS
Chairman: N. Y. Kirov

2.00 p.m.-3.15 p.m.: SECOND TECHNICAL SESSION — LEGISLATIVE ASPECTS (Cont.)
Chairman: H. K. Toner

3.45 p.m.-5.00 p.m.: THIRD TECHNICAL SESSION — SURVEYS AND PROBLEM ASSESSMENT
Chairman: H. K. Toner

Thursday, 18th July, 1974

9.15 a.m.-10.45 a.m.: FOURTH TECHNICAL SESSION — WASTE COLLECTION AND HANDLING

Chairman: J. Court

11.15 a.m.-12.45 p.m.: FIFTH TECHNICAL SESSION — TREATMENT AND DISPOSAL OF LIQUID WASTES

Chairman: J. Court

2.00 p.m.-3.30 p.m.: SIXTH TECHNICAL SESSION — DISPOSAL METHODS

Chairman: H. R. Goode

Friday, 19th July, 1974

Closing remarks by Conference Chairman. End of Formal Programme.

PAPER 1

Development of the Sydney Region Waste Management Plan

By Richard Conolly*

1. Introduction

The disposal of solid and liquid wastes in large urban communities is a major problem of ever-increasing proportions. With increasing population densities and escalating solid waste generation rates, the greater quantities of solid waste produced for a given area are placing increasingly heavy demands on existing collection and disposal systems; particularly in today's environmental climate. These demands introduce problems involving the scarcity of available land for waste disposal, particularly in the inner city areas.

Traditionally, local authorities have had the responsibility for collecting and disposing of solid waste. However, over the past decade or so there has been a growing realisation that isolated action by individual municipalities is not the best solution. The concept of a regional approach whereby technical knowledge and physical resources are pooled to mutual advantage, has been accepted and implemented by a number of major cities throughout the world.

The concept of a regional approach to waste disposal was mooted at the two previous Australian Waste Disposal Conferences in 1967 and 1971. In fact, at the termination of the 1971 Conference, the Hon. J. B. M. Fuller (now the Hon. Sir John Fuller) announced the formation of the Metropolitan Waste Disposal Authority.

The Authority has jurisdiction over the Sydney region of N.S.W., and its function can best be described by the introduction of the Act:

> "An Act to provide for the constitution of a corporation to be called the 'Metropolitan Disposal Authority'; to confer and impose on the corporation responsibilities, powers, authorities, duties and functions with respect to the transport, collection, reception, storage and disposal of waste within the Metropolitan Waste Disposal Region; and for purposes connected therewith."

With the formation of the Authority, the entire Metropolitan area can now be considered as a unit and, accordingly, the waste disposal system can be provided at least cost to the Metropolitan community as a whole. Because of its sheer size, this involves many permutations and combinations of options both in systems, location of facilities, processes and alternative directions between generation and disposal locations. Considerable economy is potentially available but the process of evaluation is complicated and time consuming.

Since its inception in June, 1971, and the appointment of senior technical staff in 1972, the Authority has been preparing the ground work necessary for the development of the Sydney Region Waste Management Plan. It should be pointed out that, in addition to solid waste, the term "waste" also covers liquid waste.

In this paper it is intended to outline the approach adopted by the Authority in developing the three key areas of the plan:

- Solid Waste Management Plan.
- Establishment of a central liquid waste treatment plant.
- Registration and licensing provisions of Part V of the Act.

The registration and licensing provisions of the Act will be the "teeth" to ensure the orderly implementation of the solid and liquid waste management plans.

2. Solid Waste Management Plan

It is highly desirable that proposed facilities and systems comprising the Management Plan be formulated on the basis of a thorough technical, financial and political understanding of all the matters involved. As it requires considerable time and effort to prepare the data base to gain this understanding, and as new facilities are required in the immediate future, the Authority resolved to attack the problem in two phases.

Phase 1: to develop a short-term plan for the period 1974 to 1978, based on the best information available to the Authority in 1973-74;

* Director and Chairman, Metropolitan Waste Disposal Authority.

Phase 2: to conduct a major study to develop a comprehensive long-term waste management plan for the period to the 1990's.

There is no dividing line between the Phase 1 and Phase 2 studies and those described as Phase 1, or "short-term", relate to the urgent first conclusions of a continuing study.

2.1. *Present Situation*

With the exception of the waste incinerated at the Waverley/Woollahra plant, all solid waste generated in the Region is currently disposed of by the landfill method. At this time, only 14 of the Council-operated depots and 9 of the commercially-operated depots have significant remaining capacity (life in excess of 2 years).

2.2. *The Problem Defined*

Based upon the best data currently available to the Authority, it is estimated that the current rate of solid waste generation per head of population is 0.68 tonnes per year. This rate is expected to escalate to 1.06 per tonnes per year by the year 2000. The increase in the per capita rate of generation, when coupled with the expected population increase to 4,400,000 by 2000, represents a disposal problem of staggering magnitude. For example,

— If landfill were the only method of solid waste disposal from now to the year 2000, an additional disposal capacity of approximately 210 million cubic metres would be required.

— If a policy of maximum incineration (i.e., the method requiring least space) were adopted now, the additional space requirements to the year 2000 (for non-combustible waste and ash) would be approximately 80 million cubic metres. (Nine incinerators each of 1000 tonnes per day capacity, double that of the Waverley/Woollahra plant, would be required by the year 2000 to bring about this result. At current prices, this would represent a capital expenditure exceeding $100 million.)

2.3. *Short-Term Plan*

At the time of preparation of this paper, the following alternative short-term solutions were under detailed evaluation:

(a) *That as Councils exhaust their own facilities they make use of regional transfer stations and/or regional landfill sites operated by the Authority.* The adoption of this solution would involve the opening of new regional facilities, both landfill and transfer stations, at an early date to operate at a small, and therefore possibly less economic, scale.

(b) *That Councils use up all existing space in their sub-region before transferring to regional facilities.* This would necessitate the direction of waste by the Authority to existing Council depots from other areas until all local facilities were exhausted. In selecting those existing sites that have a regional potential, environmental aspects must be taken into account.
The adoption of this solution would result in an increased scale of operation at existing facilities used for regional purposes, and at new regional facilities when they commence operation, with consequent economy of scale. While increased transportation costs can be expected in certain cases, overall economic benefits are likely.
The present operator (Council) of a potential regional facility will, in most instances, have invested considerable funds on the development and would seek some return from either the Authority or the other Councils who would use the facility.

(c) *That Councils use existing small depots for weekend tipping only and all other wastes be directed to other large existing depots or to regional facilities.* This is, in effect, a variation of alternative (b) that makes provision for weekend tipping by residents. Approximately 18% of the total waste received at depots in the Region is accepted during the weekend and there will be a continuing need to provide a facility for this purpose in order to reduce the likelihood of indiscriminate dumping of waste.

(d) *That the putrescible and non-putrescible waste streams be separated and directed to selected existing and/or proposed facilities.* This procedure could be adopted in a number of instances where facilities are more suited, on environmental grounds, to the receipt of one or other type of waste.
The provision of satisfactory leachate treatment facilities will prove to be expensive in the future and it is therefore desirable to minimise the number of depots requiring treatment plants.

It would be necessary in same instances to differentiate between the general non-putrescible wastes and the selected hard-fill wastes, i.e., brickbats and other demolition material. This would permit the ultimate use of certain sites, with a hard foundation, for building purposes.

As a general policy, the Authority will consider directing the various classes of waste to the most suitable available disposal facility — such facilities to become Regional depots. The criteria for such decisions will be to achieve overall long-term community benefits, with particular reference to cost and environmental considerations.

It is intended to finalise the short-term plans by mid-1974 and it will be possible at the Conference to elaborate on the details of the plan adopted.

2.4. *Long-Term Plan*

Approximately two million tonnes per year of solid waste is currently handled by Councils and private contractors at a transportation, processing and disposal cost of between $3 and $10 per tonne. This represents an expenditure of between $6 and $20 million per year. If the cost of collection is taken into account, the direct community cost is probably of the order of $30 million per year for the removal, transportation and disposal of solid waste.

When one considers that the $30 million per year solid waste management business is plagued by erratic patterns of waste movement, rapid depletion of remaining landfill sites, long lead times for major processing facilities and extensive pollution problems, the necessity for effective long-term planning and central control is self-evident.

Considerable progress has been made by the Authority in developing the long-term plan, although it is expected to take a further two years to finalise the detailed study. The main elements of the study are referred to in the following paragraphs.

In 1973, an intensive systematic search of the Region was conducted to locate sites suitable for the establishment of regional landfill depots. Over 200 sites have already been examined. After detailed assessment on economic and environmental grounds and discussion with the other affected departments or instruments of Government, most were rejected. Some 14 remaining sites are still under investigation. The Authority's first regional depot, Jack's Gully, is currently being developed in conjunction with Camden Council.

The Northern Sector is at present the subject of a particular study by the Authority's officers to evaluate cost and performance of existing disposal methods in relation to other options available in both the short and long term. Much of the information gained will be directly applicable to the remainder of the Region. While, during this study, particular attention will be given to landfill and incineration as disposal methods, other methods will also be carefully examined in developing the overall regional plan.

For the purpose of developing a solid waste management plan, the Region can be considered as a system comprising 40 Local Government areas generating domestic waste, some also generating large quantities of industrial waste; potential regional landfill sites on the outskirts of the Region; potential transfer station sites throughout the Region as well as possible incinerator (or other process plant) sites; with a variety of collection and transportation methods. The number of permutations and combinations of possible solutions to the problem is huge and a mathematical model (MADIS) has been prepared to assist in solving the problem. Using the Control Data 6600 computer, alternatives can be rapidly evaluated to generate a theoretical least-cost solution. This technique will assist in developing the most appropriate long-term system of transportation and disposal.

The Authority has gained the co-operation of the National Materials Handling Bureau to assist in studying solid waste handling methods. It is hoped that this expert group will be able to help us devise the most efficient waste handling and transportation system.

2.5. *Other Continuing Studies*

A comprehensive scheme for the disposal of obsolete motor vehicles has been prepared and furnished to the Joint Government Members Committee set up by the Government to enquire into the general question of the disposal of abandoned and derelict motor vehicles. At the time of preparation of this paper, the scheme was still under study by the Committee.

Preliminary investigations into recycling indicate that, in Sydney, the rate of recycling of solid waste is already significant and at near maximum for certain materials. Paper is being recycled at a rate higher than for most other Western Countries and with existing plant, there is virtually no additional capacity for waste paper. A similar situation exists with respect to glass, although some spare capacity for cullett is available. We are all aware of the cry for more recycling of waste. However, from a community benefit point of view, it has not been established that recycling more waste is the best answer. In a recent article, the Director of Sanitation of the City of Los Angeles analysed the situation and concluded that the potential recyclable materials in domestic solid waste are amongst the most abundant in the earth's crust. In most cases, the materials and energy needed to collect and process such materials exceed that to mine and process them from natural resources. On the other hand, those materials in short supply (e.g., copper, tin, lead) are not present in large amount in solid waste. The question of increased recycling, therefore, is one of great complexity and requires deep study.

Leachate from landfill sites is a potential environmental hazard of considerable proportions and is perhaps the major factor hindering the establishment of regional putrescible landfill sites. Design studies have been conducted into engineering design and drainage of large putrescible depots in order to minimise the formation of leachate. As a matter of urgency, a research programme has been initiated by the Authority into improving methods of leachate treatment.

3. Liquid Waste Treatment Plant

In May, 1970, the Barton Report highlighted the fact that Sydney was facing a critical situation with regard to the disposal of liquid wastes. At the time it was known that considerable amounts of liquid wastes were being placed in the Water Board sewers, both legally and illegally, and that many gallons of liquid wastes were being disposed of by generators and contractors in watercourses and bush areas. The above facts, the projected closing down of various disposal sites and the introduction by the Board of far more stringent requirements for liquid waste disposal into its sewers, made it imperative to determine the true extent of the present liquid waste situation in Sydney.

The Authority commissioned consultants to carry out a detailed survey of liquid wastes generated in the Waste Disposal Region, to prepare projections of generation into the future and to propose the type of treatment plant needed for the disposal of these wastes and the plant's location. Further, preliminary plant designs and capital and treatment cost estimates were to be provided. The consultants' report was received in May 1973 and the results of the survey are discussed in detail elsewhere in this conference (see Paper 11 by A. G. Forrester).

After detailed examination of the consultants' proposals and an appraisal of industry's reaction to the predicted treatment charges, the Authority resolved to adopt the concept of a "minimum plant". This concept envisages a plant of as simple construction as possible, designed to treat most wastes expected in the early years of operation, with the capacity of evolving to meeting future requirements.

Follow-up discussions with the major generators has vindicated the "minimum plant" concept as, faced with a substantial increase in disposal charges, many companies are now studying their waste stream with a view to minimising waste quantities, providing in-plant treatment or otherwise reprocessing their wastes. A good example is a chicken processor who, at the time of survey, generated large quantities of a high B.O.D. waste. The waste is now converted into a high protein feed for chickens and virtually no waste leaves the premises.

One aspect of the discussions with industry is of particular interest. On the advice of Authority officers, a large company which had been storing a particularly unpleasant waste for some time, could well enter into an agreement with an interstate company which would use the waste as a raw material for its process.

Design data for the plant is being finalised in conjunction with our consultants and the plant is expected to be "on line" early in 1976. Initially, our physiochemical treatment and incineration facilities will be provided, although the plant can evolve to include biological treatment, resource recovery and solid waste treatment facilities. The selection of a suitable system for the incineration of mixed liquid wastes still represents a considerable problem and a research programme is currently being conducted at the C.S.I.R.O. Division of Mineral Chemistry, North Ryde, to assist in developing the most suitable incinerator configuration.

To provide a disposal point for liquid waste until the plant is completed, an interim liquid waste disposal depot has been constructed by the Authority at Castlereagh, some 30 miles west of Sydney. This is basically a landfill operation where liquid waste is absorbed on solid waste under carefully controlled conditions to ensure proper environmental protection.

4. Registration and Licensing Provisions

Part V of the Act, which deals with registration and licensing, was excluded when the Act was proclaimed. As proclamation of this

Part is anticipated in mid-1974, it is of specific interest at the time of this conference.

The objective of Part V is to establish control over the waste stream by means of a system of registration and licensing; that is by:

- the registration of both solid and liquid waste treatment, storage and disposal depots,
- the licensing of persons who transport waste for fee or reward, and
- the registration of premises on or in which trade waste is created.

This system will give the Authority control over the generation, transportation and disposal aspects for most of the waste stream (see Fig. 1); control which will be used to ensure the orderly implementation of the regional waste management plan.

Licences and certificates of registration will be issued subject to conditions, such conditions to be set according to minimum standards laid down by the Authority for handling and disposal of waste. The minimum standards are based on good housekeeping practice and environmental considerations.

Licences and certificates of registration will be subject to an annual fee. The scale of registration and licensing fees are designed to provide only sufficient revenue to meet costs. The major cost component of operating Part V will be the inspectorate necessary to provide day to day control.

Part V also provides for the inspection of premises and penalties for breaches of conditions attached to certificates or licences. Maximum penalties range from \$1000 to \$5000 with daily penalties for continuing breaches, ranging from \$200 to \$2000.

4.1. *Depots*

A "depot" is defined as "any premises, other than a regional depot, to which waste is transported from any other premises for treatment, storage or disposal." Therefore, all landfill depots, transfer stations and incinerators must be registered with the Authority.

In general, conditions will be imposed so that the operation of landfill depots will progressively improve to comply with the Authority's

Fig. 1

Simplified Representation of the Waste Stream — GENERATION TRANSPORTATION DISPOSAL

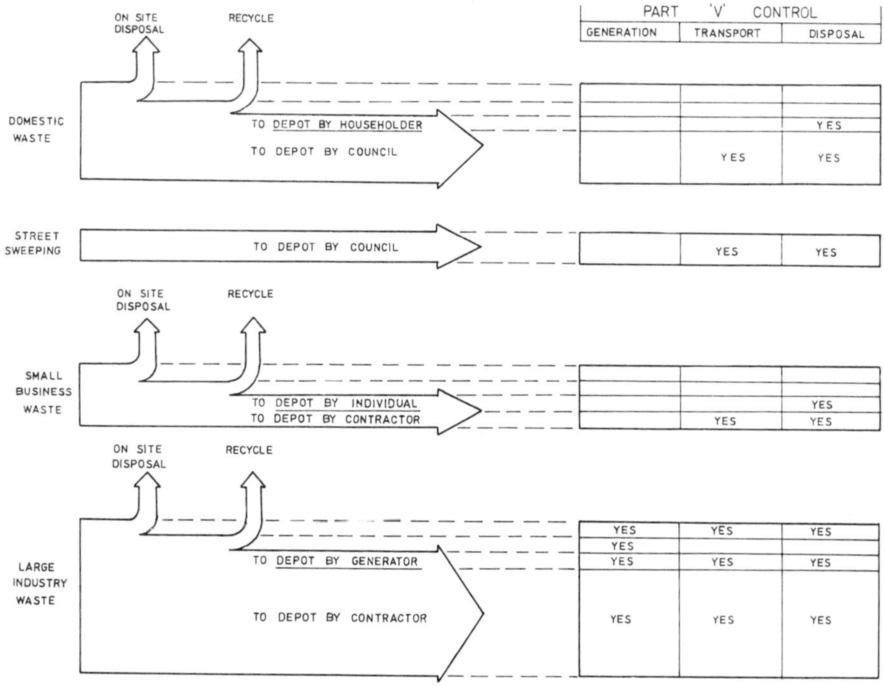

Standard Conditions of Operation S.W. 1A as applicable.

4.2. *Transporters*

A person who transports waste into, out of or within the Region for fee or reward must be licensed with the Authority. This licensing requirement is also extended by the Act to cover any public authority which transports waste and any registered generator of trade waste who transports trade waste.

In general, conditions will be imposed requiring the licensee to discharge wastes at registered depots which are authorised to accept such wastes, maintain all vehicles in a clean and roadworthy condition and, in certain cases, use such routes as determined by the Authority.

4.3. *Generators*

Premises (not being approved premises) on or in which trade waste is created must be registered with the Authority. "Trade Waste" is defined "as any matter or thing, whether solid, gaseous or liquid or a combination of any solids, gases and liquids, which is refuse from any industrial, chemical, trade or business process or operation, including any building or demolition work".

Under the above definition, some 80,000 premises in the Region would be classed as generators of trade waste; a large proportion of these generate relatively harmless wastes such as paper or packing materials in small quantities. It is not the intent of the Authority that such small businesses or shops should be registered. In order to achieve exemption of certain generators of trade waste from registration, the Act provides for "approved premises" to be defined by the Authority by order in the Gazette.

A system based on the number of employees has been adopted to classify generators of trade waste and the line dividing small and large generators has been set at 20 employees. Thus, premises on or in which trade waste is created with fewer than 20 employees are defined as "approved premises" and are not required to be registered with the Authority. However, as some small premises can generate significant quantities of toxic or otherwise hazardous wastes, all generators of hazardous trade waste must be registered with the Authority. If it is assumed that the quantity of trade waste generated on premises is directly proportional to the number of employees (generally true although other factors are also important), it can be seen from figure 2 that by adopting a cut-off point of 20 employees, the Authority gains direct control over almost 70 per cent of the trade waste stream by registering only about 10 per cent of the generators.

Fig. 2

REGISTRATION OF TRADE WASTE GENERATORS

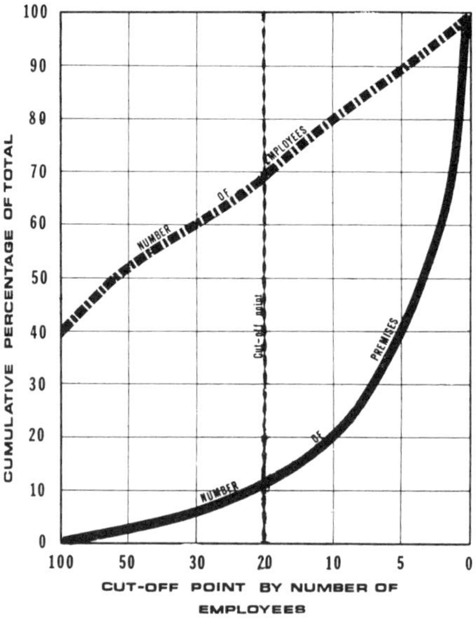

Conditions applicable to generators of trade waste will generally relate to good housekeeping practice for storage of waste, provisions to ensure that waste is removed only by licensed transporters or that on-site treatment facilities are satisfactory.

5. Conclusion

It is clear that waste disposal in all its aspects is one of the major problems confronting large cities throughout the world today. Research aimed at discovering new or improved methods of disposal and resource recovery are being conducted at many levels and must result, at the least, in producing improved technology.

Sydney has seen the introduction of the Metropolitan Waste Disposal Authority since the 1971 Waste Disposal Conference. This Authority is the first of its type in Australia, and, indeed, one of the first in the world.

In the brief period since its inception, the Authority's planning for future disposal on a regional basis, in both the short and long term, has reached an advanced level. The comprehensive studies which have been completed in the region have formed a sound base for this planning, and we look forward, as a consequence, to the operation of an efficient, well co-ordinated system of disposal/treatment.

The cost of proper disposal is, in many cases, high. Therefore, it is necessary to educate the community to accept the fact that it must be prepared to pay for environmentally-acceptable waste disposal methods.

The registration and licensing provisions of the Act give the Authority the "teeth" to ensure orderly implementation of the Sydney Region Waste Management Plan. Environmental considerations are basic to the plan and, therefore, as one arm of the State's environmental control programme, the Authority is making a significant contribution to environmental protection and improvement.

ACKNOWLEDGEMENTS

The assistance of Messrs. D. R. M. Thomas, P. Horsley and E. van den Broek in the preparation of this paper is acknowledged.

PAPER 2

Environmental Control in Queensland

By P. L. Ellis* and R. G. Hoare†

Introduction

The Queensland Government has, in recent years, passed several Acts associated with environmental control, in response to social attitudes that call for maintenance and improvement of our quality of life. While it is true that social attitudes have changed and are continuing to change, it is important to realise that recent legislation does little more than co-ordinate and rationalise environmental control measures that have long been established through many Acts in the Statute Books and in By-Laws brought down by Local Authorities. Specific legislation such as the Clean Air Act 1963 and Clean Waters Act 1971 with their Regulations did introduce specific standards for pollution control and therefore must be classed as innovative legislation. However, virtually every licence, lease, permit, or authority issued by a State Department or Local Authority attracts conditions, many of which include an environmental parameter.

Environmental matters such as waste disposal; soil, water and mineral conservation; flora and fauna protection; and forestry, agriculture and fisheries management have been subject to control through relevant Acts and By-Laws for many years.

It was the recognition that so many Acts singularly and often unilaterally affected environmental management and planning, that led the Queensland Government to develop the administrative machinery needed to co-ordinate the activities of the numerous Authorities.

Co-ordination of Environmental Control in Queensland

As mentioned above, many State Government Departments and Local Authorities have been involved in environmental management for many years. The co-ordination of their respective activities has been achieved through the State and Regional Planning and Development, Public Works Organisation and Environmental Control Act 1971-73.

This Act does two main things: firstly, it provides a systems approach to long term regional planning and, secondly, it provides within this planning framework, for environmental matters to be considered in the decision-making process. In effect, it established machinery to help the Queensland Government, with the help of Statutory Bodies and Local Authorities, to solve the complex problems of today and of the future in a co-ordinated, integrated and systematic fashion.

The Environmental Control Council

Under the above-mentioned Act, the Environmental Control Council, first established in 1970, continued in existence. The Council, chaired by the Co-ordinator-General, and including senior representatives of most State Departments, meets about four times per year. It is charged with the co-ordination of environmental-orientated works by State Departments, Local Authorities, Statutory Bodies and Independent Bodies; liaison with these groups in the fields of environmental control and improvement, reviewing on a continuous basis the state of the environment; investigation of submissions made to it; co-ordination of environmental education; making recommendations to State or other bodies concerned with environmental matters; and advising the Government on matters dealing with environmental control.

Also, the Environmental Control Council co-ordinates environmental research and makes recommendations to the Government on the basis of technical data supplied to it by its Technical Advisory Committees, Statutory Bodies and other governmental committees. The role of each of these in environmental management in Queensland is discussed below.

(a) Technical Advisory Committees

Three Technical Advisory Committees have been established to cover the fields of Waste Disposal on Land, Noise Control, and Land

* Director (Environmental Control), Co-ordinator-General's Department, Brisbane.

† Planning Officer, Co-ordinator-General's Department, Brisbane.

Use. These Advisory Committees draw considerable membership from outside the Public Service. Generally, Technical Advisory Committees have been created to examine environmental problem areas which are not adequately controlled under existing legislation.

(i) *Disposal of Waste on Land Committee*

At the first meeting of the Council in 1971, a submission was tabled which described a potential problem in solid waste disposal in the Brisbane area. It was resolved that a Committee, the Disposal of Waste on Land Committee, be established to investigate the matter. The Committee is chaired by a representative of the Co-ordinator-General's Department and draws its members from the State Departments of Health, Primary Industries, Local Government, Lands, Commercial and Industrial Development, the Chamber of Manufacturers, the Institute of Health Surveyors, the Queensland Institute of Technology, and the Brisbane City Council.

Under Queensland Statutes, control of local authority refuse disposal services is exercised under the Health Act and control of discharges to water courses is exercised under the Clean Waters Act. Because of this, the Disposal of Waste on Land Committee is examining major problems such as future planning of waste disposal services and control of dumping of poisonous waste.

Since its inception, the Committee has been collecting data on waste disposal practices in Queensland. These suggest that some cities have potential major solid waste disposal problems and that local authorities practice inefficient methods of waste disposal, to the extent that some are not getting benefits commensurate with their expenditures. The specific problems of the Cairns City Council were the subject of a study organised by the Committee under the aegis of the Environmental Control Council. This study, the first of its kind in Queensland, was completed last year and is the subject of a separate paper to this Conference.

Guidelines on the disposal of waste from intensive systems, including cattle feedlots, have been prepared by the Committee in consultation with the Water Quality Council. These will assist local authorities in their consideration of applications to construct intensive animal production farms. Also, a research project on the disposal of manure on land is currently being sponsored by the Committee.

(ii) *Noise Control Committee*

This Committee, which was set up in the latter half of 1971, is chaired by a representative of the Co-ordinator-General's Department and draws its members from the State Departments of Health, Industrial Affairs, Local Government, Main Roads, Transport, Railways, the Australian Government Department of Civil Aviation and non-Government bodies including the Chamber of Manufacturers, the Trades and Labour Council, the University of Queensland and the Australian Medical Association.

The Committee's terms of reference provide for investigation of all aspects of noise pollution in order to identify problem areas and to recommend control measures where practicable. Early work related to the examination of possible methods of noise control and the reasons for their ineffectiveness. Recognition of the ineffectiveness of overseas legislation led the Committee to direct its attention towards identifying major noise problems in Queensland and attempting to outline specific solutions to these.

A series of surveys which will enable a clearer assessment of noise problems in Brisbane has been initiated. Surveys of traffic and aircraft noise, and of community annoyance are proceeding and future ones are proposed. It is hoped that the results of these surveys will lead to practicable recommendations.

The vexed question of controlling legislation has been looked into. Whereas generalised legislation can provide a means for controlling "normal" noise sources such as motor vehicles, motor mowers, and some forms of industrial noise, it is less effective in controlling major problems such as freeway and airport noise, or noise impact on sensitive locations such as schools and hospitals. It is expected that this Committee will recommend noise legislation to the Environmental Control Council this year. It is also expected that the Committee's activities will extend to areas outside of Brisbane during the year.

(iii) *Land Use Committee*

The Land Use Committee, which was established in 1971, is currently chaired by a representative of the Co-ordinator-General's Department and draws members from the State Departments of Co-ordinator-General, Harbours and Marine, Irrigation and Water Supply, Lands, Forestry, Local Government, Mines, Primary Industries, and the independent bodies — the Australian Conservation

Foundation, Queensland Conservation Council, Australian Institute of Landscape Architects, Local Government Association, the University of Queensland and Griffity University.

The Committee was empowered, firstly, to carry out investigations and to make recommendations to the Environmental Control Council with respect to policies on the use of land, and the management of land and its resources, and, secondly, to make recommendations to the Environmental Control Council on implementation of these policies. These terms of reference are very broad and the Committee has had to face a monumental task in performing its function.

This Committee's most outstanding contribution was the drawing up of broad guidelines for environmental studies that should accompany feasibility studies for major development projects both public and private. These guidelines were accepted by the Environmental Control Council and subsequently formed the basis of a submission to Cabinet. The Cabinet decision relating to environmental policy is described below.

Other work being undertaken by the Land Use Committee includes a study of legislation pertaining to tree clearing and current problems in relation to rural subdivision of land. The Committee is currently compiling a draft procedural manual for environmental studies. A register of consultants interested in undertaking environmental studies has been compiled.

The Council has also agreed that a study to determine the principles applicable to land usage in the coastal zone be undertaken under the auspices of the Co-ordinator-General's Department. Planning for this major study is proceeding. In particular, the long term environmental problems of canal estate development will be examined in detail.

(b) Statutory Bodies

The Environmental Control Council keeps in close touch with established statutory bodies, both by exchange of observers and by direct consultation. These are the bodies with clearly defined executive function in the environmental sphere. Their legislation extends from control of part of the background to broad environmental studies and regional planning. They include the Air Pollution Council, the Water Quality Council and the Beach Protection Authority.

(i) *The Air Pollution Council*

At the time of formation of the Environmental Control Council the only agency of the State involved in environmental matters was the Air Pollution Council and the Division of Air Pollution in the Department of Health. These bodies were established as early as 1963 under the Clean Air Act. Consequently, it was not necessary for the Environmental Control Council to establish an advisory group to attend to air pollution problems; it was able to draw on the considerable expertise of the existing bodies.

(ii) *The Water Quality Council*

In 1971, the Clean Waters Act established the Water Quality Council and the Water Quality Division of the Department of Local Government. A great deal has been achieved in Queensland in this field, due to the hard work and practical approach of the Chairman of the Water Quality Council and the Director of Water Quality and his staff. Liaison with the Environmental Control Council in matters dealing with water pollution has been systematised; the Chairman of the Water Quality Control is a member of the Environmental Control Council, the Director of Water Quality attends all meetings of the Environmental Control Council as an observer, and the Director of Environmental Control, as the representative of the Environmental Control Council, attends all meetings of the Water Quality Council as an observer.

Both the Clean Air and Clean Waters Acts with their Regulations control emissions through a licensing system. This principle has been criticised as a licence to pollute, which, of course is true, but the aim is to control emissions so that they enter the environment in a quantity and at a rate which can be assimilated.

(iii) *The Beach Protection Authority*

A third statutory body that works in close liaison with the Environmental Control Council is the Beach Protection Authority which was established under the Beach Protection Acts 1968-70. The Acts are administered by the Department of Harbours and Marine. The activities of the Authority and its Advisory Board are of vital interest to the Council in its deliberations.

The coastal zone continues to feel the pressure of population growth and it is in this area that many of our environmental problems are generated, particularly in and near urban centres. The Beach Protection Authority is certainly involved in environmental control by planning the protection of existing foreshores and the rebuilding of degraded beaches, and by attempting to preserve those areas that may be rebuilt.

(c) Other Bodies

Other Departments, local authorities and specialist groups such as the Advisory and Co-ordinating Committee on Soil Conservation play an important role in environmental management. Their cumulative effort, co-ordinated through the Environmental Control Council, is quite significant. In those cases, where such authorities issue licences for a variety of development proposals, they can legally require full environmental impact studies to be made prior to the granting or rejection of the licence. Others can write restrictive clauses in the leases and agreements in order to protect the environment.

The Health Department, for example, in considering an application for use of a particular area of land as a refuse tip from a local authority, may grant approval subject to conditions. In practice, the Health Department and Water Quality Council jointly consider applications, and normally the former specifies conditions which will safeguard human health while the latter specifies conditions which will minimise water pollution from waste disposal sites.

The Australian Environment Council

This body co-ordinates environmental control at the national level. While this paper is not the vehicle to set out the structure and organisation of the Australian Environment Council, it is pertinent to note that there is close liaison between the Environmental Control Council and its Committees and the Australian Environment Council and its advisory Committees.

Environmental Impact Studies in Queensland

The State Cabinet policy decision of 27th November, 1972, is as follows:—

"(1) Environmental Studies are required for all major developmental projects and other projects where the Environmental Control Council or the Statutory Authority considers these studies necessary.

(2) Environmental Studies must be undertaken according to the Environmental Control Council's recommended guidelines.

(3) Environmental Studies constitute part of the project investigation cost.

(4) All requirements for Environmental Studies apply uniformly to proposals by State, Semi-Governmental, Local Authority and private bodies.

Initial guidelines for environmental studies required

(1) definition of the study area, e.g., its physical boundaries, the time period involved and their regional implications of the proposal;

(2) detailed definition of the scheme and its alternatives;

(3) preparation of a working brief, achieved through consultation with appropriate experts with knowledge of the environmental and social consequences of the development;

(4) preparation of a report by these experts outlining the existing situations, and anticipated changes in the study area;

(5) evaluation of requirements designed to optimise beneficial changes and to eliminate or reduce undesirable changes to a socially acceptable level. 'Acceptable levels' in many instances are defined by existing legislation.

(6) Collation of all pertinent data in the final report, which must be sufficiently comprehensive for decision-makers to be aware, as far as is possible, of the total environmental effect. An Environmental Impact Statement in the Queensland context, is intended as a summary statement of the findings of all of the various investigations."

It is the responsibility of an administering authority, for instance a local authority, to call for environmental studies in the first instance, if in the opinion of that authority a proposal shall involve significant environmental disturbance. Further, it is also the responsibility of such authority to assess the study.

To elaborate on this point, one particular authority is responsible for granting approval for, or rejecting, development proposals, e.g., local authorities for urban subdivisions and rezoning, the Mining Department for mining leases, etc., and the Department of Harbours and Marine for canal estate proposals. Such an authority has primary responsibility for—

(a) determining if a study and Environmental Impact Statement is required;

(b) advising which authorities should be consulted;

(c) advising on the general requirements for studies;

(d) obtaining expert assessment of the developer's submission prior to granting approval.

Implementation of this procedure should ensure that all pertinent aspects are covered in the preparation of the study and its final assessment. The Policy requires that normal inter-departmental / departmental statutory authority co-operation will prevail and that recourse to the Co-ordinator-General's Department or the Environmental Control Council will be minimal.

It is the responsibility of the State, firstly, through the Environmental Control Council, to issue a procedural manual for environmental studies to facilitate administration of the policy and, secondly, to engage in regional land use studies incorporating environmental planning considerations to a major degree. In relation to such regional studies, it should be noted that the first project will be an in-depth study of the coastal zone, encompassing the particular problems of canal estate development. The area under greatest pressure, that is from the New South Wales border to the Noosa River, will be studied first. Briefs for this work were prepared in the Co-ordinator-General's Department with the assistance of other interested Departments and Authorities. Specific environmental planning parameters derived from this are expected to assist greatly in future decision-making for the area, and certainly will obviate the need for duplication of studies which is inevitable when each major development proposal is considered in isolation. Further, it will lead to constraints on developments which in isolation are environmentally acceptable, but which in toto have considerable environmental impact.

The Role of Local Authorities

In outlining the administrative machinery presently operative in Queensland in relation to environmental control, the responsibilities of the "administering authorities" have been specified. It is quite clear that local authorities, as the administering authority in the majority of cases, carry a tremendous burden in the field of environmental management. Accordingly, it is the responsibility of Government, particularly those Departments administering Acts relating to environmental control, to assist local authorities in every reasonable way to implement rational environmental management.

The policy requirement for environmental studies and impact statements to accompany economic feasibility studies for major development proposals, although in the nature of a "fire fighting" mechanism at the present time, will in the short term indicate to Shire Planners just what the major local problems are. Hopefully, such recognition will lead to the preparation of well-balanced Shire and Town Plans.

Insofar as regional environmental planning is concerned, it is hoped that local authorities will take advantage of the machinery inherent in the recently created Regional Co-ordination Councils, to ensure that all local and regional problems are fully considered, and as required studied in depth.

Conclusion

In conclusion, it should be noted that environmental control in Queensland will only be as effective as the policy will allow. Controlling legislation is easily written; policing it, in a State the size of Queensland, is very difficult, but it can be done.

The procedural system in Queensland has a great deal of inbuilt flexibility that hopefully will allow decision-makers to be very informed. However, history shows that flexibility is a commodity that can often be abused. Therefore, the succes or otherwise of environmental management in Queensland will depend to a large extent on the social conscience of all persons and Authorities concerned.

If all parties observe the policy, the Co-ordinator-General's Department is optimistic enough to believe that by the end of the decade, environmental considerations will be regarded as perfectly normal components of planning and development — not new fangled and expensive constraints.

Acknowledgement

The authors gratefully acknowledge the support of the Co-ordinator-General and his staff in the preparation of this paper.

PAPER 3

Environmental Protection in Western Australia — Past and Future

By Brian J. O'Brien*

Introduction

Environmental protection in Western Australia is maintained by 18 Government Departments and statutory bodies, administering a total of some 55 statutes.

It might be thought that as a consequence there is a melange of disarray. In fact, however, perhaps because of personalities and certainly because of close co-operation between different instrumentalities, the system seems to be working rather well.

In 1970 and 1971, however, it became apparent that the increasing population, increasing development and increasing public concern for the environment, made the previous system inadequate. The Environmental Protection Act (1971) therefore established what might be regarded as an umbrella or co-ordinating and overall organising body known as the Environmental Protection Authority (EPA).

The EPA, together with its advisory Council and several committees (Figure 1), looks in the first place to the individual "service" departments to manage and control waste discharge. Thus, for example, it is the Clean Air Section of the Public Health Department which is responsible for licensing of premises with gaseous emissions. However, when a controversy arose as to whether an area north of Kwinana industry should be rezoned as urban residential, it was the EPA which recommended and organised a co-operative air pollution study by industry and State and Commonwealth Government departments (see CAPS — Coogee Air Pollution Study — Figure 1).

Table 1

MAJOR W.A. WASTE PRODUCTS

Solid Waste	Amount per year
Metropolitan waste[1]	435,000 tonnes
Gypsum into Cockburn Sound from CSBP[2]	128,040 tonnes
Laporte, Bunbury[3]	84,629 tonnes
Liquid Waste	
Metropolitan sewerage[4]	23,218,745 m^3
Swan Brewery[5]	909,220 m^3
Gaseous Waste	
SO_2 emitted by Kwinana industries[6]	48,300 tonnes

In this paper, environmental protection as relates to solid, liquid and gaseous waste control in W.A. is discussed in turn. The major waste products generated in the State are summarised in Table 1.

Solid Waste

The amounts of solid waste generated in Western Australia by comparison with those of other States are relatively small because of the relatively low population density and low industrial development of the State. Three particular examples are chosen, and management and control of each is discussed.

Metropolitan waste in the Perth area amounts to a total of 435,000 tonnes per year.[1] Local councils in general are responsible for the disposal of this waste, of which approximately 63% is household waste. The general administrative control is carried forward under the Health Acts 1911 to 1970 by the Public Health Department and local authorities. Presently some difficulty is being experienced in finding adequate space within reasonable proximity to the individual local authorities for disposal of the waste using the sanitary landfill technique, and attention is being given to co-operative crushing plants and the like.

Some trial experiments have been carried out in which the individual citizens, of their own volition, pre-sort metropolitan waste being carried into tips. To date this has been successful from a public relations point of view and reasonably efficient from a sorting point of view, but uneconomic and so far has resulted in a negligible saving in tip space.

Two out of the many industries are chosen to illustrate discharge of solid waste, one into the enclosed water of Cockburn Sound by

* Director of Environmental Protection, Perth, Western Australia.

CSBP and the second into sandpits near Leschenault Inlet by Laporte Industry.

CSBP under its enabling Act[2] can discharge over 128,000 tonnes of gypsum per year into Cockburn Sound. Initially they were discharging into slightly more than 8 fathoms of water; however, a cone of gypsum has risen up about 3 fathoms from the ocean floor and, being on the northern side of the Bulk Handling Terminal, this could present something of a navigational hazard. While it is not critical as yet, serious endeavours are now being made to find alternative methods of organising this waste discharge.

On the other hand, Laporte, near Bunbury, which treats black sands to produce almost 20,000 tonnes of ilmenite per year, produces a mixture of ferrous sulphate and sulphuric acid of approximately 85,000 tonnes per year. Under its enabling legislation the State is responsible for disposing of the waste. Various techniques have been tried[3] the latest of which,

Fig. 1: ENVIRONMENTAL PROTECTION ACT, 1971

Statutory Environmental Bodies

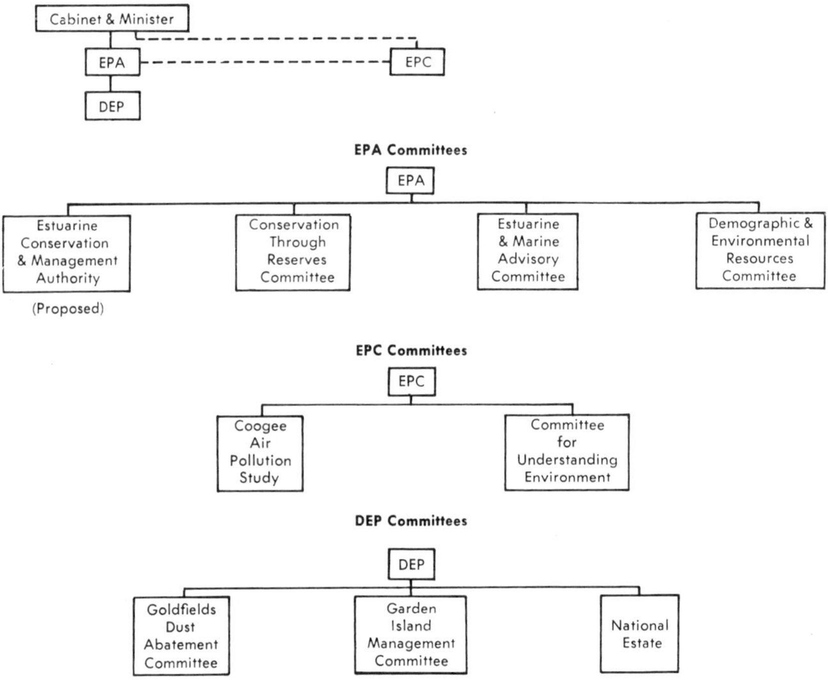

on a massive scale, has been the deposition of the liquid waste into sands on the western side of the Leschenault Inlet. This has resulted, however, in the iron stains gradually seeping through the sand after the sulphuric acid is neutralised, producing an extensive red stain on the seaward side which may reach as far as swimming beaches.

Various chemical treatment processes are under active experimentation but consideration is now being given quite seriously to the possibility that the only satisfactory solution for disposal of this waste will be to barge out to sea some 50 to 60 miles and dispose of it there. However, at this point in time (March 1974) the matter is still awaiting further tests.

The above three examples illustrate three major problems of solid waste disposal: one of community waste, and the other two of industrial waste. There is also the solid waste disposal problem common to most industrialised communities, that is disposal of litter of all types.

The fact that Western Australia is one million square miles in area with the population relatively localised in a few centres, and with very extensive road systems, means that it is essentially impractical to control waste or littering along roadside verges in the country by means other than educating the public and increasing public awareness. Excellent work is being carried out in this regard by the Western Australian Branch of Keep Australia Beautiful.

For those in the metropolitan and more heavily settled areas, legislation exists whereby fines of up to $200 can be imposed for littering. To my knowledge such a heavy fine has not yet been imposed but many smaller fines have been.

Liquid Wastes

In the matter of liquid waste, consideration will be given again mainly to the metropolitan area of Perth, although there are some rather exotic and specialised techniques of sewerage disposal used in some of the more outlying parts of the State. In the Perth metropolitan area the annual rate of sewerage discharge in 1973 was over 23 million cubic metres.[4] Approximately half the residences in Perth are sewered.

The sewerage treatment plants vary in age and in degree of treatment, and the rapid surge in population of the 1960s and early 1970s, wherein population increases rose to over 4.5 per cent per annum, resulted in a philosophy adopted by the Metropolitan Water Supply Sewerage and Drainage Board that a preference should be given to extending sewerage to the majority of the population rather than concentrating the limited sums of money on very refined, e.g., tertiary treatment plants. In some of the country areas, for example at Exmouth, the Harold Holt North-West Base, a considerable effort has been put into a sewerage treatment plant which produces potable water. On the other hand, in the mining township of Shay Gap, a vacuum system has been utilised to minimise the usage of vital water.

Perhaps of more general interest and certainly of particular interest in Western Australia, is the liquid waste effluent from the Swan Brewery. This amounts to a total of almost 910,000 cubic metres per year,[5] which consists mainly of cereals and sugars. The Swan Brewery has recently installed a treatment system which permits this waste to be discharged into the sewerage system.

The major metropolitan waterway, the Swan River, is protected against liquid waste pollution under the Swan River Conservation Act 1958 to 1966. Although waste may be discharged into the Swan River, this may be done only if authorised by permit, and the permit of course sets forth the conditions of the discharge. For example, any waste discharged must not contain: (a) sewage, (b) acidity or alkalinity of pH less than 5 or greater than 9, (c) poisons, or (d) any substance likely to contribute to the formation of sludge, scum or odours or which is likely to be injurious.

It is of interest that, in part because of the pride Perth people have in the Swan and in part through the vigilance of the Swan River Conservation Board, there have been no prosecutions under this Act.

Gaseous Wastes

Again we find that the relatively low level of industrialisation in Western Australia is such that air pollution is not a serious hazard. There are, of course, localised problems such as the blowing of fine iron ore dust from stockpiles at Port Hedland. This can become extremely serious under certain wind conditions, and the companies concerned have spent up to two million dollars endeavouring to alleviate the problem. There are also the usual problems encountered in the loading of alumina from shore to ship, and again the appropriate companies have been quite diligent in their attempts to reduce the problem to a minimum.

At present the major fuel source for industry is crude oil with a sulphur content of approximately 3 or 3½ per cent. As a consequence, from all the Kwinana industries, including the State Electricity Commission power station in the area, the total amount of sulphur dioxide emitted per year is approximately 50,000 tonnes.[6] The dispersal of the sulphur dioxide is controlled as usual by selection of an appropriate chimney height using the U.K. standards. The height of the chimney and the permit to discharge is issued by the Clean Air Section of the Public Health Department.

It would be appropriate to conclude with an illustration of what was mentioned in the introduction, namely that the high degree of co-operation which exists between industry and State and Commonwealth organisations in Western Australia, permits the Environmental Protection Authority to have a relatively small staff and to look to the service organisations to do most of the field work until they are faced with a position beyond their control.

Concluding Remarks

In the case of water pollution the situation has now been reached whereby the Fremantle Port Authority has sought the assistance of the EPA in controlling industrial effluent discharge into Cockburn Sound, which was so intensive that it was killing off seagrass. The EPA is now co-ordinating industries and State Government instrumentalities and private consultants, and beginning a concerted attack to reduce the seagrass mortality.

In a similar vein, a study of air pollution in the Kwinana area, which was again a co-operative effort, is now (March 1974) nearing completion. The study is known as the Coogee Air Pollution Study[6], and it was brought about by virtue of the fact that there was 1500 acres of Government-owned land with attractive vistas across Cockburn Sound to Garden Island which the Government was considering rezoning as residential rather than industrial.

The region is just to the north of the Kwinana industrial centre and the question arose as to whether the air pollution levels might be so high as to make rezoning undesirable. As the land was worth several million dollars, it was clear that a decision based on air pollution levels had to be credible. The Air Pollution Control Council felt intuitively that this area should be maintained as a buffer between any northern urbanisation and the southern industrial region. The Industrial Land Development Authority, which owns the land, naturally enough queried the rationale for this allocation of such a valuable piece of land.

On examination, the E.P.A. found that there were no serious air pollution levels in the region and so recommended, which recommendation was accepted, that a 12 month study of the actual air pollution levels be carried out over the 1500 acres in question and the near vicinity. The Department of Environmental Protection enlisted the aid of Kwinana industries, the Commonwealth Bureau of Meteorology, the Public Health Department, the Agriculture Department, the State Electricity Commission and the University of Western Australia and established a network of air pollution monitoring stations throughout the calendar year 1973 and early 1974.

Analysis of the results and the development of computer modelling techniques is now nearing completion (March 1974) and undoubtedly will be of great value in assessing the future usage of this particular area of land. But the Coogee Air Pollution Study (CAPS) has been of value, as it was intended to be, far and above the value associated with this immediate project. It enabled the various participants to develop techniques of co-ordination, to develop techniques and instruments of analysis and to assess the accuracy of predictive modelling. It will thus form a prototype from which air pollution monitoring and regulation throughout the entire metropolitan area and in the vicinity of future industrial developments in Western Australia can be achieved.

The techniques employed in CAPS and in the other projects mentioned are indicative of the essence of environmental protection and management in Western Australia, namely, a spirit of co-operation and co-ordination between all interested parties.

REFERENCES

1. Anon. *The First National Survey of Community Solid Waste Practices in Australia 1972-73* (Australian Consolidated Industries Ltd.) 1973.
2. *Industrial Lands (Kwinana) Agreement Act 1964.*
3. Anketell, K., *The Laporte Effluent Problem*, 1972.
4. Metropolitan Water Supply Sewerage and Drainage Board, *Annual Report*, 1973.
5. Private Communication.
6. Coogee Air Pollution Study.

PAPER 4

The Role of the EPA in Waste Management and Control in Victoria

By F. R. Meldrum*

The Environment Protection Act of the state of Victoria came in operation on 15th March, 1971, and is a comprehensive Act which brought the control of waste discharges to all sections of the environment under the jurisdiction of one body. The Act specifically provides that it shall prevail over any other Act or regulations, if there are any inconsistencies.

The Act set up three bodies, the Environment Protection Authority, the Environment Protection Council, and the Environment Protection Appeal Board. The Authority consists of three members, two of whom are required to be qualified in environmental control, and the third, an experienced administrator. The Environment Protection Council has seventeen members, appointed from specified public authorities and other organisations which are likely to be affected by the operations of the Authority or which are likely to have an impact on the thinking of the Authority. The Environment Protection Appeal Board comprises three members, one of whom must be a barrister and two experienced in environmental control or management.

The prime function of the Authority is to administer the Act, that of the Council to advise the Authority on matters relating to its responsibilities and powers, while the Board is there to deal with all appeals under the Act from decisions of the Authority or its delegated agencies with respect to licensing the discharge of wastes.

The Authority was set up on 1st July, 1971, with Mr. J. C. Fraser, an American biologist, particularly experienced in the water pollution field, as its first chairman. Following on his resignation to return to U.S.A., Mr. A. Gilpin, whose previous experience was in the clean air field, was appointed. It was not an easy process to recruit suitably qualified staff to perform its functions, and at the end of the 1st year, the total staff was 17, rising to 134 at the end of the 2nd year, and as at 1st January, 1974, 150. However, some Government and semi-Government authorities had already been carrying out certain of its functions and the Authority has chosen to continue to use such authorities as delegated agencies to carry out specific functions, mainly concerned with licencing, in some areas. Licencing of discharges to water, depending upon the area, is carried out by the Latrobe Valley Water & Sewerage Board, the Dandenong Valley Authority, the State Rivers & Water Supply Commission, the Melbourne and Metropolitan Board of Works, or the Authority itself. Air discharge licences are in the hands of the Latrobe Valley Board and the Authority itself, while licences to discharge to land are the responsibility of the Latrobe Valley Board and the Health Commission.

As at 1st January, 1974, the approximate staffing of the agencies in connection with work for the Authority was 60, but staffing overall must increase further to carry out effectively the functions of the Authority.

Under the Act, the Authority is responsible for co-ordinating all activities relating to the discharge of wastes into the environment and for protecting and improving the quality of the environment. It has power to issue licences to control the volume, types, constituents, and effects of waste discharges or deposits which are a danger to the quality of the environment, and this power is the basis of its main effort of control.

The licensing provisions of the Act became operative on 1st March, 1973, with the gazettal of regulations scheduling exemptions and the scale of fees for various types of discharge. Applications for licences to discharge wastes were required to be made within 3 months and for this reason alone, many small industrial plants making discharges to the atmosphere were exempted initially. They will be brought into the system progressively as the work-load allows.

The method of setting licence fees depends upon the type of discharge. Water licence fees are related to the probable cost of monitoring the waste for compliance with the licence

* Chief Land Waste Management Officer, Environment Protection Authority of Victoria.

conditions. There are fixed fees for such discharges as those from sewage treatment works of sewerage authorities. Trade wastes are assessed on the "rated waste load" which is calculated by summing the maximum number of units of each component or characteristic of the waste multiplied by the rating factor. The fee is then determined from the maximum daily discharge and the "rated waste load".

Air discharge licences are mainly related to the energy consumption of the plant producing the emission, either using the rated electric power installed, or the maximum fuel consumption calculated on the gross calorific value of the fuel consumed, whichever gives the higher fee. Fees for incinerators are based on grate area, and fees for stationary tanks which discharge to the atmosphere are based on capacity.

Fees for land discharge licences are related to the population served for garbage tips with a loading if industrial wastes are accepted. A set fee applies to factory licences for discharge to land, and transport operators are charged a fee related to the number of vehicles regularly used for transporting and depositing waste.

A licence issued under the Act is not a licence to pollute, but is a licence which lays down conditions for each discharge which will ensure that the environment will not be harmed. The licence application must contain details of the particular discharge, and many applicants found that they lacked specific knowledge of the composition of their discharges. Applications were not acceptable until this information was available, and many were not acceptable until the quality of the discharge had been improved substantially. This meant that each licence application required detailed consideration before appropriate conditions could be applied, and in the event, the issue of licences has proved to be a time-consuming task.

We are concerned with pollution, so that factories which deposit wastes on land, even if it is the factory's own land, require to be licensed for such a discharge as the possibilities for pollution of surface or underground waters are considerable. One of the notable examples which led to this approach concerned phenol contamination of an aquifer from phenolic wastes being deposited by a chemical factory in a quarry hole on site. On the other hand, it is not intended to licence factories whose wastes are removed by cartage for disposal elsewhere, but to licence the transport operator, to transport and deposit with the object of ensuring that wastes are deposited at sites suitable for their reception. The operation of this system has been delayed pending provision of adequate and proper disposal sites. By definition, wastes which are raw-materials for further processing, are exempt, as is deposit on land as part of an agricultural operation of waste from intensive animal husbandry such as piggeries and beef-fed lots. Whether such wastes would be permanently exempt is not yet decided, but certainly at the beginning of our operations, we did not have the proper criteria for the establishment of licence conditions. Where such operations would lead to a discharge to a stream, a licence would be required, but the condition of the discharge necessary for granting of a licence may be very difficult to achieve.

One of the problems which has caused some concern to applicants and more particularly potential applicants is the absence of standards for many discharges. The reasoning behind this is best illustrated by reference to the stream situation.

The Authority has prepared draft policies for several rivers discharging into Port Phillip Bay and for the littoral waters of the bay itself. These policies aim to define the beneficial uses of the water in the river and the quality of the water necessary for each beneficial use. The beneficial uses of the Yarra River range from potable water supply in the upper regions through the range of fishing, bathing, irrigation, cooling, to navigation at the mouth. It is important to control and limit waste discharges so that the quality of the water in any section is maintained at the appropriate standard for its beneficial uses, even in periods of low flow.

Knowledge, both of the ambient water quality, which will be obtained by a monitoring programme, and the wastes which are being discharged, is necessary before setting licence conditions for discharges which will ensure that the river will be restored to and maintained in the condition that has been laid down in the draft policy.

It is very apparent from the work to date that a large part of the problem with polluted streams flowing into Port Phillip Bay is the result of housing development in unsewered areas where soil conditions and area of land are not adequate for satisfactory septic tank effluent or sullage absorption. Issue of further building permits in some critical areas has been frozen. It is now expected that permits will be issued provided an all-wastes septic tank, grease trap, sand filter, and chlorination unit are provided, together with satisfactory arrangements for regular maintenance of all items in the system.

A similar attitude is taken as regards emissions to atmosphere where the concept of "Airsheds" will be applied. Again it will be necessary to obtain a great deal of data on ambient air conditions, and a start has been made in establishing monitoring stations in the Melbourne area. The first such station has been operating since the end of July 1973 and daily observations have been published in the press. There should be three fixed and two mobile stations operating by July 1974. The value of data from such observations is obvious when evaluating the effect of emissions from a new operation such as the projected Newport Power Station, which will make substantial additions to the atmospheric waste load over Melbourne. Conditions for operation of the station have been set by the Authority following licence applications so that design and construction can proceed.

The Clean Air Act (1958) was the legislation under which emissions to the atmosphere were controlled before the EPA Act became law, and this Act made provision for the use of the best practicable means of reducing emissions, and for the submission of plans and specifications for new installations to the administering body. To a large extent, this procedure is being followed in the licensing system. Under some circumstances, however, licence conditions may contain specific conditions, e.g., the design and operating conditions of an afterburner to reduce an odour where measurement for control is difficult.

Pollution of land and waters both surface and underground from the deposition of waste materials onto the ground is controlled by licensing the deposit of wastes onto land. Such wastes can be solid or liquid wastes generated in a factory and disposed of on the factory's own land, or household and commercial waste deposited in sanitary landfill operations.

In Victoria, by far the most commonly used method of disposal of household and commercial waste is by landfill, and in the Melbourne metropolitan area, there have been a large number of basalt, clay and sand quarries which have provided convenient disposal sites. The licensing system aims at setting conditions of operation which will raise the standard and thereby reduce the nuisance caused by pollution from wind blown debris, odour, etc. Part of the problem in Melbourne has been the realisation that substantial aquifers underlie the basalt and sand, and there has been considerable concern that liquids may leach harmful materials from the landfill and reach the aquifer, causing pollution. This possibility has led to many councils banning the deposit of liquids into tips, even the few remaining ones in clay pits which are very suitable for the disposal of liquids and hazardous materials because they are generally not connected to an aquifer and pollution of surface waters is unlikely. The EPA in conjunction with the Mines Department is conducting a drilling and sampling programme to determine to what extent leaching and pollution of an underlying aquifer actually occur under Melbourne conditions.

It had been hoped that construction of a plant to treat hazardous and difficult industrial wastes would have been well advanced by now, but at this stage (1/1/74) the proposal seems to have been shelved by the Government, even though the Melbourne Metropolitan Board of Works had recommended a tender. The plant would have included incineration, chemical, and biological methods of disposal. As at present, we are largely dependent upon one large commercially operated tip which functions as a bio-bed and which has facilities for treatment of leachate in ponds by aeration.

Consideration has been given to other methods of disposal, e.g., incineration, but the low cost of landfill, while sites are available, would seem to preclude the building of incinerators except for certain specialised purposes.

The charge (or cost) per tenement in the Melbourne area ranges from \$7 to \$14, with an average generation per tenement of about 500 kilograms per annum of waste. Collection costs generally range from \$6 to \$9 per tenement, so that disposal costs are relatively low compared with recent incinerator costs.

Little consideration is felt to be warranted at the present stage for reclamation proposals involving separation because of the very high cost of operation, and the low value of products recovered. In Melbourne, there is already a very effective recovery of glass before garbage reaches the tip, waste paper collections produce more carton and paper waste than the paper companies can consume, and other scrap is too small in amount to warrant the separation costs.

The Authority has been faced with the task of preparing a plan for the long-term disposal of Victoria's and in particular Melbourne's garbage. Data has been collected by means of a questionnaire from municipalities on generation of garbage, now and in the future, together with information on their disposal facilities. At the same time, a survey of potential landfill sites being created by extractive industries in the metropolitan area has been made, and this has shown that provided these need to adopt sophisticated high pressure comsites can be used for refuse disposal, there would be adequate space for many years, without the need to adopt sophisticated high pressure compacting techniques. Arising from this study, regional groupings of municipalities have been developed with the aim of putting forward to municipalities a plan for regional disposal of refuse. There are many municipalities which have no available landfill sites within their municipality while others have very large volumes available and being generated. Some of these municipalities are already accepting garbage from other municipalities, and it is considered that there is no reason why this practice should not be extended by the regional plan so that all municipalities have access to landfill sites on a long-term and planned basis.

The Investigations Branch of the Authority is responsible for enforcing licence conditions and investigating complaints regarding pollution. It has a force of mobile trained inspectors under radio control to do this. To date, there have been three successful prosecutions for discharges, with substantial fines being imposed in two cases. These prosecutions were for untoward or accidental discharges. There has been criticism that there have not been more prosecutions, but the Authority's success is not to be measured through the number of prosecutions, but by the improvements brought about by the licensing system. There is no doubt that many industries have improved the standard of their discharges as a result of the introduction of the licensing system.

Provision has been made in the Act for appeals against the issue of licences by third parties, and for appeals by applicants for licences either against the conditions of a licence when issued or non-issue of a licence. Depending upon the nature of the appeal, it may be heard by the Authority or by the Appeal Board. There have already been a number of appeals, both second and third party, and it seems likely that there will be a considerable number of third party appeals for some time to come, as licences are issued for some discharges which have aroused sectional public interest.

Victoria faces the usual problem of littering through careless discharging of small packages and containers as well as the roadside dumping of garbage. Although the Litter Act gives Councils power to take offenders to court, there are difficulties which have made the number of prosecutions under this Act relatively small. It is considered that public education, plus public disapproval of the litterer, will be the best means of reducing this nuisance.

The Authority has during this summer undertaken a small radio campaign against littering. Where Councils provide adequate litter bins which are cleaned regularly and keep an area clean, the public appears to be co-operating. The Authority is also stressing to Councils in its proposals for regionalisation of waste disposal, the importance of providing convenient disposal facilities for householders with special waste problems, in order to reduce illicit roadside dumping.

There can be little progress where reliance is placed on maintaining the status quo, and all branches of the EPA have a research section to deal with particular problems in their field. Many of the problems of pollution have arisen because of improper land use, and a small planning and research branch has been established to deal with the environmental and pollution aspects of changing land use.

PAPER 5

Collection, Handling and Disposal of Wastes in the National Capital

By J. G. Henty*

Introduction

Canberra resembles the other 17 federal national capitals throughout the world in that it does not have an elected local government. Consequently, it does not have any comprehensive legislation similar to the local government acts of the various states under which local administration operates within the strict framework of the law. Legislation in the Capital Territory is in the form of ordinances, or regulations authorised by those ordinances, made by the Governor-General under the provision of Section 12 of the Seat of Government (Administration) Act of 1910-1973. In addition there are Commonwealth and New South Wales Acts which apply in the A.C.T.

Section 12 provides that the Governor-General shall make ordinances which shall be notified in the Australian Gazette and shall take effect from the date of notification or such other date as is specified within the Ordinance, or by notice fixed by the Minister in the Gazette. This Section further provides that all ordinances shall be laid before the Houses of Parliament for 15 sitting days, during which time the ordinance, or part of it, may be subject to a motion of disallowance. This motion should be withdrawn, debated, or disposed of within a further 15 sitting days, otherwise the ordinance or part thereof specified in the motion is deemed to have been disallowed. Should another ordinance or a provision in another ordinance similar to the ordinance or part of the ordinance disallowed, be presented to the Parliament within six months of the date of disallowance of the original ordinance, the second ordinance (or provision of the ordinance) shall be considered as void. All regulations made under the provisions of various ordinances are also subject to the 15-day scrutiny period by the Houses of Parliament. The authority for this provision is contained in the Acts Interpretation Act 1901-1973.

Regulations differ from ordinances in that the responsible Minister may make the regulations whereas the ordinances bear the signature of the Governor-General. It is interesting to note that the definition of regulations includes rules and by-laws.

For administrative purposes the Government from time to time gazettes an administrative arrangements order which, in effect, is a list of all Acts of Parliament and the Ministers who are responsible for their administration. The administration of the Seat of Government (Administration) Act of 1910-1973 is the responsibility of the Minister for the Capital Territory. It is on this legal base that all Canberra's administration, including the control, handling and disposal of wastes, is founded.

There is one further check on delegated legislation. The Senate Committee on Ordinances and Regulations reviews all ordinances and regulations as well as those made by various Commonwealth Territories and Authorities. Primarily, the function of this Committee is to ensure that the rights of citizens are not in any way impinged upon or constrained. Over a number of years it has become increasingly apparent that the Committee frowns on administrative and ministerial discretionary powers. The Committee reports to the Senate which invariably approves its recommendations. This in turn necessitates the remaking of the disallowed section of the legislation.

From this it can be seen that it could take a considerable length of time for ordinances and regulations to come into effect.

Another aspect in which Canberra differs from other Australian cities is that the Minister for the Capital Territory is not solely responsible for the full range of administrative functions carried out by local Government authority. For instance, the Minister for Health is responsible for all health matters. In addition, the Attorney-General, the Ministers for Urban and Regional Development, Social Security, Housing and Construction, and Education all have some responsibility in the planning and administration of the Capital Territory.

* City Manager's Office, Department of the Capital Territory, Canberra City.

Amid this diverse framework there has developed the system of administering the collection, handling and disposal of wastes by administrative decision and ministerial order.

CLASSIFICATION AND PROCEDURES

Collection of Wastes

(a) Street, highway, shopping centre and public places wastes and litter — collected by Departmental staff or contractors to the Department of the Capital Territory.
(b) Household garbage — collected by contractor to the Department.
(c) Traders' wastes — collected by contractor to the Department.
(d) Industrial wastes — waste oil delivered to Departmental disposal area by generators of the waste.
(e) Automotive wastes — delivered to Departmental disposal area by the generator of the waste.
(f) Householders' wastes — delivered to Departmental disposal areas by householders.
(g) Sewerage — this matter is handled for the Minister for the Capital Territory by the Department of Housing and Construction.

Each of these functions will now be considered in more detail.

Wastes in Public Places

The Department employs a staff of 22 daytime cleaners who collect litter from all roads, highways, unfenced playing fields, parks, picnic areas and unattended public toilets under a strict rotational service schedule. In addition, a small team of cleaners attends to specialised cleaning tasks such as bus shelter sheds, highway underpasses, fountains and the lake foreshores. A further task force of 42 cleaners, working between midnight and 8 a.m., cleans all the City district and suburban shopping centres on a seven day a week basis. The tasks undertaken by this force could be briefly summarised as follows:—

- sweeping pavements, plazas and gutters,
- emptying litter bins,
- hosing all pavements and plazas twice weekly,
- sweeping and cleaning all service lanes and courtyards.

Household Garbage

The Department employs a contractor who, at this time, has some 18 months remaining of a five-year contract. He is required to collect household garbage from all households and schools within the City on a twice-weekly schedule. Householders are allowed two garbage bins for each collection. The costs of this service are borne by the Municipal Account.

Traders' Wastes

Once again this is a contract service. The existing contract has six months to run. Unlike the household garbage service, this service is supplied on request up to a maximum of six days a week. There are no limits on the amount of wastes which a trader may place for collection. Costs for this service are borne as a component of the Municipal Account.

Industrial Wastes

Waste oil — the Department requires the generators of this waste to deliver it to a special disposal area.

Automotive wastes — the same procedure as that concerning waste oil applies to all wastes generated by the automotive industry.

Housing industry — as for waste oil and automotive wastes.

Householders' Wastes

The Department operates five tips for household wastes, located in strategic positions throughout the City. To these tips householders deliver all wastes other than household garbage. The average depositing by householders in each of these tips is approximately 100 cubic yards per day. Tips are open from 7.30 a.m. to 6 p.m., seven days a week.

Sewerage

Since the Department has no professional or engineering staff working in this area, all such work is delegated to the Director, Department of Housing and Construction, who operates sewerage treatment works on behalf of the Department of the Capital Territory.

Putrescible Wastes

One major disposal area is operated for the receipt of all putrescible wastes. These wastes are deposited in trenches 100 metres long, 15 metres wide and 7 metres deep. A D8 type bulldozer is employed six and a half days a week to continuously compact and cover the deposited material with clean soil. All materials are so treated before close of work on the day of depositing. All putrescible household garbage and traders' wastes are delivered to this area.

Because of the rapid growth of the City it has been necessary to consider opening another disposal area of this type. An area of some 250 acres has been identified in the north/west sector of the City. Environmental impact, hydraulogical and geological studies have been completed and are at present being evaluated. It is anticipated that this disposal area will be available for use in 1975/76.

Householders' Tips

Six landfill areas strategically located throughout the built-up area are operated for the benefit of householders and the building industry. These tips perform a cosmetic function by ultimately improving the natural land form. They are operated either from the top or the base of the tip face depending on the physical attributes of the area. Clean soil and builders' rubble are used to cover the wastes progressively while they are being deposited. D6 class bulldozers are employed in these areas, one bulldozer alternating between two tips.

All tips are manned by an attendant for 10 hours a day seven days per week. No charges are made for the depositing of wastes.

Suburban Refuse Hoppers

Due to the shortage of land suitable for cosmetic treatment, the Department, in conjunction with the planning, design and construction authority, the National Capital Development Commission, will shortly be taking delivery of an experimental suburban refuse hopper system. In essence it is a specialised transfer station for use by householders. This station will be fitted with six 40 cubic yard transfer bins of the dinosaur type. Two-level road construction provides for depositing on the higher level and transport of the bins from the lower level. Bins will be roofed over and an automatic fire sprinkling system installed. First of these stations, located in the suburb of Ainslie, will open for public use in April 1974. It will be in full operation by the time this conference is held.

Further plans for the extension of hopper installations have been deferred until sufficient operating time has elapsed to allow for the evaluation and costing of the system.

Waste Oils

At the present moment all waste oil generated in the Territory (some 300,000 gallons per annum) is delivered to the oil disposal area, which is adjacent to the deep trench putrescible tip site. The soil in this area is quite absorbent. Oils are placed in a compartmented "turkey nest" structure which allows the sludge and water to settle before the oil moves through the next partition, and so on, whence it is absorbed into the soil. The subterranean rock structure in this area slopes away from all creeks, streams and watercourses, thus minimising danger of contamination. This practice has been used for a number of years without any adverse effects to date.

That is not to say that the Department is complacent with the system. Plans are well advanced for the installation of an oil recycling plant.

Automotive Wastes

These can be classified into their two main component parts — used tyres and reject or old cars:

Tyres — Used tyres are at present buried in the disposal area. Tenders have been called for the supply of a tyre shredding machine to reduce tyres to more manageable sizes. At the present time no use for the end product has been found. Experiments are being undertaken in this regard. Such things as drainage membranes under sports grounds, use under playground equipment, and as an additive to the bitomastic hotmix are being considered.

Old Car Bodies — All derelict car bodies in the Territory are delivered to a special car body disposal yard operated by the Department's contractors. The bodies are crushed on the site and are transported by rail to Sydney where they are processed by a fragmentiser located at Mascot. The Department, under the terms of the contract, pays all freight costs.

Cost of Services

All costs for the services mentioned above are borne in the Municipal Accounts which are balanced by the Municipal General Rate.

LOOKING TO THE FUTURE

Having regard to the manpower costs in all cleansing, waste collection and disposal operations, the Department is continually investigating the feasibility of purchasing suitable new machinery and updating its wastes handling practices.

The household garbage service, the cost of which this year will be in excess of $600,000, is the subject of considerable investigation and experimentation. Last year an experiment was conducted over an extended period with the aid of the Plastics Institute of Australia, the Contractor and the Transport Workers' Union into the use of plastic garbage bags. This year a similar experiment, using disposable paper garbage bags, is to be undertaken.

Specialist consultant reports are prepared for the Department by the National Capital Development Commission. The subjects of two recent reports were the containerisation of traders' wastes, and an investigation into the suburban refuse hopper concept. As mentioned above, the refuse hopper concept has been accepted by the Department and a trial plan will be in operation during April.

Any viable methods of altering the household garbage service or the containerisation of the traders' wastes service must of necessity be phased in with new contracts.

CONCLUSION

A paper on this topic would be incomplete without reference to recycling of re-usable materials. It is proposed to conclude, therefore, with a brief description of the Department's present program and future plans for recycling.

For many years provision has been made at all departmental tips for the depositing of bottles in areas separated from the main tip face. The Boy Scouts' Association has the franchise for all tips. A departmental contractor some years ago commenced collection of paper. His operation has grown over the years from a sideline to a full-time occupation in which he employs staff and uses a number of vehicles.

Early in 1972, as a result of discussions held with the Research Re-cycling Group of the Society for Social Responsibility in Science, the Department opened an experimental recycling depot for paper, glass and tins in the car park at Waramanga Shopping Centre. Records were kept of weekly depositings and an evalution in December 1973 showed the following averages of:—

1. Paper .... 2 tonnes
2. Re-usable bottles .... 82 dozen
3. Cullet glass .... 2/3 tonnes
4. Cans 1500-1800

The experiment was enlarged to include two older district shopping centres in February and to the five remaining suburban shopping centres in the Weston Creek area in March this year (1974).

After suitable period of time the results of all experimental plants will be evaluated and a decision made on whether to provide recycling facilities at all existing and proposed shopping centres.

The Department realises that the greatest saving of material for re-use would be achieved if separate household collections were made for glass, paper and tins. However, costs for this type of service would be prohibitive. The present experimental program at shopping centres is aimed at encouraging the householder to deposit re-usable material at the depot before embarking on the weekly shopping; in other words, the object is to have people with full shopping bags both to and from the shops.

PAPER 6

Waste Management in Tasmania

By Brett E. Struck*

INTRODUCTION

This paper reviews the waste management problem in Tasmania and outlines legislation and administrative organisation for the control of waste disposal within the State.

Climate and Topography

The State of Tasmania is a group of islands with a total area of 68,300 km^2 lying some 240 km south of the Australian mainland at approximately 42° south latitude and 146° east longitude.

Of the islands, Tasmania is the largest and most populated. It is mostly mountainous and cut by steep river valleys, particularly in the west and north-east, although a few flat, open areas exist along the north-west and east coasts and in the northern Midlands.

As the island lies in the latitudes known as the "roaring forties" weather can be described as variabl[illegible]e more sustained periods of inclement w[illegible]er are likely to occur in the late autumn, winter and early spring. Rainfall is heaviest in the west and north-east of the island, falls of 250 mm and 150 mm a year respecti[illegible]y being common; eastern areas of the islan[illegible]ceive approximately 75 mm a year on the average.

Snow, alt[illegible]h not uncommon in winter, is seldom se[illegible] low elevations where the majority of people live.

Population

The population of Tasmania is about 400,000, with the main centres being located at Hobart in the south, Launceston in the north, and Burnie and Devonport in the north-west. The overall population density is high compared with other States, but over 90 per cent of the State is uninhabited bush or agricultural land. The north-west coast is the most highly settled area other than the metropolitan areas of Hobart and Launceston, the population of which is 150,000 and 62,000 respectively.

Estimates for the next 30 years suggest that the State's population will approach 600,000 by the year 2000. At that time, due to increasing urbanisation, a projected 90% of the population will live in urban areas compared with 73% in 1970 and 50% in 1930. An interesting fact is that the present urban population occupies less than three per cent of the land.

Waste Production

The major sources of wastes in the State are mining and metallurgical operations, and mullock and tailings from mining activities in the west and north-west make up the bulk of these wastes. The remainder of the 10 million tonnes a year generated by these industries is produced as waste from refining operations on the north coast and in the Hobart area.

Other significant sources of waste are wood product industries such as logging operations, sawmills and paper manufacturers; food processing industries; orchard pruning activities; operations involving intensive raising of livestock and poultry; commercial enterprises such as car wrecking firms, tyre sales firms, and service stations and other waste oil producers; various institutions such as hospitals and schools; and finally, domestic sources (which contribute some 110,000 tonnes a year to the waste stream).

Problems in Collection and Disposal

The types of wastes which must be handled in the State vary considerably, ranging from insoluble, inorganic solids such as some mullock and tailings through highly putrescible organic wastes such as food processing wastes and from inflammable liquids such as solvents or oil to corrosive liquids like acids and caustics.

Many of these wastes present unique problems in collection and/or disposal. For instance, rural disposal areas are often littered with discarded corrugated iron water tanks and broken farm implements. These take up a disproportionate amount of space in relation to actual numbers and present problems in crushing.

The trend toward increased concentration

---

*Waste Management Officer, Department of the Environment, Tasmania.

and quantities of manure from intensive raising of livestock and poultry and toward the use of easily handled and readily available artificial fertilisers is resulting in a growing problem with disposing of manure without causing pollution.

The growth of the oil disposal problem is proportionate to the increased use of motor vehicles. Many councils refuse to take waste oil into their disposal sites and most methods of disposal other than incineration or reclamation are unsatisfactory because they pollute soil and underground water. To help combat this situation the oil companies are setting up a reclamation project which, though initially covering only the Hobart area, will eventually be extended to the rest of the State.

Water pollution problems in the form of silting, heavy metal pollution and changes in pH arising from the disposal of mullock and tailings are considerable. The very bulk and permanency of this waste is also troublesome because disposal can take up large areas, leaving ugly blots on the land. The increasing use of properly engineered tailings dams, disposal of tailings and mullock into worked-out sections of mines and a regard for landscaping will help eliminate these problems.

Some wastes, such as jarosite, a thixotropic product of zinc refining, are difficult to handle and have no adequate disposal sites available within a reasonable distance. In fact, with some of these wastes the question is whether any disposal sites are really safe. At present, as an alternative solution, when there is easy access to the sea the waste is barged to deep water and dumped. This method is used by two Tasmanian companies to dispose of large volumes of materials. But in view of the impending restrictions under the International Convention on Ocean Dumping, applications to establish other such operations are unlikely to gain the support of the Department of the Environment.

Domestic and commercial wastes present much the same problems in collection and disposal as are encountered elsewhere in Australia. The small scale of operations, even in the larger centres, would seem to preclude more sophisticated refuse handling and reduction equipment such as incinerators and pulverisers in the foreseeable future. In rural areas, distance, too few people and insufficient funds impede disposal.

Finally because radioactive and pathological wastes are generated in small quantities they present no disposal difficulties.

Legislation

Tasmania has two Acts which cover waste management — the Litter Act 1973 and the Environment Protection Act 1973. A third Act, the Statute Law Revision (Environment Protection) Act 1973 amends several earlier Acts to bring them into line with the Environment Protection Act 1973.

The Litter Act 1973

This Act provides uniform anti-litter legislation for the whole State and uses officers appointed by local government and police officers as its principal enforcers. Two philosophies prompted this arrangement:

(1) Justice, and provision of an atmosphere of uniform respect throughout the State toward the control of litter, inviting a need for uniform legislation which could best be achieved by State action.

(2) The need for sections of government most immediately concerned with the problem to be the principal enforcement agencies. These parties are the police and local authorities, who have grass roots contact with people.

In order to achieve a cleaner environment, officers authorised under the Act are charged with the responsibility of apprehending persons violating the Act or its Regulations.

In turn, the Act charges all persons with the responsibility of not causing litter. Persons selling commodities in no-deposit or refund, non-returnable containers for consumption in or near the place of sale have to provide readily available receptacles for containers used in or near their premises. Persons issuing tickets or other documents enabling entry to any place for admission to which a charge is made have a similar responsibility. These regulations further specify conditions for the containers supplied, their placement and emptying.

As stated previously, council police officers enforce the Act, but other persons also have these powers. These persons are: (1) all authorised officers within the meaning of the Environment Protection Act, 1973; (2) all bailiffs of Crown lands appointed under Section 13 of the Crown Lands Act, 1935.

Under the Litter Act an authorised officer has these powers: Where there is a contravention of Section 4(1) of the Act (prohibits littering except in certain conditions listed) he can: (1) order the offender to remove the litter or to place it in a receptacle intended for that purpose; (2) issue a notice indicating the offence and requiring a payment of 10 dollars within 14 days; (3) issue a complaint which could lead to fines of up to $500.

The Act also gives courts the power to recover expenses for a person who has to spend money to clear offenders' litter from his property. The court has an alternative of ordering the offender to remove the litter. Breach of the Regulations carries fines of from $15 to $50.

The Act also provides that the Governor may make other regulations, with fines up to $200, which would regulate the use and disposal of: (1) boxes, bags, packets, bottles, tins and other containers; (2) papers; and (3) things of no use or further use to their owner, so that their use or disposal will not contribute to the increase nor the pollution of the environment within the meaning of the Environment Protection Act 1973. No regulations of this nature are contemplated in the foreseeable future.

The regulations came into effect on 18th December, 1973, and the various municipalities have appointed authorised officers. As yet no prosecutions have occurred under the Act or the regulations, but notices of infringement were not available until late January, so this is understandable. As the Act is more widely applied councils with particularly bad litter problems are expected to apply its powers vigorously, while others may seldom serve a notice.

Environment Protection Act

Scope. — The Environment Protection Act was proclaimed to provide adequate powers to control air, water and noise pollution, and waste management problems.

The need for the Act became evident after the 1969 Australian Senate Committee Inquiry into the Environment which pointed out, among other things, that: (1) Tasmania was the only State without formal clean air legislation; (2) effective pollution control was impossible in Tasmania because of (a) divided authority (anti-pollution legislation was scattered through numerous Acts) and (b) a lack of scientific standards on which to base prosecutions.

This is not to say that there was complete chaos in the State. Good efforts were being made, which set up a code of practice for operation of waste disposal sites and worked hard to bring about a change in the existing practice of burning, loose tipping, and sparse covering. As a result of the Senate inquiry, the Tasmanian Government decided to rectify the situation by placing the responsibility for most environmental problems in the hands of one Department, the Department of the Environment, which would act under one Act, the Environment Protection Act 1973. This Act supplants all previous legislation in conflict with it. However, two other Acts which prevail over the Environment Protection Act where conflicts exist are: (1) the Oil Pollution Act 1961, which gives the Marine Board power to deal with oil spills in a marine environment; and (2) the Public Health Act 1962, which charges the Division of Public Health with the responsibility of looking after matters concerning public health.

Both agencies continue to function in areas of previous interest: the Marine Board by itself, as before, and the Division of Public Health in conjunction with the Department of the Environment where environmental and health matters interact.

In this context, the Division of Public Health retains responsibility for health matters and the Department of the Environment sets comprehensive conditions relating to environment protection and enhancement including health protection. But the two departments liaise to ensure that the conditions set by the Department properly state the requirements of the Division of Public Health.

The Act begins by charging the Director of Environment Control with these broad responsibilities: (1) protection of the environment in Tasmania; (2) ensuring the control or prevention of any act or emission which causes or is capable of causing pollution; (3) coordinating all activities, governmental or otherwise, as are necessary to protect, restore or improve the environment.

The Director has the following additional responsibilities in enforcing the Act: (a) To consider means and initiate steps for the protection of the environment and for preventing, controlling, abating, or mitigating pollution; (b) to investigate problems of environmental protection; (c) to obtain the advice of persons having special knowledge, experience, or responsibility on environment protection; (d) to keep under review the progress made in attainment of the objects and purposes of the Act, and to publish reports and provide information to increase public awareness of such progress and of the problems and remedies that exist in relation to the protection of the environment; (e) to advise on standards and criteria and methods of sampling and testing to be used for any purpose; (f) to promote, encourage, co-ordinate, and launch short- and long-term plans and projects in environmental protection together with or separate from other State Departments and instrumentalities; (g) Generally, to administer and give effect to the provisions of the Act and to perform such other functions as may be prescribed.

Potential polluters are charged with responsibilities under the Act. Some of them are: Industries, manufacturers, producers and facilities such as mines, chemical works, abattoirs, refuse disposal sites, sewerage treatment plants, etc., which are listed in Schedule *1 of the Act, must obtain a licence issued by the Director in order to operate under the Act. All such scheduled premises must operate within the conditions of the licence and inform the Director of proposed changes in plant or operations which could lead to emission of a pollutant or substantially increase the emission of a pollutant. All persons, industries, etc., must obey the conditions of the Act and its regulations, which include compliance with standards of operation of waste disposal sites, e.g., no open burning to be carried out on waste disposal sites; compliance with standards relating to acceptable emissions of pollutants into the environment with regard to the air and water pollution sections and the noise control section of the Regulations.

The Act provided for an Environment Protection Advisory Council whose responsibilities under the Act are to: (1) advise the Minister when se seeks its advice; (2) make recommendations on matters where the Act provides for such a recommendation; (3) make recommendations to the Minister or Director on any matter affecting the purposes or administration of the Act.

Also set up under the Act is an Environment Protections Appeals Board whose responsibilities are to hear appeals concerning: (1) conditions, limitations or restrictions on licences; (2) refusals to grant or renew licences; (3) revocations of licences; refusals of the Director to approve of changes in processes or construction installation alterations, etc., changes in materials dealt with or used or increases in the quantities of materials dealt with or used which may lead to emission of a pollutant or increases in the emission of a pollutant; or (4) notices served by the Director or by a municipality requiring abatement of a pollutant or noise being emitted from land occupied by the appellant.

Enforcement

The Act is enforced by authorised officers drawn from the Police and from State and Local Government Departments.

The enforcement of the Act is divided into two sections: (1) enforcement of the Act concerning unscheduled premises, which is the area of operation of local government appointed officers and the basic area of operation of police officers, particularly with regard to motor-vehicle exhaust and noise emissions; (2) enforcement of the Act with regard to scheduled premises, which is the basic area of operation of Department officers.

Officers enforcing the Act on unscheduled premises will be concerned primarily with determining that the levels of emissions of pollutants do not exceed those set by the Regulations and that any changes directed by the municipality are made. These officers will liaise with the Department for help where they need it.

The Department officers will be enforcing both the Regulations and the conditions set on licences for operation of premises scheduled under the Act.

In enforcing the Act the Department uses the following staff of 10 technical personnel in addition to the Director and the Executive Officer:

Testing and Analysis Branch — made up of five personnel including the senior chemist, an analyst, two monitoring officers and a laboratory assistant. This branch does routine air and water pollution monitoring and analyses as well as spot tests and special analytical work to support the Research and Enforcement Branches.

Research Branch — consisting, at present, of a Research Officer — evaluates environmental impact statements required by the Department and acts as the Department's adviser regarding these. He also does such research as is needed by the other two branches, or by the Director, whether it is of the literature or involves test apparatus and extensive studies.

Enforcement Branch, itself, is comprised of an Air Pollution Officer, a Noise Control Officer, a Water Pollution Officer and a Waste Management Officer. These officers investigate complaints, act as advisers for the Department, other Government bodies, industry and the general public. They sit on technical subcommittees appointed by various State and National Councils which are oriented toward environmental protection, e.g., Australian

Environment Council, Australian Water Resources Council, National Health and Medical Research Council, and Environment Protection Advisory Council (Tasmania). Finally, they inspect premises, hold discussions with licencees regarding enforcement and set the conditions of licence for operation of premises scheduled under the Act.

The Regulations (Air, Water, Noise and Waste) which form a basis for prosecution under the Act, have been drawn up and are currently under review by the Environment Protection Advisory Council before recommendation for adoption by the Government. The licensing of scheduled premises is limited at present, because the Act provides that only premises which came into being on or after 26th July, 1974, can be licensed before 1 January, 1975.

Powers

The Act sets forth certain powers to assist authorised persons with its enforcement. These powers are wide ranging and cover the followign areas: (1) Entry at any time with or without assistants, vehicles and equipment to any land from which waste, pollutant or noise is emitted or likely to be emitted; to (a) examine and inspect equipment, (b) take and remove samples, (c) make such examinations, tests or enquiries to determine whether the Act, Regulations or licences and conditions thereof are being complied with. (2) Power to take any photographs he considers necessary. (3) Power by notice in writing, to require: (a) the occupant or person in charge to produce reports, books, maps or documents relating to emission of pollutants or noise, processes being used on that place, and data from monitoring equipment giving information on waste discharges, etc.; (b) any person or body to produce any books, reports, etc., in their possession or custody relating to any equipment, works, etc., used for the discharge, deposit or emission of waste.

(4) Power to: (i) quesion alone, or in the presence of another, with respect to matters under the Act, any person he finds in a place he enters, and require him to answer any questions and sign a declaration as to the truth of the answers; (ii) require any such person to tell who the occupier of the site is.

(5) Power to: (i) by notice in writing, require owners of motor vehicles to bring them in for inspection; (ii) stop any motor vehicle and inspect it; (iii) require the driver of a motor vehicle to stop the vehicle for purposes of inspection; and (iv) inspect any motor vehicle to ascertain whether it complies with the Act.

(6) Power to board vessels and proceed as in (i) above with respect to premises on land.

The Director has certain powers under the Act as well as in his function as an authorised officer. He may: (1) require applicants for licences to give such plans, specifications and descriptions of emissions and other information as he may require; (2) grant an applicant a licence to operate: (a) unconditionally, or (b) subject to such conditions, restrictions or limitations as he sees fit; (3) refuse to grant a licence to operate; (4) renew a licence (a) as originally issued, or (b) with added and/or varied conditions, etc.; (5) set conditions for implementation at a future date; (6) give or withhold approval for changes in processes, nature of materials used, quantities of materials or in structures or equipment which may give rise to the emission of a pollutant or noise; (7) require applicants for transfer of licences to produce evidence of the applicant's right to operate; (8) give consent to such a transfer.

The Act, in addition to the powers given to the authorised officers and the Director, gives the courts the power to levy fines and impose prison sentences under the Act. These fines range from \$500 plus \$50 per day for noise pollution to \$5000 plus \$1000 per day for allowing the discharge of a pollutant into the air or water or onto or into land where a standard is not being met and/or the best practicable methods are not being used to control the pollution (where standards do not exist). Breaches of confidence gained under the Act are punishable by \$5000 or two years' imprisonment or both.

Powers also exist for declaration of a wide range of regulations as the need arises.

Progress

The Department's efforts to date have been directed mainly to an understanding of the problems to be encountered in the State and familiarisation with individual premises preparatory to licensing when the time comes.

In the field of waste management, the Department has circulated a survey of the waste collection and disposal practices of the State's 49 municipalities in an attempt to obtain a more accurate picture of the situation. To date, close to 50% of the authorities have replied and it is hoped that the remainder will answer soon. The results of the survey, with the results in years ahead, should supply the necessary data for analysing trends in collection and disposal of wastes, such as (1) needs for regionalisation of collection or disposal or for co-operative efforts by council in this regard; (2) the effectiveness of the Department in promoting proper waste disposal and collection; and (3) the rise in volumes and types of wastes over the years.

The future may yield projects, or possibilities for projects, along the following lines:

(1) **Amalgamation of waste collection facilities** and/or equipment either cutting across municipalities to form a regional authority or as a co-operative of municipalities. In Tasmania, where the majority of the 49 municipalities are rural and are under-financed for waste management, this could be a blessing. Many municipalities have limited or non-existent collection of waste, and disposal sites are given maintenance only when funds become available. By the pooling of resources the same amount of money should be available to provide more regular collection and disposal services, in addition to the equipment to perform the services. A route could be worked out to ensure that every disposal site would receive maintenance once every two weeks or once a month, etc., and every town could have collection once every week (or so).

(2) **Creation of all waste disposal sites as closed sites.** Although it is argued that this procedure will not work, authorities which have adopted it report that it permits good control once an initial spate of illegal dumping has been weathered. In conjunction with this concept, perhaps opening hours could be introduced in the evenings and on weekends, with fewer daytime openings to allow better access by the homeowner.

(3) **Resource recovery** — this is another area in which the Department is interested, and as time goes by it should assume an increasingly important role.

While it is uneconomical to ship such waste products as paper to the mainland for recovery, there is the possibility that locally based industries selling to local demand (and elsewhere when economics allow) could use locally produced wastes for feedstock and produce competitive products. Much research into markets, technology, costs, fluctuations in supplies, demand and their possible effects would have to be done before this could be proved (or disproved). Some of the possibilities are rubber tyres — hoses for cars, carpet underlay, boiler feedstock, insulation, etc.; aluminium — castings of a non-critical type; paper — new newsprint, shredded paper for packing goods, inclusion in paper made on the island for use by the Government, manufacture of egg cartons, etc.; glass — use in highway construction, re-use for bottles; oil — reclaim and use for engines or boiler feedstock; and

(4) **Education** — production of a series of films, lectures and reading material dealing with aspects of waste management and resource recovery, instilling the desirability of well-engineered and well-managed projects to obtain the optimum results, in addition to acting as an aid to teaching operators proper techniques.

Much lies ahead under this Act that can be a challenge and an exciting chance to influence the patterns of waste management for the State and the nation for years to come.

Statute Law Revision (Environment Protection) Act 1973

Rather than reviewing all the revisions, most of which are of a minor nature, comment will be limited to a revision of the Local Government Act 1962, which states that henceforth municipalities may license "persons contracting to remove trade wastes from places in the municipality" and have the power in the case of licences under paragraph 6A of this part (the previously stated revision) to impose conditions as to the places, within or without the municipality, to which trade wastes may be removed. The responsibilities, enforcement and powers are readily appreciated, but as will be seen, the effects of this Act could not come under the heading of Progress.

An operator collecting trade wastes from the Hobart hinterland could well be required to license his vehicles or his business in 18 municipalities, and an operator working out of Launceston could require at least 12 licences. Compared with the possibility of one licence, with standard conditions set by the State and enforcement by local government, the revision is a step backward. An amendment of this Act is clearly in order to prevent misuse of a worthwhile device — i.e., licensing of waste haulers to ensure clean operation.

CONCLUSIONS

The various Acts mentioned in this paper have yet to make their full impact felt, and it is hard to predict what that will be. However, it is safe to say that with the Environment Protection Act at least a cleaner environment will result as all sections of government and business begin to put their houses in order. Whether the Litter Act accomplishes its aims will depend largely on the willingness of local councils to exercise the powers the Act affords them.

Much needs to be done to bring about a cleaner environment in Tasmania; it will only be achieved through co-operation, open communication and thorough planning. The Department is committed to this and invites others to join it in meeting the challenges ahead.

REFERENCES

1. Department of the Environment (Tasmania): "Environmental Pollution in Tasmania", Government Printer, Hobart (1972).
2. State of Tasmania: "Environment Protection Act and Regulation 1973", Government Printer, Hobart (1973).
3. State of Tasmania: "Litter Act and Regulations 1973", Government Printer, Hobart (1973).
4. State of Tasmania: "Statute Law Revision (Environment Protection) Act 1973, Government Printer, Hobart (1973).

PAPER 7

Towards Effective Solid Waste Management in the U.K. —Legislative Developments

By N. Y. Kirov*

1. Introduction

Growing public concern in the U.K. about the environment has focused attention in recent years on the need for vigorous Government action. In consequence, substantial administrative and legislative measures have been introduced to improve the environmental services and to provide a more efficient and rational basis for the management of wastes from all sources.

In 1971, the Ministry of Housing and Local Government was reconstituted into an enlarged Department of the Environment.

In the same year, in the Government's White Paper on Local Government Reorganisation[1], far-reaching proposals were outlined for the first comprehensive Local Government reform of this century, set for implementation in April 1974. Its objectives were to establish "a new pattern of areas in which the major services can be efficiently administered and developed in the future", and at the same time "by retaining important responsibilities at the more local level, give power of decision and action to those who will be immediately affected" . . . This reorganisation is expected to go a long way towards meeting the need for a scale of operation for waste disposal on a County or even Regional level and thus, through the benefits of an Environmental Service Grouping, ensure greater efficiency and overall economy. One effect of such environmental grouping would be to replace the existing 1300 county boroughs, boroughs and district councils of England and Wales by a much smaller number of new district councils. Responsibility for refuse disposal will be assumed by the new county authorities, whereas the district councils will be charged with the duty of refuse collection.

In keeping with these current changes, the Institute of Public Cleansing was renamed in 1973 the "Institute of Solid Wastes Management".

Effective waste management calls for new legislation in order to overcome the deficiencies of existing outdated legislation, and appropriate steps have been taken or are being taken to provide it. Developments in this area will now be considered.

2. Early Legislation and its Shortcomings

The Public Health Act of 1875 was the first national legislation to deal with public cleansing. It enabled authorities to provide a necessary service for the cleansing of streets, the removal of refuse and the provision of means (receptacles) for the deposit and disposal of rubbish; it provided penalties for failure to do so. Such legislation became necessary, following the growth of the great industrial towns and cities of the 19th century and public concern at the widespread disease and grave sanitary problems resulting from lack of sewers and drains and lack of provision for the adequate removal of refuse. However, this legislation did not extend to Scotland, Ireland or London.

The Act was updated in 1936 by a new Public Health Act which has provided the basis for the operation of the collection and disposal functions of solid waste management by local authorities up to the present time. This new Act empowered the local authority to undertake the collection and removal of domestic and trade wastes, to remove or require the removal of obnoxious matter from premises, and to make by-laws regulating these services or the carriage and transport of noxious matter or liquid through their district. It prohibited the interference with, or the disturbance of, deposited refuse by any other person than the person employed by the local authority. The provisions of the Smoke Abatement Act of 1926 are also included in the Act.

It is only in recent years that grave deficiencies in the 1936 Act have become apparent. For example, the existing legislation is too

---

* Dr. Kirov reports on an aspect studied during a six months' sabbatical leave during the latter part of 1973.

narrow, it is concerned mainly with domestic refuse and does not make provision for the handling and disposal of waste. There is a lack of any definition or classification (commercial and industrial) of wastes and this has given rise in practice to many problems. While the Act provides legal powers for the local authorities to carry out the various services associated with the collection and disposal of wastes, it does not demand that these essential services be carried out to any definite standards. It fails to provide minimum standards for waste disposal, and hence recommended codes of practice, which are not enforceable, are frequently being ignored. It does not demand, and fails to provide for, overall control and co-ordination in the management of all waste operations.

In other words, what had previously been regarded as a minor branch of public health administration, has in fact now become a major and important problem of waste collection, processing, utilisation and disposal, requiring its own new, specific and coherent legislation and control.

3. Recent Legislation

Conscious of these defects in the legislation and aware of the urgency of some of these growing problems, the Government introduced more recently a number of Acts, all of which have some relevance to the problem of Waste Management, namely:—

(i) The Town and Country Planning Act, 1962.
(ii) The Civic Amenities Act, 1967.
(iii) The Removal and Disposal of Vehicles Regulations, 1968.
(iv) The Clean Air Acts, 1956 and 1968.
(v) The Litter Act, 1971.
(vi) The Deposit of Poisonous Wastes Act, 1972.

3.1. *The Civic Amenities Act*

This Act originated as a private member's bill, and amongst legislative measures for the protection of ancient buildings and trees, includes provisions for the disposal of bulky household waste and unwanted vehicles. Part III of the Act makes these two matters a main obligation imposed upon local authorities by which they are duty-bound to appoint suitable places where unwanted articles can be taken for disposal by residents. It also provides for increased penalties for dumping.

Various methods have been adopted in providing this service and in some cases serious problems have arisen on a number of sites from lack of proper control. There has been much criticism about such "do-it-yourself" refuse disposal and the general opinion among local authorities is that no real benefits can be said to have accrued from the Civic Amenities Act as it stands; that such a system is unworkable and that high standards of hygiene cannot be maintained without full-time supervision.

The widespread nature of these problems prompted the Department of the Environment to set up a Working Group to study the problems of disposal of awkward wastes arising from households, with the following terms of reference:—

> "To identify wastes other than normal household refuse which because of their capacity to pollute, their poisonous or otherwise dangerous nature or because of their bulk or difficulty in handling, create problems for householders in disposal; these will include used sump oil, poisons or pesticides surplus to householders' needs, and aerosols. It is expected that the group will examine the suitability of existing arrangements and facilities for the collection and disposal of these wastes and possible alternative methods, and will also consider the working of Section 18 of the Civic Amenities Act 1967 as regards the disposal of refuse. The object will be to formulate advice which local authorities should be recommended to make available to householders and ways in which such advice could be given publicity."

The report of this Working Group is shortly to be released.

3.2. *Air Pollution Control — the Clean Air Acts of 1956 and 1968*

Air pollution nuisance in solid waste handling and disposal in the U.K., and in particular the control of emissions from incinerator plants are, in the main, administered by the local authorities, through their Environment Health staffs. To minimise and control pollution they apply the provisions of the Clean Air Acts 1956 and 1968, the Public Health (Recurring Nuisances) Act 1969 and the Noise Abatement Act 1960 (concerned with problems of excessive noise).

Chemical Incineration Works (defined in the Alkali, etc., Works Order 1966) are controlled by the Alkali and Clean Air Inspectorates. (**Note:** The Alkali Act, 1863, came into force on January 1, 1864.)

For smoke control purposes, an incinerator unit is treated in the same way as any other furnace. Section 1 of the Clean Air Act 1956 prohibits (except in exceptional circumstances) the emission of dark smoke (smoke as dark or darker than shade 2 on the B. S. Ringelmann Chart) from a chimney.

Section 1 of the Clean Air Act 1968 is concerned mainly with the prevention of excessive smoke emission caused by the indiscriminate burning of industrial or trade waste in the open, and prohibits the emission of dark smoke from industrial or trade premises, as distinct from chimneys.

The Clean Air Act 1968 (Section 6) requires that in the case of a furnace which burns or is to burn at a rate of 100 lb/h or more any solid matter, an application for chimney height approval should be submitted, on a prescribed form, to the local authority. The authority must give a written decision within 28 days of receiving the application. If it fails to do so, the approval applied for is deemed to be granted without qualifications.

The Regulations of 1971 under the Clean Air Act on emissions of grit and dust from furnaces do not apply to incinerators burning refuse or waste matter (solid or liquid) whether or not the resulting heat is used for any purpose. However, the technical panel of a working party set up in 1972 has now reported with recommendations for admissible levels of emission from incinerators which become progressively more stringent with increasing unit size, namely emissions equivalent to 0.4 grain/ft^3 for very small units and 0.1 grain/ft^3 (0.23 g/m^3) at 50 million Btu/h or larger, i.e. for incinerators burning over 4½ tons/h of solid waste. These emissions are to be standardised to conditions of 10% CO_2 at N.T.P. dry combustion gases. Until such recommendations are legally implemented, emissions of grit and dust may be controlled by the local authority by the application of the "practicable means" clause.

For any new plant installed after October 1st, 1969, Section 3 of the 1968 Act states that if it is to be used to burn at a rate of 100 lb/h or more any solid matter, it must be equipped with an approved grit and dust arrestment plant. In general, local authorities do not favour the installation of small incinerators.

The present legislation in the U.K. relating to air pollution control from incinerator units seems to be somewhat vague and outdated. Inevitably, legislation will become more clearly-defined and more stringent, in line with that already existing in many countries in Western Europe.

3.3. *Control of Odours and Gaseous Wastes*

There has been a steady increase in recent years in Britain of public complaints about the malodorous gases and vapours arising from various industrial processes. In consequence, the Government appointed in June 1971 a Working Party "to examine the problems of unpleasant odours emitted by offensive and selected other trades, and to make recommendations about the best practicable means for their minimisation and suppression".

The first report of this Committee — Part 1: Assessment of the Problem in Great Britain — was published towards the end of 1973. It analysed the problem of unpleasant odours, described how they arise in the various relevant industries and trades, and surveyed the known methods of measurement, analysis, abatement and counteraction.

The Committee concluded that complete answers to many industrial odour problems do not exist at present, that reliable technical and design data are seriously lacking and that an intensive programme of research and investigation is needed on abatement methods. More detailed guidelines for the design and operation of abatement processes and plant are promised in the second part of the report, expected to be completed by the end of 1974.

3.4. *Deposit of Poisonous Waste Act, 1972*

The much-publicised dramatic incidents of children's exposure to cyanide wastes in 1971 and early 1972 led to the introduction of interim legislation in the form of the above Act. This is concerned with safeguarding the environment and public health from problems which may arise in the collection and disposal of the relatively small quantities of poisonous, harmful and noxious wastes, i.e.

> "Subject to the provisions of this section (Section 1 (1)), no person shall deposit waste on land, or cause or permit waste to be deposited on land, where the waste is of a kind which is poisonous, noxious or polluting and its presence on the land is liable to give rise to an environmental hazard."

The environmental hazard implies subjection of persons or animals to risk of death, injury or impairment of health, or to the threat of pollution or contamination of any water supply.

This legislation needs now to be amplified:

(a) to provide for a proper system of registration, authorisation, inspection and control;

(b) to define the standards and limits for safe disposal;
(c) to specify the standards required for the transport of dangerous materials.

4. Reports of the Major Working Committees on Wastes and Their Disposal

In order to assess the extent and nature of the new legislative measures needed, the Government established three major working groups, i.e.

(i) *Refuse Storage and Collection,* appointed in May 1963 with the following terms of reference: "to examine the facts of refuse collection; to what extent it is unsatisfactory, what the difficulties are, and which methods are proving most successful; and to consider what advice can be given to local authorities on how to obtain the best results."

(ii) *Disposal of Toxic Solid Wastes,* appointed in July 1964, "to consider present methods of disposal of solid and semi-solid toxic wastes from the chemical and allied industries, to examine suggestions for improvement, and to advise what, if any, changes are desirable in current practice, in the facilities available for disposal and in control arrangements, in order to ensure that such wastes are disposed of safely and without risk of polluting water supplies and rivers."

(iii) *Refuse Disposal,* appointed in May 1967, "to consider the various methods available for the disposal of refuse and to ascertain how satisfactorily they are being operated; to evaluate these methods with particular regard to the increasing quantities and changing character of refuse; and to consider what advice should be given to local authorities and others concerned on the choice of methods best suited to local conditions and on the practical aspects of refuse disposal generally."

This Committee did not consider radioactive wastes because the disposal of materials of this nature is covered by special legislation, nor did it consider the disposal of waste products from the electricity and mining industries, which come under the control of separate organisations.

The reports of these three Working Committees, which have been published,[2,3,4] provide a comprehensive survey and relevant information on problems associated today with wastes and make specific recommendations as to how these problems are to be overcome.

Implementation of the recommendations is inevitably linked with the overall reorganisation of local government currently taking place; nevertheless, the Department of the Environment has already initiated action to implement the recommendations of the working parties.

5. Proposals for New Legislation — The Consultation Document[5]

As a first step in implementing the recommendations of these three working parties, the Department of the Environment in conjunction with the Scottish Office and the Welsh Office issued in February 1973 a Consultation Document in which they outlined proposals for new legislation in relation to waste disposal. So far this document has had only a limited circulation, seeking the views of those most likely to be involved with its implementation.

The document sets out a system for the planning and control of waste disposal in England, Scotland and Wales and points to the legislative requirements for such a system. It stresses the need to improve standards of collection and of disposal, and of the need for properly qualified and trained people to ensure adequate control and execution of these operations. It emphasises the importance of providing a comprehensive review of all disposal problems, of the classificatoin of waste products, and of the necessity to provide an overall plan for satisfactory disposal of wastes in an area. The setting-up of a proper administrative and executive structure and the granting of power to the appropriate authorities to implement the plan is implied.

In an annex to the document it is suggested that there should be a bringing together of all existing legislation, "so as to lay down in a single place an up-to-date code on all refuse and waste disposal matters."

The implication of the Consultation Document is that new legislation is to be expected in the fairly near future, and that it will not only go further than present legislation but also take the opportunity to consolidate in a single Act the existing legislation. It is expected, too, that the new legislation will lay down an absolute minimum standard which must be observed by all who are involved in the various aspects of waste collection and disposal.

The new legislation is expected to be enacted in 1975.

6. Research Into Refuse Collection, Storage and Disposal

The Report on Refuse Storage and Collection[2] stressed the need for a continuous programme of research in this area and recommended the setting-up of a small research committee. Such a committee was appointed in 1969, its terms of reference being:

"to assess research projects which are in progress, recommending co-ordination where desirable; to study new techniques and processes; and to advise what new research, including collection of data, would be useful and by whom it could best be undertaken."

The first report of this Standing Committee, covering the period September, 1969, to December, 1972, was published in 1973.[6] The report stressed the need for a nationally co-ordinated or sponsored arrangements for research. While much of the work of the Committee is still in progress, assessment of some projects has led to certain conclusions and recommendations.

For example:

(i) owing to increased costs and maintenance, the Committee "cannot now recommend dustless loading systems";

(ii) on-site compaction systems have been found to work satisfactorily and a full report on the comparative study of available compaction equipment is expected soon;

(iii) "the Committee are quite unable to support the idea of installing small incinerators in individual houses", as "the resulting multiplicity of incinerators would not be acceptable on clean air grounds"; "on-site incinerators require skilled maintenance to ensure that an acceptable performance is maintained";

(iv) in the light of recent American progress with large-scale compaction methods, the Committee considers that this method is "worth investigating further".

The Committee commissioned various research organisations and consulting firms to investigate and report on a number of projects such as:

(i) a feasibility study of pipeline collection and transportation of refuse;

(ii) appraisal of design factors of different types of treatment plant;

(iii) the carrying out of tests on analytical methods for calorific value, noxious gaseous emissions from incinerators and the assessment of the organic content of incinerator residues;

(iv) a survey of disposal of scrap tyres;

(v) pyrolysis of refuse, salvage and reuse of ferrous scrap, reuse of plastics waste and the extraction of non-ferrous metals from incinerator residues.

Of particular interest, too, is the contribution by the Committee towards the cost of the Reigate Borough Council's multiple hearth incinerator, which should provide information of general interest on the combined incineration of refuse and sewage sludge.

7. Conclusions

Efficient waste management needs a strong legislative framework within which to operate. Its task is no longer seen in the U.K. as a straightforward service of the collection and disposal of refuse but as an integral part of the provision of a safe, healthy, clean and well maintained environment.

The three major reports of the working parties leave no one in doubt that public cleansing and waste disposal would benefit from reform and modification under one comprehensive enactment, and of the need for a scale of operation for refuse disposal on a County or even Regional level.

The Consultation Document on Waste Disposal follows closely the working parties' conclusions and recommendations and affirms the Government's determination to ensure that an adequate framework of powers and legislation is provided to support the implementation of a progressive nation-wide waste management programme.

---

8. References

1. Local Government in England. Government Proposals for Reorganisation. (Cmd. 4584) February 1971.
2. Refuse Storage and Collection. (Report of the Working Party.) H.M.S.O., 1967.
3. Refuse Disposal. (Report of the Working Party.) H.M.S.O., 1971.
4. Disposal of Toxic Solid Wastes. (Report of the Technical Committee.) H.M.S.O., March 1970.
5. Consultative Document — Waste Disposal: Proposals for a New Framework, Dept. of the Environment, the Scottish Development Dept. and the Welsh Office; February 1973.
6. First Report of the Standing Committee on Research into Refuse Collection, Storage and Disposal, Dept. of the Environment, H.M.S O., 1973.

PAPER 8

The Activities of the WHO International Reference Centre for Waste Disposal (IRCWD) and the Federal Institute for Water Resources and Water Pollution Control (EAWAG) Swiss Federal Institutes of Technology

By U. Bundi*

1. Introduction

The WHO International Reference Centre for Wastes Disposal (IRCWD) is an annex to the Federal Institute for Water Resources and Water Pollution Control (EAWAG). The EAWAG is divided up into different departments, one of which is concerned with the problems of solid wastes management.

In the following we will therefore not only present the IRC's activities but also briefly discuss the main activities of the solid wastes unit at the EAWAG.

2. The Activities of the IRCWD

2.1. *Introduction*

The staff of the IRCWD consists of seven persons. These are: a manager, an assistant manager, two project engineers, a translator, a documentalist and a secretary. While the manager devotes only part of his time to the IRCWD, the other collaborators are employed on a full-time basis. Despite limited personnel the IRCWD is engaged in the following three main activities: (i) documentation on solid wastes; (ii) training of WHO staff and other persons; (iii) setting up research projects of international interest.

2.2. *Documentation*

Since the establishment of the IRCWD in 1968, 2300 abstracts of scientific papers have been collected. A thesaurus with about 500 key words was prepared and published in the spring of 1973. Copies have been and will still be sent to all interested institutions.

Aside from abstracts, the IRC also intends to review specific problems in the field of solid wastes management.

The "IRCWD News", a bi-annual publication, deals with problems of environmental pollution. This news bulletin is sent to all of the 45 collaborating institutions with the purpose of generating contacts between them and the IRCWD; owing to a lack of initiative on the part of the collaborating institutions this aim has been only partially accomplished.

2.3. *Training*

In 1972 the IRCWD had 80 visitors from 23 countries, mainly scientists, fellows and students working in or looking at different departments of EAWAG or touring plants in different regions of Switzerland.

Every year a field training for the students of the "Centre International de Génie Sanitaire" (Rabat) is carried out. The training program focuses mainly on planning, design and operation of wastewater and solid waste disposal plants.

These activities will be continued although the IRCWD would prefer fewer longer traineeships to many short-term visits, being convinced that the visitors can benefit more from traineeships of a longer duration.

Every year staff members of the IRCWD give a course on solid wastes management at the above-mentioned "Centre International de Génie Sanitaire" in Rabat.

2.4. *Research projects*

Since its foundation the IRCWD has carried out a number of projects. A few are listed below:

— Survey of Solid Wastes Management:
A manual, which is clearly intelligible to the layman, discusses the important aspects of solid wastes management.

— Compilation of Legislation on Solid Wastes Management:
A study of legislation and regulations governing solid wastes disposal in countries around the world.

— Segregation of Refuse:
A public opinion poll carried out in a small Swiss town to find out whether the inhabitants would be willing to segregate glass and plastic bottles from the rest of household wastes.

— Inquiry on Industrial Wastes:
An inquiry into all aspects of the waste

* Swiss Federal Institute for Water Resources and Water Pollution Control, Dübendorf.

problems of the main industrial branches in Switzerland, except the chemical and watchmaking industries.

— Methods of Sampling and Analysis of Solid Wastes:

A compilation with a view to recommending standard methods that will permit comparative analysis of data.

— Comparative Cost Accounting in Refuse Treatment:

A comparison of incinerator and sanitary landfill operations.

Another finalised project deals with the establishment of a basis for international regulations concerning industrial wastes. A preliminary report has already been completed and distributed to interested institutions.

Besides the work on solid wastes, the IRCWD is active in the field of water pollution control. The IRCWD, for example, participated in a study on the development of the quality of Swiss natural waters by the year 2000. In this study the inter-relationship between the activities of modern man, industrial and agricultural activities, possible water pollution control measures and the quality of natural waters is shown. It is hoped that the methodology of this project will prove generally useful in establishing priorities for water pollution control and thus constitute an important step in water resources management in developed as well as developing countries.

As already mentioned under Item 2.1, the IRCWD intends to review specific problems in solid wastes management in the near future.

A project which even goes beyond this objective is the development of concepts for recycling. The work for this project is now under way and will be finalised by the summer of 1974.

In November 1972 a meeting of directors of about 20 collaborating institutions was held at the EAWAG organised by WHO HQ. International collaboration in solid and liquid wastes management was the main topic. Research projects and other developmental work to be carried out within the framework of the WHO environmental program were identified and priorities recommended. Among the many suggestions, some were qualified as being of great importance:

— Seminar on ecological aspects of waste management.
— Research leading to the development of techniques and basic criteria for regulations for the management of hazardous and toxic wastes.
— Toxicology and epidemiology of wastewater for domestic use.
— Alternative methods for the collection of human excreta.
— Alternatives to water carriage methods for the disposal of human excreta.
— Methods for the design and operation of a sanitary landfill for developing countries.
— Seminars on rural sanitation.
— Training courses in solid wastes management, water pollution control and soil pollution control.

If WHO could make available the necessary funds, some of these projects could be carried out by the IRCWD or a collaborating institute in the next few years.

3. The Activities of the Solid Waste Unit of the EAWAG

3.1. *Introduction*

At present the solid waste unit of the EAWAG consists of 16 collaborators and five doctoral candidates. The overall objective of the solid waste unit is to find ways of handling wastes economically and with the least detriment to the environment. Thus all aspects of solid waste management — collection, treatment, final disposal, recycling and recovery of solid wastes and sludges from household, the trades and industry — are covered. The solid waste unit carries out research projects on its own initiative and on a contractural basis. It also acts as a consultant on behalf of the authorities responsible for wastes management. Some of the collaborators give lectures at the Swiss Federal Institute of Technology.

3.2. *Present and future investigations*

Investigations on the following subjects are under way or will be in the near future:

— Technical, chemical and biological problems arising from waste treatment processes such as incineration, composting, sludge dewatering and conditioning, and from the final disposal of wastes such as controlled tipping and land disposal.
— Effects of the different treatment processes and final disposal methods on the natural environment.
— Examination of newly constructed or dedeveloped plants and processes by means of physical and chemical measurements.
— Recovery and recycling of solid wastes (household refuse, industrial wastes) and treatment residuals (compost, slag).
— Segregation of refuse at source to avoid the mixing of different types of wastes, thus permitting recovery or recycling.

PAPER 9

Some Overseas Developments in Waste Management

A summary of reports submitted by the recipients of the travelling grants awarded by the 1971 Waste Disposal Conference Committee.

By G. D. Sergeant*

From the proceeds of the 1971 Waste Disposal Conference, the Committee awarded five travelling grants available on application to any member of the organisations associated with the Conference, i.e., The Institute of Fuel, The Local Government Association, The Clean Air Society of Australia and New Zealand, and the Australian Institute of Health Surveyors. The conditions of the award were:—

(1) That the successful applicants could spend at least ten days of their time overseas studying a problem of interest to the Waste Disposal Conference Committee.

(2) That applicants were to be overseas between May 1972 and June 1973.

(3) That the successful applicants submit a full written report on an approved topic of study and that an outlet for a wide publication of the report would be sought, e.g., at a Conference or in a technical journal.

The value of the travelling grants is perhaps best illustrated by an extract from the introduction of one of the submitted reports:—

". . . I also wish to acknowledge the Committee of the 1971 Australian Waste Disposal Conference and extend my appreciation for the scholarship awarded by them which assisted in initiating the tour and defraying some of the costs."

The grants served to initiate overseas visits in some cases and to extend them in others.

The recipients of the travelling grants were:—

Mr. T. J. Somers, Health Inspector, City of Salisbury, S.A.

Mr. H. Healey, Deputy Health Surveyor, Woollahra Council, N.S.W.

Mr. G. F. Sparks, Health Surveyor, Brisbane City Council, Q'ld.

Dr. D. Iverach, Air Pollution Control Branch, Department of Health, Sydney.

Dr. G. D. Sergeant, Fuel Department, University of N.S.W., Sydney.

The following statements are extracted from the reports submitted by the recipients and are self explanatory in terms of subject matter studied.

1. Mr. T. J. SOMERS studied the disposal of glass, paper, tins and used car tyres in the U.K. and France.

Plastics constitute about 1.56% of Britain's refuse and it is suggested that an increase in this commodity by 10% to 15% annually can be envisaged. This material is ideal for "one trip" containers, but if this trend is allowed to continue, the amount of plastic scrap presented for disposal could perhaps vie with tyres as a difficult item to dispose of.

Glass. 9% of refuse in England and France consists of glass. There are various means used to recycle this material and it is not established whether any of them are economically viable propositions.

Pulverised glass is being used for various useful commodities, including road surfacing carried out experimentally in the U.S.A.

If the volume of discarded glass is controlled the disposal problem becomes that much easier. At present the public appears to have little alternative but to accept industry's extended use of "one trip" bottles.

Tyres. About 300,000 tons of scrap auto tyres are disposed of each year in Britain. Closer to London there appears to be no awareness of a problem existing, due to accumulations of tyres for disposal. In the industrial north-west, what could have been a serious problem is being dealt with most satisfactorily and with the minimum of fuss. The attitude — specially in Britain, appears to be "the problem exists — it must be solved". Whether tyre manufacturers were inveigled or forced to combat the problem arising as it does from their production, was not established. In France an attempt was made by one tyre company to label tyre disposal as a "social problem".

Cans, like other "one trip" containers and paper could constitute "excessive packaging" in that alternative means do exist, which result in less refuse for disposal. Where drinks are

* Senior lecturer, Department of Fuel Technology, University of New South Wales.

concerned the returnable glass bottle is one answer. However, tinned cans have been with us for a long time and did revolutionise the Food Industry at their inception.

It is the excessive use of this packaging device that presents the problem. Recycling of the aluminium container appears to be a viable proposition and would certainly reduce the drain on bauxite deposits.

Steel cans are not in much demand as scrap — other means of disposal are found. A reduction in the use of this type of container is considered desirable; recycling of the non-ferrous containers would reduce the volume of refuse.

Disposal Methods. The various means of disposing of domestic refuse — including glass, paper and tins, includes the following:—

(i) *Controlled Tipping:* A tried and approved method when controlled by informed personnel, this means if refuse disposal is valuable in land reclamation. However, for the disposal of plastics — which do not "breakdown" and tyres, this method is not ideal. Dumping of tyres in land reclamation schemes could result in the re-claimed area being unsuitable for building.

(ii) *Pulverisation* provides an aesthetically acceptable way of running a dump. Greater compaction of the refuse is achieved and less cover is required. However, the problems of plastics and tyres remain.

(iii) *Composting* has lost favour over the years and except for perhaps some "die hard" authorities is not seriously considered for disposing of refuse from large towns.

(iv) *Incineration* would appear to be the complete answer for today's and future (15-20 years) refuse disposal problems.

Specially designed furnaces are being used for tyre disposal but some authorities dispose of large quantities of tyres in their direct refuse disposal incinerator.

Maintenance problems do exist, but appear to be well controlled by a programme of preventive maintenance. Running costs of the various disposal methods are as follows:—

Method	**Cost per Ton** 1969 Figures (Britain) including Land Costs
Controlled Tipping	50p to £1
Pulverisation	£1.25p to £2
Composting (fully mechanised)	£2.50p to £3.50p
Incineration	£2.50p to £3.50p

Capital costs of incinerators are high and the units observed appeared to be worked intensively. Attempts to recover some costs are being made by utilising the heat generated for production of steam or/and electricity. The recovery of iron from incinerator residue depends on the demand at any given period.

A Royal Commission on Local Government in Greater London began its work in 1961. A recommendation that a unified organisation be established to carry out systematic research into methods and costs of refuse disposal eventuated. This recommendation was accepted by Parliament in 1963 and the Greater London Council became the authority responsible for the disposal of all refuse throughout its administrative area.

It is the concensus of opinion in Britain that the disposal of solid wastes of all types should be co-ordinated and controlled over reasonably large areas of Solid Waste Disposal Authorities. Suitable local authorities for this function would be of a size similar to the Metropolitan (London) Authority.

While it is conceded that there is no universal solution to the problem of refuse disposal in the writer's considered opinion, a solution of major importance is direct incineration. In the case of tyre disposal there appears to be no ideal alternative.

2. Mr. H. HEALEY made the following recommendations on solid waste handling in residential flats based on U.S.A. and European experience.

2.1. *Waste Programming in Residential Flat Buildings*

The importance of a well designed solid waste disposal provision within multi-occupied buildings cannot be over emphasised.

The choice of systems currently practised are:—

(i) Chute Feeding to Containers.
(ii) Chute Feeding to Compactor Equipment.
(iii) Chute Feeding to Incinerators.
(iv) Chute Feeding to Pneumatically Operated Underground Trunking in an Estate.
(v) Water Carried on an Estate Basis.

The latter two schemes have little to offer in this country as rarely is the one developer engaged in a project of sufficient magnitude to make this practicable.

Schemes (i), (ii) and (iii) are possible and practical and some local experience is being gained in dry chuting as opposed to chute-fed incineration which has a long history in this country. While on-site incineration is losing favour with many environmentalists, the standard of efficiency demonstrated locally is far ahead of any overseas installations viewed by the writer.

In dry chuting, the American design is preferred to the European as the metal liner seems to offer a more impervious tube which can be flushed for cleaning if required.

The minimum standard for chute construction, in the writer's opinion, is covered in the Woollahra Draft Code. However, after viewing a number of installations overseas, additional requirements should be included, such as:—

(i) Increase of chute wall thickness in the lower section of the tube from 16 gauge to $\frac{1}{8}''$ thickness.

(ii) Increase fire protection by the addition of a sprinkler immediately below the lowest chute hopper. This sprinkler should be set to actuate at 160 degrees Fahrenheit while the fire damper linkage should be set at 130 degrees Fahrenheit.

(iii) Provision of vermin spray automatically actuated at container or compactor level.

(iv) Physical separation of all electrical controls from the main chute and/or compactor device to reduce restoration after fire damage.

2.2. *On-Site Waste Treatment*

Incineration in the past has offered a worthwhile volumetric and weight reduction in solid waste at its source. If this practice is not to be continued, an alternative reduction method should be sought. In the writer's opinion, in any development greater than 15 occupancies, volumetric reduction should be demanded (Ordinance 71 Clause 62).

At present there is little choice in equipment and the advertised performance varies considerably. The writer has no knowledge of any local research on the recommended performance specification that should be sought, and for this reason only the New York performance specification might well provide a temporary standard until local work is attempted.

The New York standard is "Input Density". "The weight of uncompacted refuse used for test should be within the range of 150 to 230 pounds per cubic yard."

"Output density". "Compacted refuse should be within the range of 450 to 700 lbs. per cubic yard."

These figures relate fairly well to local densities and the treated waste should incinerate or land-fill satisfactorily.

It is considered that existing incinerators should be permitted to remain at present, and that new buildings be compulsory equipped with refuse chutes and compactors or modern fuel-fired incinerators. This policy would assist in developing a larger number of equipment suppliers and build up reliable local experience, prior to a full scale conversion of existing incinerators should this become necessary through poor performance or unpredicted changes in refuse composition (rendering the material unsuitable for combustion on a small scale).

2.3. *Solid Waste Collection*

It is strongly recommended that Local Authorities implement mechanical lifting of containerised waste. There are obvious economics to be gained, together with aesthetic improvement.

Compaction of waste results in an increased weight per unit and mechanical lifting is a basic necessity in certain types of systems.

Generally, councils should provide a variety of vehicles in their fleet so that sufficient flexibility is available to the consultant when designing the most efficient waste handling programme for a particular building, whether of residential or commercial type.

2.4. *Research and Statistics*

That research should be carried out into such fields as:—

(i) The volumetric measure of waste produced by persons living in various dwelling styles.

(ii) The volumetric measure of waste generated by various commercial and industrial undertakings.

(iii) The most advantageous compaction density related to the haulage and disposal means.

(iv) The chemical and physical composition of waste in various regions.

The result of established data in these areas would greatly assist consultants in designing suitable long term solid waste provisions in buildings and further assist approving authorities in analysing the worth of proposals in new structures or the conversion requirements in existing buildings.

3. Mr. G. F. SPARKS studied trade effluent control in England and Wales.

With any legislation, the Government of the day envisages that its Act will be effective and capable of withstanding legal scrutiny as well as coping adequately with present-day problems and those that could arise in the forseeable future. This applies to Water Pollution Control legislation as regards England and Wales, and over the years there has been considerable progress made in achieving a wider scope and power in dealing with this aspect of environmental pollution.

The introduction of water-borne waste accompanying the industrial revolution in Britain brought the country's attention to focus on the various problems, from the public health aspect, of water usage and this resulted in the institution of a series of Public Health Acts. The fact that polluted air is not as conveniently channelled away from our immediate environment and is water-borne waste in the majority of cases, most probably caused public and hence Government reaction and consequent action to prove far more active, positive and as history can now recall, effective.

Industrial progress together with great achievements in the technological field have necessitated the interpretation of the scope of water pollution from a broader viewpoint than the public health aspect although this remains the essential concept from the "short-term" point of view. Today, Britain realises that she has the technical ability to satisfy the public health and recreational amenity demands together with the requirements of industry. To this effect, the Government intends spending some £St.300,000,000 over a five-year period on water pollution control activities in England and Wales according to a recent ministerial statement.

The water pollution control structure as it exists in England and Wales today will be completely re-organised by the present Government's proposal, which comes into operation in April 1974, to institute ten regional water authorities to control the country's water supplies. This proposal has brought rather caustic criticisms from various people actively engaged in the water pollution control field. One notable critic is Sir Desmond Plummer, leader of the Greater London Council which is responsible for the River Thames. This body has achieved quite a remarkable restoration of this river in recent years and is justifiably proud of the accomplishment.

The present structure, however, relies upon a "chain of command" in which the River Authorities, set up under the 1963 Water Resources Act, lay down and enforce standards of purity for waste liquids discharged to watercourses under their jurisdiction. This applies to both industrial effluents and to discharges from Local Authority sewage treatment works.

In 1903, the Royal Commission on Sewage Disposal suggested in its third report "that purification of trade effluents by the Local Authority is, in the greater majority of cases, practicable, purification by the manufacturer is, in some cases difficult, if not impracticable, while purification by the manufacturer would be more costly than purification by the Local Authority"!

Today, this is still the general opinion held by those concerned in this field and most sewage purification bodies consider it their duty to receive and treat all trade wastes which can be conveyed via the sewers. However, there is always a proviso to the effect that these wastes shall not include any substance of a nature and quantity likely to injure the sewers into which it is discharged or to interfere with the free flow of the contents thereof or cause any injury to persons engaged in work in such sewers or to cause damage to sewage purification plant and equipment or to interfere with any of the processes of purification to such a degree that statutory requirements cannot be complied with, i.e., those standards laid down by the River Authority as acceptable for the discharge of effluent from a treatment works to a river outfall within its Area.

To protect its plant and equipment and for reasons of human safety, the Local Authority must then advise the trader on what the maximum standards or levels of impurity that they consider can be adequately dealt with by their treatment works. Hence having being so directed by the Local Authority as second body in this "chain of command", the trader may find it necessary to undertake and implement some form of pretreatment prior to the discharge of his effluent to the Local Authority's sewers.

It is generally acknowledged that the admixture of trade effluent with domestic sewage will normally result in the most economical treatment of these wastes, since they then receive considerable dilution and the presence of nutrients may be beneficial to their treatment and, furthermore, the Local Authority in most instances employ purification specialists, a factor which cannot be claimed by most industrialists.

It is only fair to assume that the Local Authority's costs incurred in treating the trade waste to an acceptable standard should be borne by the traders involved, taking into account the degree of difficulty encountered in the treatment of the respective discharge. Most Authorities base their charges on a formula similar to the well established "Mogden" type formula, originated by the Middlesex County Council, which relates the pollution load of the trade effluent to that of sewage in assessing the additional cost involved.

Having established a method of charging both fair and acceptable to the trader and Local

Authority alike, it is necessary to have an efficient and qualified personnel to undertake the work of trade effluent control. In addition to taking samples of the trade effluent for analysis to ensure the compliance with consent conditions as laid down by the Local Authority, these officers must be capable of gaining the trust and respect of the industrialist as well as offering advice and noting alterations in plant design and process changes. It is quite obvious that only a conscientiously implemented trade effluent control programme together with the support of the trader will achieve satisfactory results in the control of the pollution of waterways.

The role of the industrialist and the necessity for him to "pull his own weight", was admirably explained by the following quotation from "Disposal of Trade Effluent in England and Wales 1962", a publication of the Confederation of British Industry (Handbook 49):—

> ". . . it will be seen that the provisions of adequate sewage works capacity and the development of new methods for dealing with intractable effluents will be required if the pollution of rivers by trade effluent is to be satisfactorily reduced. These requirements will impose on industry a considerable financial burden, but though it may not be productive in the ordinary sense, it will contribute in no small measure to the preservation of underground and surface water supplies of this country. The successful implementation of such a programme will be largely dependent on the co-operation of government, local authorities, river boards and industry, particularly in planning the economic grouping and operation of treatment plants in agreeing to mutually acceptable conditions for trade effluent discharges."

It is acknowledged that for Britain to maintain a healthy "Balance of Trade", the interest in industrial progress and hence, whilst not giving way to pressure from industrial sources, the standards for purity of effluents as laid down by the River Authorities should not be more restrictive than absolutely necessary.

This present structure of water pollution control has been operating efficiently from the viewpoint of minimising the pollution of waterways and in particular the method of formula charging as outlined, whilst being considered the most reasonable method of charging for both parties concerned, also provides the trader with a significant financial incentive to decrease both the volume and strength of the discharge from his premises.

4. Dr. D. IVERACH travelled extensively in the United States and studied on-site incineration standards.

The 1971 Australian Waste Disposal Conference discussed many aspects of the waste disposal problem. On my recent visits to the United Sates I had the opportunity to add another dimension to my understanding of some of these aspects. For this I thank the Executive Committee in their generosity of awarding me a Travelling Grant. Some of the more important impressions gained on these trips are discussed below. Incineration is discussed in some detail because the experience gained overseas played a major part in the promulgation of new design standards by the Health Commission of N.S.W.

Development of Design Standards for Small (<300 Kg.hr^{-1}) On-Site Incinerators

In most large American cities the incineration of small quantities of refuse on-site is discouraged. In most cases it is discouraged as a result of the local Air Pollution Control Authorities requiring expensive, sophisticated units which have a reasonable chance of meeting the applicable emission standards. These standards vary quite widely from state to state and further, the zeal with which they are enforced also varies. Some cities require an emissions test on completion of the incinerator but this is always done under controlled conditions and may not be representative of the emissions under normal operation (or maloperation).

The following points were held important by most authorities:—

(i) Auxiliary firing (usually natural gas) was was considered essential. Sometimes as primary burners but more usually as a secondary burner sited either near the end of the primary chamber, in the mixing chamber (if one existed) or in the secondary combustion chamber. After-burners were also accepted on some designs.

(ii) Much attention was usually paid to the loading provisions of the incinerator — automatic doors designed to prevent an operator from charging the incinerator and leaving the doors open were not uncommon.

(iii) The use of low-medium pressure drop water scrubbers to control particulate emissions was quite widespread. Some authorities considered the disadvantages of this type of control equipment — maintenance requirements, power requirements, by-pass flues used when they shouldn't be and steam plumes — to outweigh the advantages.

(iv) For apartment house incinerators, requirements were becoming even more stringent — possibly because compaction has proved an attractive alternative.

In general, industrialists and apartment owners in the United States appear to be prepared to pay more for efficient incineration than their counterparts in Australia. Some units looked more like package boilers than incinerators. Keeping in mind the status of incineration in Sydney in 1972, many of the designs seen in the U.S.A. were found to be too expensive and complex. However, it was recognised that some upgrading of incineration in Sydney was warranted.

In American cities, as in Australian cities, incineration is regarded as playing a minor role in the overall air pollution problem. However, the potential for irritating local problems always exists and the objective of many control Authorities is to minimise the development of such problems. This has been the objective of the Health Commission of N.S.W., but the effort in attaining such a goal has been substantial in terms of man-hours spent on checking proposed designs. Discussions with American agencies revealed that this problem did not arise to the same degree either because approval of the unit depended on an emissions test during the commissioning period or because only those designs manufactured by an established and well respected company were approved. Although both these mechanisms might have a place in N.S.W. neither was seen as the answer to the overtaxed evaluation system.

The problem of how to simplify and accelerate the approval (or disapproval) process has been solved in a unique fashion. First it was agreed that of all the designs which find acceptance in N.S.W., the Los Angeles type multiple chamber incinerator had certain advantages. Foremost among these is that such an incinerator performs noticeably better than other types under conditions of mal-operation such as the charging doors being left open, vigorous and prolonged raking of the firebed and low burning rates.

Then, based on American experience and to a large degree our own, it was agreed that secondary burners were essential on all general waste incinerators. This raises questions of cost and safety and the Commission's Guidelines stress the need for properly designed, fail-safe burners and also require the tenderer to estimate the annual fuel bill for the auxiliary burner. Not surprisingly, some prospective buyers have decided against incineration when confronted with this significant operating cost.

Finally, the sizes of the respective chambers were related to a single variable — the burning rate. It was determined that this could be done without any loss of accuracy in designing for minimum and maximum velocities, temperatures and heat release rates, etc. The calculation process has been reduced to simple arithmetic and the evaluator or checker need have no basic knowledge of combustion principles. This has meant that local councils have taken over the job of approving applications to build incinerators. Since the promulgation of the guidelines encouraging developments have occurred, the standard of incinerator design has improved and there has been a growing demand for virtually pollution-free units. Some industries, too, have found it more economical to have their refuse collected than to replace an old incinerator with one designed to the new standards. In either case the result is a net decrease in air pollution and happier neighbours.

5. Dr. G. D. SERGEANT studied the reuse of otherwise waste materials from the coal industry and domestic waste in the U.K. at the National Coal Board Research Establishment and at the Warren Spring Laboratory.

5.1. *The Coal Industry*

Plastics from Colliery Waste

The main waste disposal problems from collieries are those of large tonnages of washery tailings and wet, fine materials, the utilisation of which obviously requires large tonnage applications. However, the development of a wide variety of applications, some quite small, is of value both in consuming some quantity of otherwise troublesome waste material and in exploring in general the possibilities of utilisation. The Stoke Orchard Coal Research Establishment has been successful in developing several commercial products via this approach.

Conveyor Belts — An early success, and now a commercial venture, was the utilisation of worn, disused or discarded PVC conveyor belts to form "Hyload". As the market for the product expanded, the amount of available waste conveyor belt proved insufficient and other textiles have been used.

Washery Tailings — The coal content of the tailings is converted into a plastic binder by digestion with a heavy coal tar oil, water is evaporated off, mineral matter acts as a filler and the resulting mixtures can be moulded, pressed or extruded into a wide variety of artefacts, including drainage pipes, cables, ducting, etc.

Other binders — for example, silicates, bitumen and synthetic latexes — could be used and another possible binder source that has been proposed is waste rubber automobile tyres.

Colliery Waste: Unburnt and Burnt Spoils

Highway Construction — Burnt spoil has been widely used for some years in highway construction and has been found to be both good and economical. In many areas where highways are being constructed there is abundant unburnt spoil, but the objection has always been raised that such material is liable to spontaneous combustion and is therefore unacceptable.

To check the acceptability of this material the C.R.E. constructed an embankment using 8000 tons of weathered spoil, containing about 20% coal equivalent (on a calorific value basis), and having 180 thermocouples buried in it. After three years no temperature has been recorded above the environmental temperature. The compaction and drainage have been found to meet the required standards.

Skid-Resistant Road Surface Dressing — Tests of a range of road surface dressings have shown that natural stones do not have lasting skid resistance, due to ease of polishing. C.R.E. has shown that a satisfactory stone can be produced from burnt spoil. Some burnt spoils have been found to be suitable as a concrete aggregate, especially for precast concrete blocks.

Disposal of Colliery Tailings by Fluidised Combustion

Fluidised combustion offers the possibility of processing reject wet fines to produce a stable, disposable product, with the further possibility of utilising the ash to offset the cost of treatment. Fluidisation is also considered to be potentially a cheap way of drying wet fines as an alternative to filter pressing or centrifuging.

5.2 *Domestic Refuse*

Recovery of Metal Values from Incinerator Residues

At the Warren Spring Laboratory, processes have been developed to recover metal values from the clinker of domestic refuse incinerators. After the initial magnetic separation of ferrous metals, heavy-media fluidised-bed separators have been used to separate the non-ferrous fraction into light (aluminium rich) and heavy (copper rich) fractions. The processes work well with feed-stocks larger than 10 B.S. mesh. For finer particles the pinched sluice, also developed at W.S.L., may be used. With this equipment the mixture is fed along a sloping porous deck through which air is blown. At the lower end of the deck a splitter plate separates the heavy and light fractions.

There is always a problem in the efficient separation of mixed metal scrap originating from such sources as domestic refuse incineration and car fragmentation plant. The W.S.L. process described in the previous section depends on suitable physical properties of the metals, e.g., density, for their separation; but the use of such physical differences is not always possible. Fortunately, in many cases there are wide differences in melting points of various metals or alloys, and selective melting in fused salt media offers a method of separating alloys based on zinc, aluminium and copper. Fused salt media enable scrap to be heated without oxidation and with more even temperature control.

Pyrolysis

Pyrolysis (i.e., heating in absence of air) can provide the following advantages over incineration for treatment of domestic refuse:—

(1) It can provide high-grade fuels which can be stored and therefore used as and when required.
(2) There is no oxidation of metals.
(3) The solid residue is a friable char from which metal values can easily be separated.
(4) The temperature of pyrolysis can be regulated to control the nature of the products.
(5) There are no moving parts in the reaction zones of the reactor.
(6) The process can be designed to minimise pollution.

W.S.L. has a program of investigation under way to examine the possibilities of this process in detail.

Ferrous and non-ferrous metals can be recovered from the solid products by the processes previously described, and these systems will have the advantage of treating non-oxidised, non-fused metals.

The sysem is applicable also to other solid organic wastes — for example, rubber tyres, from which a high oil yield is obtained.

This study has been reported in the September 1973 issue of Process and Chemical Engineering.

PAPER 10

The Cairns Regional Solid Waste Disposal Study

By R. G. Hoare* and N. Y. Kirov†

1. Introduction

The Environmental Control Council of Queensland, at its first meeting in May, 1971, considered a report outlining possible future refuse disposal problems in Brisbane. As a result of the submission, the Disposal of Waste on Land Committee was formed to investigate the problems of Waste disposal in Queensland. To assist this Committee in its deliberations, a questionnaire was prepared on waste disposal which was circulated to all local authorities in the State.

At approximately the same time the City Engineer's Department, Cairns City Council, being conscious of the problem of waste disposal in the City, brought it to the attention of the Cairns City Council which in turn approached the Co-ordinator-General's Department requesting assistance with a study of the waste disposal problems of the City. The Disposal of Waste on Land Committee and the Environmental Control Council agreed that such a study would lead to a better understanding of the problems of waste disposal in the area.

The aim of the study was to define the present problems encountered with waste disposal in Cairns, to define the factors likely to affect waste disposal, both present and future, and, using these inputs, to determine the most practicable method of disposal for the area. Prof. N. Kirov was engaged by the Co-ordinator-General's Department as a consultant to advise on the planning and organisation of this study, particularly in respect of collection and evaluation of the waste generation data. Data was collected between March and November, 1972, and a report — "Cairns Region Waste Disposal Study" — was presented to the Council in February, 1973.

Throughout this paper Cairns means urban Cairns, that is, the area administered by the Cairns City Council and the urban areas about the City of Cairns administered by Mulgrave Shire Council.

2. Nature of the Problem

Unlike most areas of the world experiencing problems with waste disposal, the major problem in the City was related to the *cost* of the disposal operation rather than to the use of unsanitary disposal methods or a critical shortage of land for refuse disposal. In actual fact, from the health aspect alone, waste disposal practices in Cairns are amongst the best in Queensland.

Currently, 6 ha (15 ac) of land are filled annually with the refuse generated by the City's 31,000 persons (1972 Census population). This rate of land filling, on a per capita basis, is much greater than that in other Queensland cities, e.g., Brisbane 18 ha (45 ac), population 700,000 and Townsville, 4 ha (10 ac), population 72,000.

After reclamation with refuse, the land is ultimately developed as open playing fields but, as demand for playing areas in the City falls far short of supply, reclaimed areas become a liability to the Council. The actual cost of disposal is also very high and is related to the available covering material which costs the Council approximately \$1.50 per m³. This includes the cost of quarrying, transport and distribution of the covering material at the tip face. The total cost of maintaining the waste disposal service in the City in the 1972/73 financial year was approximately \$90,000. Use of large amounts of covering material is necessary to prevent fly breeding and water pollution.

3. Factors Affecting Waste Disposal in the Cairns Region

3.1. *Climate*

The combination of high temperature, average 31°C to 32°C between November and March, and high relative humidity, average 66%, is conducive to fly breeding. Under these conditions it is common for flies to develop from eggs to adult in five days. Because much of the refuse entering the tip contains flies at the larval or maggot stage of development, refuse must be treated with at least 300 mm (1 ft) of compacted covering material.

---

* Planning Officer, Co-ordinator-General's Department, Brisbane.

† Head, Department of Fuel Technology, University of New South Wales.

3.2. *Rainfall*

It is necessary to prevent water entering refuse disposal areas as the leachate from disposal areas may cause water pollution. As the area is subject to particularly intense rainfall, it is important that the covering material be compacted so that water would drain from the area rather than penetrate the covering layer to the refuse. The behaviour of the available covering material, when moist, increases drainage problems at the disposal area. Maintenance of adequate supplies of suitable covering material during the wet season also creates temporary problems. The high rainfall leads to local flooding which increases the problems of waste management in the area.

3.3. *Topography*

Cairns is situated on a narrow coastal plain which is 7 km (4.3 miles) in width at the City and which characteristically has low relief features. At Cairns, the lowlands form ridges separated by swamps. Many of these swamps, particularly in the West Cairns area, have been filled.

In recent years, refuse has been used to fill remaining low-lying areas within the City. As the swamp areas are tidal and subject to local flooding, and as it is advisable to keep water out of the disposal areas, disposal sites have been isolated by bund walls fitted with tide gates.

3.4. *Ground Water Level*

Ground water levels in the area are high during the wet season, reaching the surface in places, and, following the wet season, the ground water level drops. It is unusual for the ground water level to fall below 1.6 m (5 feet) below the surface. The result of the high ground water level is that refuse may be dumped into water which is undesirable. In addition, it limits the excavation of an area and therefore trench and sanitary landfilling methods of disposal are not practicable.

3.5. *Development of the Area*

The Trinity Bay area was first settled as a port for the developing goldfields in the hinterland; currently the City provides goods and services for all of Cape York and much of the Gulf of Carpentaria areas. Because of this function, a large percentage of the workforce is engaged in the retail and wholesale trade and in the transport industry. These industries generate large quantities of waste, generally in the form of bulk packaging.

The recent growth of population, particularly through translocation and immigration of older groups (over 40 years of age) has led to a major increase in building activity. Increases in building and demolition wastes have been associated with this development.

Cairns is the centre of a major sugar cane growing area, mills are located at Mossman, Hambledon, Gordonvale, Babinda, Goondi, South Johnstone and Mourilyan. The magnitude of this industry can be judged from the fact that almost all of the non-urban land in the coastal plain area has been developed for sugar growing. Alienation of this land by sugar growing has accelerated the development of wetlands and left local authorities with no alternative but to use wetlands for refuse disposal.

Examination of population changes in Cairns over the last ten years gives a misleading impression of the City's development. During this period, there has been loss of unskilled and semi-skilled labour, principally displaced by the conversion from manual to mechanical sugar harvesting. Concomitantly, an increase of skilled labour has occurred which has been related to the development of the City as a tourist and service centre. This trend has markedly affected the total amount of refuse handled by Councils in the area as the per capita rate of refuse generation of affluent socio-economic groups is greater than that of less privileged sections of the community.

Over the last ten years tourism has developed as the major industry in the area and Cairns has become a dormitory city for visitors touring the attractions of the area. The influx of large numbers of tourists, particularly in the winter months, increase the rate of generation of refuse in the City.

4. Generation and Composition of Cairns' Refuse

The survey was carried out over three phases during the months of March, July and November to determine the effects of the seasonal conditions on the nature and quantities of refuse generated. Reliable quantitative data were obtained by weighing all vehicles entering the dump to deposit refuse. From this, information was obtained on generators of refuse, namely trade, commercial, Council, and domestic. Variations in domestic refuse generation by area were determined through the random sampling of 600 households.

The following table, Table 1, lists the generators of waste in the City —

Table 1

AVERAGE WEEKLY REFUSE GENERATION IN CAIRNS
(Weights in Tonnes)

	March	July	Nov.	Average
Refuse collected by Council	72.36	84.2	78.0	78.99
Domestic refuse delivered to dump by private persons	105.4	98.7	111.8	104.7
Trade, Industrial, Commercial, Council, Government waste delivered to dump	244	223.7	241.2	234.6
TOTAL Refuse Generation	421.8	406.6	431	418

Trade, industrial, commercial, Council and Government refuse collected by Councils is delivered over the five-day period, Monday to Friday.

It is interesting to note that the refuse brought to the dump by private persons is greater in weight than the refuse actually collected by the Council. This is probably due to a number of factors: firstly, convenience of the service and location of the refuse dump close to the bulk of the population; secondly, the fact that there is only one refuse collection service per household per week; and thirdly, the fact that Cairns people regularly clean up garden areas. (It is suspected that throughout Queensland, refuse taken to local authority dumps by the domestic population is a major percentage of the total waste generated.)

Taking each of these factors in turn —

(i) It is estimated that 25,000 people resided within 5 km (3 miles) of the disposal site. A related factor is knowledge of the existence of the service provided by the presence of the dump. Now that the present site is located well to the north of the City, the observed rate of land filling has continued although the tip is located over ten miles from regular users. (It is interesting to note here that in Townsville prior to cyclone Althea on Christmas Eve 1971, the rate of filling of land was about 3 ha (7.5 acres) per annum. Immediately after the cyclone, emergency disposal areas were created on the sites of former tipping sites. Refuse generated as a result of the cyclone was burnt and subsequently buried. An after effect of Althea has been the increased awareness of the disposal service provided by the Council and it is understood that the rate at which land is being filled has doubled within the last two years.)

(ii) In a questionaire to householders which accompanied the data collection, residents were asked whether they would prefer one or two refuse collection services weekly. Approximately 45% indicated a preference for two services and the remainder (55%) preferred one service. It was noted that many people took their domestic refuse to the tip during the week. Estimated rate of generation of this extra domestic refuse is 20 tonnes per week.

(iii) The long growing season in North Queensland results in exceptional growth of vegetation. Pruning of shrubs monthly and mowing of lawns twice a week are not uncommon. Much of this material is dumped and during the year, garden refuse is generated at a rate of 25 tonnes per week. On a fine weekend during the wet season garden refuse generation may be as high as 35 tonnes but this would have a relatively high moisture content. Building waste from home improvements, including waste new material and demolition material, account for a further 20 tonnes per week.

From the random sample of domestic refuse collected by the Council the weight of refuse generated per person per day throughout the year was relatively constant, varying from 0.31 kg to 0.35 kg. The average weight of refuse generated per person per day was approximately 0.33 kg.

Throughout the year, refuse generated by hotels, motels, etc., was approximately 3.7 tonnes per day in March, 6.8 tonnes per day in July, and 6.2 tonnes per day in November. The higher figures in July and November can be attributed to the larger tourist population.

In the March survey the refuse was weighed only whereas in July and November, the survey was extended to include composition of refuse. Refuse collected from the sample households was sorted into its individual components. The results are indicated in Table 2 —

Table 2

PERCENTAGE COMPOSITION OF HOUSEHOLD WASTE IN CAIRNS

Component	July (Weight) %	November (Weight) %
Glass	22.9	22.9
Paper	25.6	31.1
Metal	8.5	8.4
Food Waste	37.2	34.1
Miscellaneous	5.8	3.5
	100.0	100.0

Compared with other surveys conducted in this country, the high percentage composition of glass is significant. Most of the glass comprises non-returnable stubby bottles (370 ml; 13 fl. oz. capacity).

The refuse taken to the dump by private householders was also sampled and sorted into its components. Space does not permit the presentation of separate tables showing the composition of trade, domestic and Council collected waste.

The composition of all refuse generated in Cairns, including an estimate for refuse generated in March is presented in Table 3.

Table 3

COMPOSITION OF REFUSE GENERATED IN CAIRNS
(Weight in Tonnes)

	March		July		November		Average for Year	
	Weight	%	Weight	%	Weight	%	Weight	%
Garden Refuse	55	13.0	58.8	14.5	54.5	12.6	56.6	13.6
Soil/Mud	97	23.0	30.1	7.4	88.5	20.5	65.6	15.7
Building Waste	30	7.1	74.1	18.2	41.3	9.6	52.0	12.4
Wood	35	8.3	36.6	9.0	50.6	11.7	40.2	9.6
Food Waste	39	9.2	43.6	10.7	39.2	9.0	41.0	9.8
Paper	75	17.8	77.4	19.0	67.6	15.7	73.9	17.7
Metal	32	7.6	30.4	7.5	37.1	8.6	32.8	7.9
Glass	32	7.6	32.3	7.9	31.0	7.5	31.8	7.6
Rubber/Lino/Tyres	2	0.4	2.0	0.5	2.0	0.4	2.0	0.5
Miscellaneous	25	5.9	21.3	5.3	19.2	4.4	21.8	5.2
TOTAL REFUSE GENERATED PER WEEK	422	100.0	406.6	100.0	431.0	100.0	417.7	100.0

To obtain these data, estimates of the composition of all large loads of domestic refuse and all trade, commercial and Council refuse were made as the refuse was weighed.

Although these data are estimates, errors are not considered to be large. Data contained in this table have been used in the consideration of alternatives for future waste disposal in the City.

From Table 3 it is apparent that throughout the year, waste generation is relatively constant and also that the weight of individual components does not vary significantly. The major exception to this is soil and mud which is low in the dry season, increases as drains are cleared later in the year, and remains high during the wet season as mud is removed regularly to maintain flow.

Building waste generation is normally high during the dry season but decreases markedly during the wet season.

5. Factors Affecting Future Refuse Generation in Cairns

5.1. *Population Growth*

Over the next twenty years the population is expected to rise from the 1973 figure of 34,000 to 65,000.

The current age structure of the population is expected to continue to vary from the State average, with a larger proportion of the population aged over forty-five years than in the rest of the State. This is due to the fact that Cairns is developing into a retirement centre.

Over the next ten years, the proportion of the population in the eighteen to thirty-two years age group is expected to increase to approach the State average as employment opportunities increase with the growth of the City. It is anticiated that future employment in Cairns will continue to be concentrated in the tourist and service industries.

5.2. *Urban Land Growth*

At present, 1200 to 1300 ha of land have been developed for residential and industrial uses in the City. Assuming that the residential density will increase slightly, approximately 500 ha of land will be developed as urban land within ten years and approximately 1200 ha, or an area equivalent to the existing urban area, will be developed by the end of the century.

Urban development will encroach upon land currently under sugar production and will also affect mangrove areas at Trinity Inlet, the mouth of the Barron River and Yorkeys Knob (about 13 km north of the City). The total area of mangrove swamp in the district is slightly over 4000 ha but of this, more than 2000 ha is unsuitable for development due to its inaccessibility. Also, most of the swamp area adjacent to the mouth of the Barron River (approximately 450 ha) is unsuitable for urban development because of its proximity to the airport and its very low relief features. This area, however, may be developed as rural land although it is subject to regular flooding (one day in a hundred days). Therefore, in the remaining (approximately 1500 ha) agricultural and urban development will result in the filling of a considerable proportion of this area.

Because of the shortage of available land in the area, the future use of all land in the district must be properly planned. This, by implic-tion, includes planning future land requirements for refuse disposal.

5.3. *Growth in Waste Generation*

In line with general world trends, it has been assumed that weight of waste generated per head will increase and that density of waste will decrease. Hence, the volume of waste generated will tend to increase at a greater rate than its weight.

The role of Cairns as a service city for Cape York and the Gulf of Carpentaria areas is expected to continue. With the continued development of bauxite mining and the development of a major beef industry in the region, considerable quantities of packaging waste will be generated as a result of trans-shipment of goods.

For the purposes of the Cairns Waste Disposal Study, a growth in waste generation of 35.9 kg per person per annum was assumed. This figure is five percent of the total refuse generated per person per annum. With this rate of increase, waste generation per head doubles every twenty years.

The rate of increase of individual components will vary depending on changing patterns of living and increase in population. Garden refuse generation will be dependent on population increase but its rate will probably be slightly lower than the overall rate as it is expected that a larger proportion of the population will reside in flats and home units. Soil and mud generation vary throughout the year and are dependent upon the construction work being undertaken and drain clearing. Usually building waste generation is low during the early months of the year, increases after the wet season and drops towards the end of the dry season with the approach of the holiday and wet seasons. It is estimated that over the next ten years the accelerated growth of the City will result in an increase of building wastes at a rate greater than five per cent.

With the trend towards prepackaged food, fresh food scraps will probably decrease, but this may be complemented, in part, by an increase in processed food waste. Packaging materials, glass, metal and paper will increase in line with this trend.

The growth in total waste generation in Cairns, calculated by taking expected population and waste generation into account, is expected to rise from 418 tonnes per week in 1972 to approximately 1600 tonnes per week in 1990.

6. Alternatives for Refuse Disposal for Cairns

In general, the cost of maintenance of one disposal area is lower than the cost of maintenance of two or three separate disposal areas. Consequently, it has been recommended that Cairns City and Mulgrave Shire Councils should co-ordinate their refuse disposal services.

At this stage of the City's development very little low land remains to be filled, and the low land requirements for flood retention basins have yet to be determined. Therefore, it was assumed in the study that substantial areas of low land in the City would not be available for filling with refuse.

The following options, available for future refuse disposal for the Councils, were examined:—

(a) Continuation of present method.
(b) Modification of present method of disposal.
(c) Pulverisation.
(d) Incineration.

Composting was not considered for the following reasons:—

(a) only thirty per cent of the refuse would be degraded and therefore the remainder would require disposal
(b) composting is expensive and therefore the final product is expensive
(c) it was considered that a market for the compost did not exist in the area.

6.1. *Continuation of Present Method of Disposal*

At present rates approximately 6 ha per annum are filled with refuse, and this land is later used for sports grounds and open space. By 1990, over 28 ha would be filled annually and total land requirements for this period would be approximately 300 ha.

Because the filled area contains bulky objects, broken glass and metal fragments, industrial or urban development, other than for open parkland, would be very costly. As a land shortage exists in the area, creation of large areas of land which are expensive to develop would exacerbate this shortage. Currently, the cost of waste disposal is associated with the cost of covering material. From an estimated expenditure of $75,000 in the financial year 1971/72, the cost of maintaining the current standard of refuse disposal is expected to rise to $345,000 per annum in 1990.

These are conservative costs, based on 1972/73 values, and assume that the cost of excavation, transport and application of covering material would remain constant, that no major engineering works such as bund walls would be required, and that no new plant will be obtained.

6.2. *Modification of Present Method of Disposal*

The survey indicated that approximately 30% by weight, and 25% by volume, of the refuse deposited does not require immediate covering, e.g., car bodies, tree trunks, and bulky objects, and may be covered by soil or mud which is unsuitable for covering other refuse. At present all this refuse, including waste soil, is treated with covering material.

An alternative method of disposal would be to separate bulky waste at the disposal site, dump the waste in predetermined areas of the site and cover with waste soil and/or mud. This may be augmented with quarried covering material when necessary.

As bulky objects would be confined to particular areas of the tip, development costs of the filled land would be significantly reduced. However, the land would be useful only for industrial purposes as the presence of glass and metal residues would make residential development costly. Land filled with the bulky refuse (approximately 58 ha to 1990) would be suitable only for parkland.

Establishment costs for a refuse tip run on the above lines would be higher as the final use of the land must be planned before filling commences, but running costs would be approximately 25% lower as much less covering material would be required.

The annual cost of this alternative would rise from $75,000 in 1973 to $275,000 in 1990.

6.3. *Pulverisation*

In Cairns, land required for disposal of pulverised refuse over the next twenty years would not exceed 120 ha. If the height of the reclaimed area was maintained at 2 m, less than 100 ha would be required in the period to 1990. Also, after consolidation, the pulverised refuse could be excavated and used as a solid fill. If this practice were encouraged, reclaimed land may be filled after excavation.

The capital costs of a plant sufficient to handle refuse generated in the City would be approximately $300,000; maintenance and operating costs of such plant would be $50,000-$60,000 per annum; and the total cost of owning and operating the plant would be $70,000-$80,000 per annum.

6.4. *Incineration*

The results of the survey indicate that over one-third of the refuse generated is not combustible. Waste in this category includes soil/mud, glass, metal and some building waste and miscellaneous material.

The remaining refuse is a good fuel, with a calorific value of 1800-2000 kilojoules per kilogram.

If a high temperature incinerator were constructed, the residual ash and non-combustible refuse would require 70-80 ha of land for disposal over the next 20 years.

Capital costs associated with the establishment of an incinerator in the City would be at least $700,000. This would include the costs of electrostatic precipitators and associated gas cleaning equipment as well as the costs of other works. Annual capital and operating expenses including the costs of labour and plant maintenance would be approximately $150,000.

7. Conclusions

It was apparent that a land shortage exists in the Cairns area and that the requirements for open parkland would be much less than the predicted generation rate for this type of land. Irrespective of the cost of the current method of land filling and the modified method of land filling described above, the land requirements for refuse disposal, when considered in terms of available land and requirements for open parkland, render conventional land filling an unacceptable form of disposal on a long-term basis. Consequently, in the future, refuse should be either pulverised or incinerated prior to disposal.

As the area is developing and solid fill is required for land reclamation projects, the adoption of pulverisation or incineration would have the advantage that useful filling material would be obtained.

Because of the cost and other advantages, it is concluded that pulverisation offers the most practicable long-term solution to the City's waste disposal problem.

The survey has also provided a wealth of data as a result of which various recommendations relating to collection of refuse, salvaging possibilities and improvements in waste management have been made.

ACKNOWLEDGEMENTS

The authors acknowledge the support given by the Co-ordinator-General and his Department, and the staff of the Cairns City and Mulgrave Shire Councils for their valuable assistance and co-operation.

Permission of the Co-ordinator-General to present this paper is gratefully acknowledged.

REFERENCE

Co-ordinator-General's Department, Queensland — Cairns Region Waste Disposal Study, February 1973, pp. 82.

PAPER 11

The Liquid Waste Survey in the Metropolitan Sydney Area

By A. G. Forrester*

Introduction

The closure, in December 1969, of most local authorities depots for the reception of liquid wastes precipitated the largely hidden problem of their disposal. Waste hauliers were left with considerable gallonage to dispose of and a crisis situation developed with the strong possibility that less reputable contractors would dispose of liquid wastes into streams, rivers, sewers, manholes and in open country. The Barton report submitted in May 1970 highlighted this disposal problem and for the first time an idea of the scale of the problem was generally available. The situation revealed a potentially dangerous environmental situation. The report recommended the establishment of one authority to control waste disposal and recommended legislative procedures, including registration and licensing of waste generators, hauliers and disposal points, which would be desirable for complete pollution control. Subsequently in June 1971 The Metropolitan Waste Disposal Authority was formed.

Background

The Authority, after studying the Barton report, arranged for preparation of an interim report aimed at updating the Barton report of 1969. The results of this study indicated that the quantity of liquid waste being produced in Sydney was to some degree a matter of speculation. Discussions were held with industrial firms who had an interest or expertise in the treatment of liquid wastes, regarding possible solutions to the disposal problem. Resulting from these discussions with industry, in November 1971, by advertisement, submissions were invited from organisations with major technical and financial resources capable of designing, erecting and operating a Liquid Waste Treatment Plant. Twenty-six submissions were received. Subsequent detailed discussions over the next few months with these organisations revealed that not sufficient detailed information was available for design purposes and these organisations were reluctant to proceed. The need then for a comprehensive liquid waste generation and disposal survey was established and early in 1972 the possibility of the Authority combining with the Metropolitan Water Sewerage and Drainage Board, in conducting such a survey was investigated. Due to the differing needs of the two Authorities it was decided by the Waste Disposal Authority in May 1972 to engage consultants to carry out a comprehensive Liquid Waste Survey within the Metropolitan Area.

The Survey

(i) *Objectives*

The objectives of the survey were:—

(a) To determine the total volume and characteristics of liquid wastes generated in the Metropolitan Waste Disposal Region.

(b) To determine the future anticipated growth patterns of industry with particular reference to planned future disposal facilities.

(c) To determine the future growth of population and transportation.

(d) To investigate overseas trends in waste treatment and disposal methods.

(e) To investigate possible plant sites.

(f) To investigate the possibility of resource recovery.

The Consultants established a programme for the survey and survey evaluation which extended over a period of 12 months, an objective which was subsequently fulfilled.

The survey area, being the Metropolitan Waste Disposal Region, approximates the area covered by the Sydney Outline Plan prepared by the State Planning Authority and by the Sydney Area Transportation Study. The statistical data thus available upon population, industrial and transportation trends was of immense use in evaluating the survey data and in preparing future focecasts.

Not surprisingly, no single listing of potential waste generators suitable for the survey was available. The principal source for the preparation of the master list was the record of factory registration applications maintained by the Department of Labour and Industry. This basic listing was supplemented by information obtained from the Barton Report, the M.W.S. & D.B., waste disposal contractors, trade

* Deputy Director, Metropolitan Waste Disposal Authority, New South Wales.

directories and various types of institutional establishments.

Once the master list had been prepared and recorded on a card index, a system developed by the Commonwealth Bureau of Census and Statistics, known as the Australian Standard Industrial Classification (A.S.I.C.), was adopted. All companies on the master list were classified according to their A.S.I.C. code. A further identification required was their geographical location and the postcode was selected for this purpose. To cover the case where there were two similar concerns with the same postcodes two further digits were added.

Further consideration clearly indicated that the master list required additional stratification based on the anticipated or expected waste generation of specific individual industries and the master list was divided into three groups. Of these groups:

Group one contained 675 establishments thought most likely to generate major quantities of liquid waste.

Group two contained 6784 establishments from the main range of manufacturing industry and it was expected that considerable variation in waste generation would occur within this group.

Group three contained approximately 17,000 establishments which because of their activity or scale were unlikely to generate any significant volume of liquid waste.

Each establishment was thus individually identified by an eleven digit number.

(ii) *Questionnaire, Pilot Survey, Survey Techniques*

The questionnaire required a format that was easily followed, and that both a large technically oriented organisation and a single craftsman could understand. With this in mind a draft questionnaire was developed and a pilot study launched in August 1972. The pilot study showed that while the information required was generally available from industry, the questionnaire format was too complex, and that the response rate to a particular question was directly related to the ease of answering or understanding that question.

The final format of the questionnaire which was developed and published in two editions was attractive and comparatively easy to understand and complete. The first of the editions was applicable to the major generators in group one and was more searching than the second edition which was used for the generators of groups two and three. Both editions of the questionnaire were divided into five parts requesting the following information.

Part A — General information regarding the industry.
Part B — Details of the liquid waste(s) generated by the industry.
Part C — Details of "acceptable discharges" generated by the industry.
Part D — Details of grease trap wastes.
Part E — Preliminary details of solid waste generation.

The majority of the information requested was in Part B and to help in classifying the wastes they were divided into 12 broad streams which were further classified into 29 waste categories. This classification was later modified to 22 "feedstocks".

Sections 49 and 50 of the Waste Disposal Act

The Authority has powers under Sections 49 and 50 of the Waste Disposal Act to require any person to furnish information in regard to the creation, transportation, reception, treatment, storage or disposal of waste. Such information, however, if the person furnishing it wishes may not be admissable as evidence which might tend to incriminate him. Further the Authority or its staff may not disclose any information so obtained. Substantial penalties are provided for in both cases should a breach of either Section occur.

Survey Techniques

As outlined previously, the master list was stratified into groups one, two and three, which were established on the basis of the waste generative potential of A.S.I.C. classifications or specific establishments. Each of the groups then created was considered as a separate entity for the purpose of the survey and differing techniques were employed in the method of data collection and its processing. It is therefore convenient to consider the techniques employed for each group separately:—

Group One of the Survey Strata

It was decided that each establishment included in the group one list was significant to the entire survey and that a person to person approach should be made to each establishment to ensure the maximum accuracy in the response. For this reason a team of up to eight personnel, experienced in engineering or related disciplines, was utilised to distribute, discuss and assist in the completion of group one returns.

The surveyor's first action was to establish telephonic contact with the representative of the establishment responsible for the disposal of trade waste and make an appointment to discuss the survey. The questionnaire was delivered and explained to the respondent. A tentative appointment was made to collect the completed questionnaire. Prior to the pre-arranged collection date, the surveyor contacted the respondent and confirmed that the questionnaire had, in fact, been completed and a firm date for collection was fixed. The surveyor then re-visited the establishment and discussed the completed questionnaire with the respondent. The questionnaire was returned to survey headquarters where it was vetted and coded by either the surveyor or another senior member of the projection team before being sent to the data processing agency for card punching.

The field work on group one was carried out during the period mid-September to late December 1972, and the main volume of the group one questionnaires was in the hands of the data processing agency by the end of December 1972.

Group Two of the Survey Strata

It was decided to handle the group two strata of the master list as a 100% postal census utilising the basic questionnaire. After allowing a reasonable period for response, approximately a fortnight, a phone call was made to the establishment to ascertain whether they had received the questionnaire, whether there were any problems in completing it and the probable date of its return. To undertake this phase of the operation a team of 5 telephonists was employed.

Questionnaires were returnable in a reply paid envelope and were processed by a team of editors on their return. Queries arising from the completed questionnaire were referred to a senior member of the project team who contacted the establishment and clarified doubtful points of discussion.

The telephonists conducted a second round of follow-up calls on non-respondents after working through the section of the group two survey population in their records. Control was maintained over the system and follow-up calls to persons who had already replied were avoided by the establishment of a system which involved an action sheet recording the current status of each respondent in the process.

The despatch, expediting and coding of the group two strata of the master list was commenced on 6th October, 1972. The telephone follow-up and the coding of returned questionnaires was substantially completed by the end of December 1972, with only a limited number of returns being received early in January 1973.

Group Three of the Survey Strata

By hypothesis, this group contained establishments expected to be minor or non-generators of liquid waste. Approximately 17,000 establishments were allocated to the group on the basis of their scale of operations or their identifications as members of an A.S.I.C. classification unimportant with respect to liquid waste generation potential. One hundred per cent contact with members of this group would clearly be associated with considerable cost and, if the stratification of the universe of establishments into groups was accurate, then the value of results achieved by this approach would not be commensurate with the costs of collection of the data.

Following this philosophy it was decided to adopt a sample survey approach to data collection for this group. Survey methodology for group three was therefore quite different to that used for other groups of establishments in which all establishments in the group were contacted.

Evidence of the incidence of liquid waste generation was required for each A.S.I.C. classification within the group. For this reason, members of each A.S.I.C. classification in the group were sampled independently. The sample size necessary to be drawn from a particular A.S.I.C. classification was dependent on a number of factors such as the prevalence of the attribute of interest in the classifications to be sampled and the degree of accuracy required to be associated with the estimates to be derived from the sample observations.

In this instance, it was agreed that the attribute of primary interest was the occurrence of liquid waste as an establishment. That is, for an A.S.I.C. classification, emphasis was to be placed on estimating the proportion of members of the classification that generate liquid waste. For a specific random choice of members this proportion could easily be determined.

General Observations on the Survey Results

The results obtained from the survey represent the most accurate possible picture of liquid waste generation in the Sydney Metropolitan Region. The response rates to the survey were of a high order and the pattern confirms the validity and accuracy of the division of the universe into three groups.

The "face to face" interview technique adopted for group one produced an exceptionally high response rate of 99%.

A very reasonable response of almost 75% was received from the establishments forming group two which was the 100% postal census.

The lowest response rate was obtained from group three which was the sample survey conducted of establishments which it was anticipated would not be significant waste generators. In this case a response rate of approximately 63% was attained. Generally the standard of the responses was high and respondents did not appear to have encountered any substantial difficulty in completing the questionnaire.

Some difficulty was experienced with Part C of the questionnaire which dealt with "Acceptable Discharges". During the currency of the survey it became evident that many respondents were not certain whether or not an effluent discharged to the sewer or to a stormwater channel was within the standards specified in the Board's regulations.

Many respondents discharge large volumes of waste contaminated by proprietary brand products, i.e., dyes, printing inks, etc., into the sewer or to a stormwater channel and have declared such effluents as "Acceptable Discharges". In many cases it is doubtful whether this declaration has been made on the basis of analysis or on the assumption that, since the M.W.S. & D.B. had not so far acted to prevent the discharge, it could be regarded as acceptable. Any errors of classification which have been committed under this confusion of definition do not have any significant bearing on the value of the survey results.

The results disclosed that a volume of liquid waste of more than 1,505,000,000 gallons of liquid waste are generated annually of which 96.6% is disposed of by discharge to the sewer or stormwater, discharge to the ocean or by use for irrigation purposes. To prevent distortion of the results by the large volumes of waste which would not be transported to a treatment plant, a net volume was derived by deducting waste disposed of by the methods indicated from the gross volume of 1,505,314,000 gallons. Any erroneously declared acceptable discharges to sewer or stormwater would have been similarly deducted from the gross volume had they been declared as wastes.

It is important to stress that it was recognised that many wastes currently discharged to the sewer or the stormwater may require additional treatment to comply with the Board's standards. The treatment process will give rise to a sludge or a residue in many cases and allowance was made for this factor in the calculation of design volumes for the plant.

The survey indicated that, of the 9,256 establishments forming the survey universe, the 371 respondents with liquid waste in group one account for 1,297,564,000 gallons. Thus, approximately 4% of the universe is responsible for over 87% of the total wastes generated.

Survey Summary

The Survey determined the volume of liquid waste (defined as "A liquid slurry or sludge which is not reprocessed and which may not be discharged (a) into a sewer or where a sewer is not available (b) into a stormwater channel, under the gazetted M.W.S. & D.B. regulations") which is currently being generated within the Waste Disposal Region. Following analysis, the gross volumes were modified by deduction of those volumes of waste which could continue to be discharged to the sewerage or stormwater drainage system and a net volume was determined.

The following is a list of the 29 Waste Categories used in the Survey:—

Definition of Waste Categories

Waste Stream	Description	Waste Category
PAINTS	acryl, alkyd or vinyl base, printing inks	01
RESINS	phenolic, alkyd or vinyl base	02
SOLVENTS	chlorinated — trichlorethylene	03
	non-chlorinated — benzene, alcohols, esters, ketones	04
OILS	waste oils (lubricating, hydraulic mineral)	05
	oil/water mixtures (cutting oils, soluble oils, oil/water emulsions)	06
	oil sludges (high dirt content, tank bottoms, sludge pit residue)	07
	vegetable oils and tallow derivatives (fatty acids, soaps, drawing compounds)	08
EMULSIONS	rubber, latex	09
	bituminous based	10
ORGANIC WASTES (from living matter)	animal waste	11
	bacterial sludge	12
	settling pond sludge	13
	vegetable wastes	14
OTHER ORGANIC CHEMICALS	chlorinated — D.D.T., chlorbenzene	15
	non-chlorinated — xanthates, alkyl sulfonates	16
ACID WASTES	sulphuric	17
	sulphuric + others	18
	nitric	19
	nitric + others	20
	phosphoric	21
	chromic	22
	hydrochloric	23
	other acid wastes	24
ALKALI WASTES	caustic soda, lime	25
NEUTRAL SALTS	chrome, iron, ammonium and various metal salts, not acidic or alkaline	26
PLATING WASTES	cyanide wastes	27
	other plating wastes not included in any other category	28
OTHER INORGANIC CHEMICALS	to be specifically detailed by the respondent	29

The survey found that the following quantities of liquid waste are generated. These quantities, designated Survey Gross Volumes, are tabulated as nine general waste types:

SURVEY GROSS VOLUMES

Paints, inks, solvents	34,515,000
Oils	25,730,200
Latex/water emulsions	190,500
Organic (food) wastes	1,111,397,900
Organic (chemical) wastes	56,365,100
Acids	9,961,200
Alkalis	119,753,700
Neutral salts and neutralised Acids	72,135,700
Other inorganic wastes	75,264,700
TOTAL	1,505,314,000

The survey volumes were analysed and deductions made to compensate for the quantities of wastes which should continue to be discharged to the sewers or stormwater channels in accordance with the M.W.S. & D.B.'s policy of accepting above-limit wastes at extra cost in some cases, or are used for irrigation, or are discharged to sea by pipelines or otherwise disposed of by acceptable methods. These volumes known as the Survey Net Volumes are tabulated below under similar categories to those given for the gross volumes:

SURVEY NET VOLUMES

Paints, inks, solvents	1,804,700
Oils	5,575,300
Latex/water emulsions	189,400
Organic (food) wastes	16,026,200
Organic (chemical) wastes	934,200
Acids	2,436,400
Alkalis	5,403,200
Neutral salts and neutralised Acids	8,329,200
Other inorganic wastes	9,655,000
TOTAL	50,353,600

The net waste volume likely to be treated in a liquid waste plant may be further reduced because a number of wastes are reprocessed by the manufacturer, or other industries; process changes in hand or contemplated will further reduce or eliminate some wastes entirely, some wastes are satisfactorily disposed of on land owned by the producer, some industries will cease operation and some manufacturers are installing their own treatment plant.

When these reductions are taken into consideration the undermentioned estimated flow of wastes to a treatment plant would be:

Total net waste volume likely to be treated at a liquid waste treatment plant	20,903,000 gals.
Physiochemical section of plant	3,932,000 gals.
Incineration section of plant	11,215,000 gals.
Biological section of plant	5,756,000 gals.

In order to arrive at a final figure on which to design the treatment plant suitable allowances were made to cover survey accuracy, expansion of waste generation up to the year 1980 and an engineering design factor.

1980 total design volumes	31,014,000
Physiochemical section of plant	5,821,000
Incineration section of plant	16,612,000
Biological section of plant	8,581,000

Survey Recommendations

1. *Site selection*

Upon receipt of the early results from the survey, preliminary treatment plant design was carried out to ascertain site area and service requirements. The pattern of liquid waste generation was studied relative to municipal area and future industrial growth in the Sydney Region. Comparisons were made of the haulage mileage from the load centres of generation to likely available sites in the Sydney Region which satisfied the site area and service requirements.

The possibility of establishing a number of small treatment plants close to the various load centres of waste generation was investigated. This treatment was finally discarded in favour of a single central treatment plant for the following reasons:

(a) Higher capital cost.
(b) Higher operating cost.
(c) Lack of sufficiently large load centres generating wastes that may be used as treatment agents for other wastes.
(d) Lack of operating economies that may be gained by bulk storage and blending of many wastes at a centralised plant.

Two interim reports were received during the course of this investigation, one detailing particulars of likely available sites, and the second proposing the most favoured sites.

The final site selected is currently subject to negotiations and by the time this paper is presented further details will be available.

(ii) *The Process Plant*

It is envisaged that liquid wastes be brought to the plant in road tankers by the generator or a contract carting service. Facilities would be available for quick discharge of the waste, its categorisation and despatch to the specific storage facilities associated with the necessary process stream. Generally, there would be three main process streams:

(a) Incineration.

(b) Physiochemical.

(c) Biological.

The wastes going to the incinerator section would comprise all the combustible material such as oils, resins, paints and solvents together with high sulphur spent caustics and chlorinated hydrocarbons. Extensive gas cleaning equipment would be fitted to ensure that the emissions of gas and particulate matter are within the limits prescribed by the Clean Air Act.

Inorganic waste streams consisting mainly of acids, alkalis and neutral salts would be processed in the Physio-chemical Section to produce an effluent acceptable to the M.W.S. & D.B. and an inert solid suitable for landfill. Treatment would include neutralisation, precipitation of heavy metals, oil and solids removal and detoxification of cyanide. In addition, facilities would be available to treat special wastes that may occasionally arrive at the plant.

Organic material such as tannery, gelatine and food factory wastes would be dewatered in centrifuges and sludge obtained burnt in the incinerator. The liquid phase would be pumped to the biological plant where it would successively pass through sections for oil removal, biological filtration, activated sludge treatment and sulphide removal in order to produce a clear effluent acceptable for discharge via the sewer.

The plant would be completely self-contained with administrative offices, laboratory, workshops, stores and personnel amenities and the site would be fully landscaped.

The process plant criteria were studied under the following headings:

(i) Choice of processes available.

(ii) Residue uses and recycling.

(iii) Emission standards.

(iv) Engineering standards.

(v) Utility services.

(vi) Ground conditions.

(vii) Operational philosophy.

(a) Receival and storage.

(b) Incineration.

(c) Physio-chemical treatment plant.

(d) Biological treatment plant.

(e) Residue handling and disposal.

(viii) Costing criteria.

In particular the operating cost for treating the wastes was examined and estimates made of the actual cost of treatment of the waste to be received. The costs were examined for two primary proposals, i.e., proposal A and proposal B. Proposal A being the plant design for the projected volumes of waste for 1980, whilst proposal B was a modified plant, eliminating biological processes.

Final Proposal "Minimum Plant"

The Authority, after an exhaustive study of the Consultants' Report, and after examining the reaction of industry to the number of regulatory and cost constraints which were having and would continue to have a bearing upon industry's profitability, decided that the concept of a "Minimum Plant" was the most suitable at present.

The "Minimum Plant" is a further modification of the Consultants' "Proposal B" and involves the concept of a plant of as simple construction as possible, designed to treat most wastes expected in the early years of operation with the capacity of evolving to meet future requirements.

The incineration section will process biological wastes in addition to those for incineration, and whilst the plant will have somewhat less flexibility and storage capacity than the Consultants' proposal, pollution control standards will not be compromised. The selection of the incinerator for the Minimum Plant has posed a considerable problem. It is proposed that selection be based upon pilot scale tests currently being conducted by the C.S.I.R.O.

Therefore, the plant proposed can evolve to include biological treatment, resource recovery and solid waste treatment facilities.

The Liquid Waste Survey Summarised

The problems encountered in defining the liquid waste crisis in the Sydney region led to the decision of the Metropolitan Waste Disposal Authority to commission consultants to carry out a comprehensive Liquid Waste Survey. The objectives and design of the survey, the development of the questionnaire format and the survey methodology adopted to obtain the data necessary to quantify the extent of the problem, have been discussed in some detail.

The generally good response rate and the

minimal problems encountered during the survey, allowed considerable reliance to be placed on the results obtained. Modification of the gross survey volumes returned was necessary to allow for acceptable disposal methods and likely future in plant treatment facilities. The net volumes thus obtained were used as the basis for the preliminary design of a regional liquid waste treatment plant. An exhaustive study of the preliminary design by the Authority led to the concept of a minimum plant which is at present being further developed.

ACKNOWLEDGEMENTS

Thanks are due to the Authority's officers Mr. D. R. M. Thomas, Technical Manager, and Mr. K. J. Cleland, Chemical Engineer, for their assistance in the preparation of this paper.

PAPER 12

A Survey of Community Solid Waste Practices in Australia

By J. J. Varjavandi* and T. J. Fischof†

Introduction

The generation and disposal of waste are inherent to life itself and have presented problems to the human community for as long as mankind has existed.

In the past, up to and including the first half of this century, it was a relatively easy matter to collect and remove the waste products and either dispose of them on dumps or destroy them by fire in simple furnaces.

Since the advent of the industrial revolution, however, the variety of industrial and municipal wastes have been increasing and the quantities of refuse of different kinds, generated by both industries and individuals, have now reached alarming proportions.

It has now become essential to stimulate the growing awareness in every member of the community of the need for the protection and conservation of the environment. It can no longer be a question of relegating the responsibility for a pollution-free environment to any particular governmental or municipal body. It is each and every member of the community who contributes either directly or indirectly to the rapidly increasing mountains of waste. Therefore it is the community as a whole which must deal with this situation which is already affecting our way of life and could seriously threaten the health and welfare of future generations.

Today the problems of controlling pollution are world-wide, transcending urban and even national boundaries. In many other industrial nations, in America, Europe and Asia, during the past two decades a number of different processes have been developed and implemented to cope with the problems of effective waste disposal.

Here in Australia we are in the early stages of such development. In the past, the conventional "garbage tip" has in general been considered adequate for our needs. However, the steadily increasing population, its concentration in limited high density urban areas, and the alarming growth in the "per capita" generation of refuse which is occurring throughout the world, make it essential that we now rapidly conceive and implement more effective means of dealing with these avalanches of waste.

In order to obtain a comprehensive background of information in the field of solid wastes, Australian Consolidated Industries carried out this present survey, being the first on a nation-wide basis. This survey is part of a broader program within the Group on the general problems related to solid waste disposal and recycling.

It was the objective of the study to provide, as accurately as possible, a quantitative appreciation of the present and future situations resulting from the generation and disposal of municipal wastes, with emphasis on metropolitan areas. With the information contained in the final report it is hoped that the Federal and State Governments, Municipal Authorities and professional planners will be assisted in taking effective action, possibly on a regional basis, in their endeavours to provide adequate and lasting solutions in this essential aspect of our community life.

Field Survey

The survey was conducted in co-operation with Intercont Development Corporation Pty. Ltd., Environmental Consultants. Discussions were held with government agencies at Federal, State and Local levels prior to conducting the survey. In most cases ministerial approval and support were obtained in writing, to help secure the co-operation of the municipal authorities.

A questionnaire, as shown in Appendix I, was sent to every municipal office in metropolitan regions throughout Australia and to provincial urban centres of significance, generally with populations of 10,000 and over. A total of 305 questionnaires was sent out, each with a copy of the relevant Minister's letter supporting the survey.

The response varied in the major metropolitan areas between 60% and 70% on a strictly numerical basis. However, the actual informative value was even higher when judged

* Manager, A.C.I. Environics.
† Manager, Development Department, A.C.I. Technical Centre Pty. Ltd.

Appendix 1

FIRST NATIONAL SURVEY of WASTE PRACTICES – AUSTRALIA 1972

WASTE COLLECTION

1 What is the major problem faced by your company in the area of waste disposal and collection?

Briefly describe . . . .

. . . .

. . . .

. . . .

If more space is required please attach a separate sheet.

2 How much waste was collected by your company during the fiscal year 1971/72? (ie. July to June) boundaries during the fiscal year 1971/72?

Please tick the appropriate square to indicate if this is a measured or estimated value.

If a type of waste is not collected please write "NC" in the appropriate space.

If an individual quantity is not known please tick the "Don't Know" column for the specific waste type.

	QUANTITY COLLECTED									
TYPE OF WASTE	Don't Know	Cubic Yds. (loose)	Meas-ured	Estim-ated	Tons	Meas-ured	Estim-ated	Imperial Gallons	Meas-ured	Estim-ated
Municipal Garbage										
Residential Rubbish										
Gully Wastes										
Street Refuse										
Industrial Solid Wastes										
Industrial Liquid Wastes										
Other (specify)......................										
PLEASE TOTAL COLUMNS										

———oOo———

WASTE DISPOSAL

3 How much waste was received during 1971/72 at the disposal facilities owned by your company?

If no facilities are owned please tick ☐ and proceed to question 5.

Include quantities received from other municipalities, contractors, firms and individuals.

If some types of waste were not received please write "NR" in the appropriate space.

If an individual quantity is not known please tick the "Don't Know" column for the specific waste type.

	QUANTITY RECEIVED									
TYPE OF WASTE	Don't Know	Cubic Yds. (loose)	Meas-ured	Estim-ated	Tons	Meas-ured	Estim-ated	Imperial Gallons	Meas-ured	Estim-ated
Municipal Garbage										
Residential Rubbish										
Gully Wastes										
Street Refuse										
Industrial Solid Wastes										
Industrial Liquid Wastes										
Other (specify)......................										
PLEASE TOTAL COLUMNS										

4 What waste disposal facilities are owned by your company? (see table over page.)

(Examples of facilities include open dump, controlled tip, sanitary landfill, incinerator, transfer station, pulveriser, composting or compacting plant).

If no facilities are owned please put "NA" in the left hand column of the table below and proceed to question 5

Estimated Capacity refers to the remaining capacity of landfill sites at 30th June 1972.

Design Capacity refers to the theoretical design capacity of a mechanised facility.

Total Operating Costs should include all costs incurred in running a disposal facility including capital, operating and maintenance charges during 1971/72.

If an individual charge is not known please write "DK" in the appropriate space on the table over the page.

Appendix 1 (contd.)

TYPE OF FACILITY	ADDRESS	Estimated Capacity (Cu. Yds.)	Design Capacity (Ton./Hr.)	OPERATING COSTS 1971/72	
				Total Costs ($)	Capital Investment Expenditure ($)
			PLEASE TOTAL COLUMNS		

5 What other waste disposal facilities were used by your company during 1971/72?

Include both privately and publically owned facilities.

If no other facilities were used please write "NA" in the left hand column of the table below and proceed to question 2.4.

If an individual charge is not known please write "DK" in the appropriate space in the table.

TYPE OF FACILITY	ADDRESS	Owner of Facility	Charges levied on your Municipality for use of Facility ($)
		TOTAL CHARGES	

What planned or future possible waste disposal facilities are available to your company?

TYPE OF FACILITY	ADDRESS	Owner of Facility	Estimated Capacity (Cu. Yds.)	Design Capacity (Ton./Hr.)

What was the average man-power and equipment used by your company for the collection and disposal of waste during the fiscal year 1971/72?

MAN–POWER & EQUIPMENT	COLLECTION	DISPOSAL	TOTAL
Collectors & Drivers (man years)			
Other Labour (man years)			
Open Trucks (number)			
Bulk-Liquid Road Tankers (number)			
Compactor Trucks (number			

Name of Company ....

Postal Address , ... Post Code........

Telephone Number ....

Officer for Further Contacts ....

Signed .... Position .... Date ....

Thank you for your co-operation

—————oOo—————

Please use the reply paid envelope and return to :-

Australian Consolidated Industries Limited,
Box 5094 B.B., G.P.O. Melbourne, 3001.
Telephone: 60-0441
Attention: N.J. McCann

on the importance of the individual area, e.g., the councils of all city areas comprising the central business district in each state capital returned carefully completed questionnaires, which allowed a realistic evaluation of the effect of the high number of commuters on the generation of municipal refuse in these areas.

Additional information from other documentation and local sources and the consideration of known development trends abroad provided sufficient data for a realistic appreciation of the present situation and possible future developments in the field of solid waste practices in the Australian communities.

Survey Zones

For ease of reference the country was divided into eight zones comprising the six States and two Territories. Data from all the questionnaires was fed into a computer program, and the results were tabulated for several variables from the year 1966 to the year 2000. A typical computer read-out chart is shown in Appendix II.

CONSOLIDATED SURVEY RESULTS (AUSTRALIAN COMMONWEALTH)

Population Projections

Since waste generation is a function of population, population projections were used as the basis of the waste quantity projections.

The following comparative population projections are based on official forecasts, published by the Commonwealth Bureau of Census and Statistics, Canberra, in 1971.

The lower figures are based on average annual net migration during the four-years-period from 1966 to 1970. The higher figures are estimates allowing for an annual net migration of 140,000.

	1972	1980		1990		2000	
		low	high	low	high	low	high
N.S.W.	4,696,000	5,311,000	5,400,000	6,214,000	6,470,000	7,178,000	7,897,000
VIC.	3,577,800	4,098,300	4,140,000	4,859,200	5,060,000	5,712,700	6,284,000
QLD.	1,898,200	2,144,000	2,160,000	2,562,000	2,660,000	3,023,000	3,329,000
S.A.	1,196,500	1,376,500	1,385,000	1,627,300	1,690,000	1,887,000	2,075,000
W.A.	1,065,800	1,386,500	1,420,000	1,853,800	1,950,000	2,384,000	2,623,000
TAS.	395,600	436,000	442,000	493,000	576,000	550,000	605,000
A.C.T.	165,300	230,000	242,000	346,000	373,000	479,000	550,000
N.T.	96,200	139,000	139,000	215,000	215,000	310,000	310,000
AUST.	13,091,400	15,121,300	15,324,000	18,170,300	18,994,000	21,523,700	23,373,000

MOVEMENT from 1972 to 2000

Low 8,432,300 persons	Growth 64.5% over 28 years
High 10,281,600 persons	Growth 78.6% over 28 years

Recently published population figures indicate that by 1972, despite the slowdown of immigration and an increase of emigration over the last two years, the actual figures exceeded the lower average figures of the 1971 forecasts in all states and territories, except in Tasmania, where the projected figure was reached but not surpassed.

Already, official attempts are being made, for economic reasons, to reverse the migration slow-down trend and to re-invigorate a selective migration program. It is possible that during the next three decades even the high forecasts of this report may be exceeded.

This increase in national population up to the end of the century will make it mandatory to abandon the present piecemeal approach to effective waste management and to find long-range and lasting solutions to this vital community problem.

Metropolitan Areas

Since the potential growth centres in Australia, and consequently also the problem areas in the sector of solid waste management, are the capital cities of states and territories, only these have been analysed in detail.

POPULATION
(projected to the end of the century)

	1971	1980	1990	2000	movement	1971/2000
Sydney Region	2,922,356	3,366,500	4,183,000	4,907,000	1,984,644	67.9%
Melbourne	2,498,023	3,191,760	3,949,700	4,757,000	2,258,977	90.5%
Brisbane Region	968,363	1,206,000	1,509,000	1,885,000	916,637	94.8%
Adelaide	842,613	1,085,300	1,422,000	1,777,800	935,187	111.0%
Perth	701,392	979,000	1,315,000	1,643,000	941,608	134.2%
Canberra	141,575	263,000	408,000	548,000	406,425	287.3%
Hobart	153,024	202,500	260,000	298,000	144,976	94.8%
Darwin	36,828	68,000	118,000	180,000	143,172	389.0%

Appendix 2

Sequence Number	Name of Community	Postcode	Statistical Division	Survey Zone	Status Code	Area in Acres	Area in Sq. Km.	1971 Density Persons/Sq. Km.		1971 No. Dwellings
2026	Mosman	2088 Sydney		II N.S.W.	M	2,150	9	3,358		11,251
	Year		1966	1971	1975	1980	1985	1990	1995	2000
	Population		28,141	29,220	30,000	30,500	31,000	32,000	33,000	34,000
	Total Waste Tonnes/Year			21,171	23,214	29,056	36,887	47,538	61,429	78,802
	Domestic and Municipal Wastes Tonnes/Year			12,692	14,454	17,144	20,933	26,514	34,087	44,055
	Loose Volume Cubic Metres/Year			95,268	108,235	140,819	185,113	246,789	329,813	437,979
	Progressive Landfill Consumption including Cover in Cubic Metres				204,000	508,000	892,000	1,384,000	2,020,000	2,837,000
2027	North Sydney	2060 Sydney		II N.S.W.	M	2,585	10	5,058		22,508
	Year		1966	1971	1975	1980	1985	1990	1995	2000
	Population		51,756	52,920	53,000	54,000	55,000	55,000	56,000	56,000
	Total Waste Tonnes/Year			38,342	41,011	51,443	65,444	81,705	104,244	129,794
	Domestic and Municipal Wastes Tonnes/Year			22,986	25,535	30,353	37,139	45,570	57,645	72,562
	Loose Volume Cubic Metres/Year			172,538	191,216	249,319	328,426	424,169	559,683	721,377
	Progressive Landfill Consumption including Cover in Cubic Metres				364,000	902,000	1,583,000	2,438,000	3,521,000	4,881,000
2028	Parramatta—City	2150 Sydney		II N.S.W.	M	12,190	49	2,244		31,230
	Year		1966	1971	1975	1980	1985	1990	1995	2000
	Population		107,047	110,717	112,000	115,000	117,000	118,000	120,000	122,000
	Total Waste Tonnes/Year			80,217	86,666	109,555	139,218	175,293	223,380	282,765
	Domestic and Municipal Wastes Tonnes/Year			48,090	53,962	64,641	79,004	97,769	123,954	158,081
	Loose Volume Cubic Metres/Year			360,977	404,079	530,956	698,652	910,035	1,199,320	1,571,571
	Progressive Landfill Consumption including Cover in Cubic Metres				767,000	1,909,000	3,357,000	5,187,000	7,509,000	10,456,000
2029	Penrith—City	2750 Sydney		II N.S.W.	M	100,700	408	147		15,630
	Year		1966	1971	1975	1980	1985	1990	1995	2000
	Population		46,391	60,242	78,000	112,000	136,000	175,000	189,000	205,000
	Total Waste Tonnes/Year			43,647	60,356	106,697	161,826	259,971	351,823	475,138
	Domestic and Municipal Wastes Tonnes/Year			26,166	37,580	62,955	91,834	144,996	195,227	265,628
	Loose Volume Cubic Metres/Year			196,411	281,412	517,106	812,108	1,349,628	1,888,929	2,640,756
	Progressive Landfill Consumption including Cover in Cubic Metres				492,000	1,495,000	3,085,000	5,596,000	9,180,000	14,023,000
2030	Randwick	2031 Sydney		II N.S.W.	M	8,495	34	3,589		40,209
	Year		1966	1971	1975	1980	1985	1990	1995	2000
	Population		113,731	123,404	124,000	126,000	128,000	130,000	131,000	132,000
	Total Waste Tonnes/Year			89,409	95,951	120,034	152,307	193,121	243,856	305,943
	Domestic and Municipal Wastes Tonnes/Year			53,600	59,743	70,825	86,432	107,711	135,316	171,039
	Loose Volume Cubic Metres/Year			402,341	447,373	581,743	764,337	1,002,581	1,309,258	1,700,389
	Progressive Landfill Consumption including Cover in Cubic Metres				851,000	2,107,000	3,692,000	5,703,000	8,246,000	11,444,000

The population projections for metropolitan areas are based on recent and detailed information from regional transportation studies and other demographic surveys.

Local demographers admit that the accuracy of their forecasts could vary between ± 20% by 1990 and even ± 30% by the year 2000.

Waste Generation

For the determination of current "per-capita" waste generation figures the "recorded as collected" waste volumes, provided by the surveyed communities, have been used. An additional allowance of 30% for solid wastes disposed of by the individuals themselves, in domestic incineration units, on tips or elsewhere has been added to the figures to give total generated volumes.

The projected figures have been derived from official survey results and forecasts in affluent nations abroad.

The most extreme assumption is that in the two biggest Australian state capitals, namely Sydney and Melbourne, the ultimate "per-capita" figure by the year 2000 would be in the vicinity of *present* officially established figures in the U.S.A.

The Environmental Control Administration in the U.S. Department of Health, Education and Welfare in its "Interim Report", covering The National Solid Waste Survey 1968, published the following figures:

Average Solid Waste "Recorded as Collected" by Municipalities and their Private Contractors in Urban Areas — per person per day

5.72 lbs. or 2.600 kg

Estimated Generation of Household and Municipal Wastes

7.00 lbs. or 3.170 kg

The above figures are increasing at an annual rate of 4% by conservative estimates.

These figures represent a nation-wide urban average, whilst the present report mainly analyses the situation in large capital cities where, by experience, the per-capita waste generation is considerably above the country's average.

GENERATION OF DOMESTIC AND MUNICIPAL REFUSE IN METROPOLITAN AREAS

in kg per person per day — Calculated and Assumed Values:

	1971	1975	1980	1985	1990	1995	2000
Sydney	1.190	1.320	1.540	1.850	2.270	2.830	3.550
Melbourne	1.010	1.150	1.435	1.795	2.245	2.815	3.550
Brisbane	0.960	1.065	1.270	1.540	1.890	2.320	2.870
Perth	1.085	1.180	1.370	1.620	1.980	2.240	3.000
Adelaide	0.835	0.950	1.175	1.460	1.800	2.200	2.750
Hobart	0.818	0.920	1.090	1.300	1.580	1.980	2.500
Canberra	0.700	0.780	0.900	1.080	1.350	1.750	2.300
Darwin	0.700	0.780	0.900	1.080	1.350	1.750	2.300

Based on above "per-capita" generation figures the following Total Quantities of Domestic and Municipal Wastes have been computed:
(expressed in metric tonnes per year)

	1971	1975	1980	1990	2000
Sydney	1,269,000	1,479,000	1,892,000	3,466,000	6,358,000
Melbourne	921,000	1,202,000	1,672,000	3,236,000	6,164,000
Brisbane	339,000	412,000	559,000	1,041,000	1,975,000
Perth	278,000	348,000	490,000	950,000	1,799,000
Adelaide	257,000	329,000	465,000	934,000	1,784,000
Canberra	36,000	57,000	86,000	201,000	460,000
Hobart	46,000	59,000	81,000	150,000	272,000
Darwin	9,500	15,000	22,000	58,000	151,000

Progressive Cumulative Land Consumption
(expressed in cubic metres)

For the computation of progressive cumulative land consumption the requirements are based on the assumption that all solid wastes generated by households, public cleansing, commercial and industrial premises would be disposed of by "sanitary landfill" methods.

Whilst the density of loose refuse has been assessed at 4.5 m^3/tonne an annual linear growth rate in specific waste volume of 1.45% has been used in computations to allow for the expected decrease in density of domestic and municipal refuse only.

The computed figures for land consumption allow for a specific volume of 1.75 m^3 per tonne for solid refuse after two years' consolidation in a tip, plus 30% of the compacted volume for inert cover material. These figures resulted from long-range tests carried out on sanitary landfill operations in the U.K., U.S.A. and Europe and represent operational standards typical for this type of disposal.

	1975	1980	1985	1990	2000
Sydney	20,730,000	53,432,000	98,346,000	160,975,000	363,679,000
Melbourne	18,400,000	49,000,000	92,000,000	151,000,000	345,000,000
Brisbane	5,747,000	15,094,000	27,846,000	45,206,000	101,196,000
Perth	4,670,000	12,681,000	23,985,000	39,576,000	89,518,000
Adelaide	4,669,000	12,404,000	23,392,000	39,018,000	90,891,000
Canberra	682,000	1,985,000	3,900,000	6,766,000	16,962,000
Hobart	814,000	2,161,000	4,054,000	6,663,000	14,639,000
Darwin	168,000	498,000	1,023,000	1,846,000	5,076,000

Waste Handling Costs (in Dollars)

The waste handling costs have been computed and extrapolated from figures provided by communities covered in the field survey.

These also include the annual capital costs for plants and facilities.

	Annual Total 1972			
	$	$/pers.	$/m^3	$/tonne
Sydney Region (S.S.D., O.S.S.D.)	19,100,000	6.53	4.46	20.08
Melbourne (M.S.D.)	10,250,000	3.97	3.06	13.75
Brisbane Region (B.S.D., M.S.D.)	3,170,000	4.53	3.72	16.80
Perth (P.S.D.)	3,217,000	4.58	3.14	14.12
Adelaide (A.S.D.)	2,000,000	3.47	2.81	12.63
Canberra (C.S.D.)	1,035,000	6.77	4.47	20.10
Hobart (H.S.D.)	595,000	3.89	4.60	20.70
Darwin	n.a.	no cost figures available		

Conclusions

In Australia, waste management practices strongly emphasise the collection and transportation of solid wastes, investing on an average between 77% and 89% of the total public cleansing budgets in this sector.

Disposal activities are, with very few exceptions, restricted to rudimentary landfill methods, which rarely reach the standards of a modern "sanitary landfill" operation. Only between 11% and 23% of the total amounts spent for waste handling are allocated to disposal activities.

Street sweeping absorbs a forbiddingly high proportion of the public cleansing funds, fluctuating between 39.2% and 18.75% in the various metropolitan areas. The exceptionally low figure of only 6% for Metropolitan Perth appears to be completely out of proportion. Dwindling landfill sites within reasonable distances from the waste sources, rapidly growing haulage costs, traffic congestion and the enforcement of more stringent environmental legislation will generally cause substantial increases in waste handling costs and a noticeable shift towards disposal expenditures.

Improved collection methods and the implementation of large regional disposal facilities will characterise future developments in this field of environmental protection.

Note: Details of this survey are available in a report "The First National Survey of Community Solid Waste Practices — Australia 1972-1973", published by ACI Ltd., July 1973.

ACKNOWLEDGEMENT

The authors are grateful to ACI Management for permission to publish this article.

PAPER 13

The Cringle Dock Pulverisation and Transfer Station, London

By O. G. Hughes*

As a result of the passing of the London Government Act 1963, the Greater London Council was created from 92 various-sized authorities, ranging from the City of London to rural and urban districts. These were reconstituted into 33 authorities known as London Boroughs, but apart from boundary changes, etc., one of the most important aspects was the transfer of the function of refuse disposal to the Greater London Council in 1965, whilst "collection" remained with the new London Boroughs. As no organisation existed prior to that date, a Refuse Disposal Branch was set up within the Public Health Engineering Department; it had two main functions—that of (a) arranging to continue to dispose of almost 11,000 tons of solid waste per day, and (b) investigating improved methods and designing new stations to deal with the waste being created then as well as in the future.

A number of older substandard stations have now been closed, some improved and some replaced by entirely new ones. The two largest in the latter category are the Edmonton Incineration Plant and the Cringle Dock Pulverisation Plant. There are five stations operating on Thameside where 25% of London's solid waste is transferred into barges for conveyance to landfill sites in Essex. These five stations, of which Cringle Dock is the largest and newest, have capacities varying from 250 to 800 tons per day and the barge journeys vary from 28 to 62 miles one way. It is the purpose of this paper to describe briefly the Cringle Dock station and its conception.

An examination of records and of the content of solid waste (carried out by a small analysis team) indicated *inter alia* a steady decrease in its density. This meant that more and larger conveyances would be required—be they for road, rail or river. Alternatively, some form of volume reduction was necessary, and a number of methods were investigated — incineration, pyrolosis, compaction, containerisation, high-density baling and pulverisation. As far as river-borne solid waste was concerned it was considered highly desirable to fit any new method into that already in use.

Pulverisation was finally decided upon; because loading could be done cleanly and continuously, it could roughly double existing payloads, would cause minimal disturbance to unloading arrangements and barges of pulverised material could be "mixed" en route with other barges conveying crude solid waste. Incineration was a contender offering greater volumetric reduction but was excluded particularly due to difficulties of obtaining suitable sites, at acceptable cost, in Central London.

A small site of two acres became available on the south side of the Thames in 1968 and, following site investigation, construction commenced early in 1970 of a modern, totally enclosed, nuisance-free pulverising and transfer station, with throughput of 800 tons per day. Storage was to be provided for 850 tons of refuse accepted from 270 collection vehicle loads, the material to be processed in two 50-tons-per-hour pulverisers to a product with dimensions of 80% (by weight) below 1.75 in. and the remainder below 6 in. "Clean" delivery into barges having payloads up to 200 tons was essential.

The Plant and Operating Details (see also Fig. 1 and 2)

Collection vehicles entering the 200 ft. x 400 ft. site are weighed on one of two 40 tonnes, 40 ft. x 10 ft. weighbridges. The weights, vehicle number and originating Borough, etc., are recorded in a raised signal box type office by means of electronic data capture equipment which produces punched tape for processing through County Hall computer. Vehicles then proceed into the main building, access being governed by automatic traffic control equipment which also actuates safety barriers located in front of the eight tipping bays. Having reversed to a tipping bay and discharged its load the vehicle, in moving

* Divisional Manager (Ops. and Maint.), Public Health Engineering Dept., Greater London Council.

CRINGLE DOCK — LONDON

Fig. 1: Side Elevation of Plant

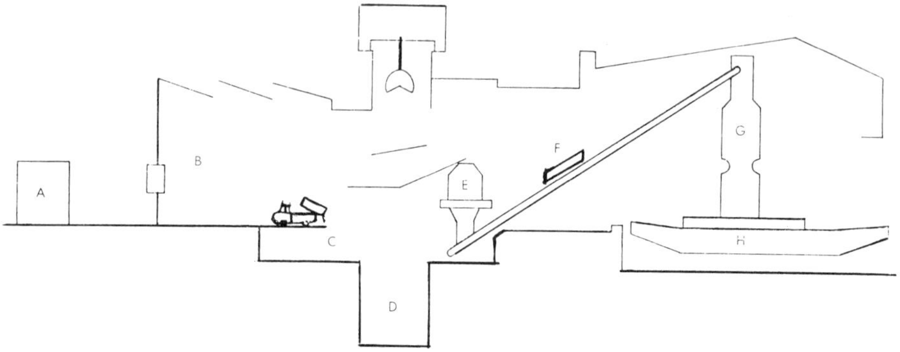

Fig. 2: General Plan Layout of Plant

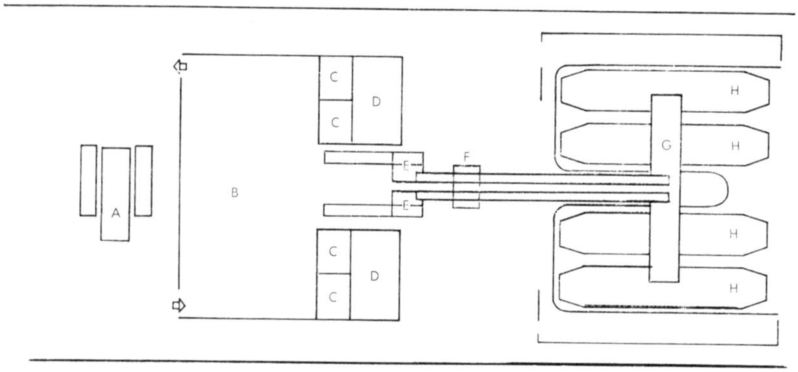

MAIN FEATURES

A: Weighbridge.
B: Tipping apron.
C: Tipping bays.
D: Bunkers.
E: Pulverisers.
F: Magnetic extractors.
G: Barge loading machine.
H: Barges.

away, actuates sonar heads fixed above it, which in turn raise the safety barrier and operate the hydraulic ram which transfers waste into one of two bunkers. Each tipping bay has a capacity of 70 cu. yd. and is 13 ft. deep, whilst each bunker is 48 ft. long, 25 ft. wide and 50 ft. deep.

Crude waste is then lifted by two 9-ton overhead travelling cranes fitted with 8 cu. yd., 4 rope, clam-shell grabs and operated from fixed control cabins high in the bunker wall. The waste is deposited on to one of two 34 ft. 6 in. x 8 ft. 6 in. vibratory feeders in order to level out the batch delivery of the grabs and produce an even layer of material on two 35 ft. long by 8 ft. 6 in. wide plate conveyors below. These are under constant observation by two attendants housed in a soundproof, ventilated cabin. Their duty is to stop the plate conveyors, when necessary, and remove any material which would cause either damage or blockage in the pulverisers. The plate conveyors deliver the waste to two 70 in. x 90 in. hammer mills, each weighing 65 tons and containing 40 hammers. The mills are driven by 400-h.p. 3.3 kv motors at 525 r.p.m. and pulverise the waste by forcing it against breaker bars and through and over heavy grids having various sized openings in them. Of the four grids presently in use in each mill three have apertures of 8 in. x 5 in. and one of 14 in. x 8 in. The hammers are reversible, four edged, double bossed and weigh 136 lb. each. Hard facing material is applied to the wearing surfaces of the grids and to the wearing corners of hammers. Each pulveriser stands on its own platform, which in turn is on anti-vibration spring-type mountings in order that it is isolated from the main building as well as from the other machine.

The pulverised waste then falls on to two output vibrating feeders and thence on to two 57 in. wide 154 ft. long conveyor belts which travel to the barge loading machine mounted on a 120 ft. long central jetty—with two barges each side of it in a tidal dock, each barge being 24 ft. 6 in. wide and 120 ft. long. A horizontal boom on this machine contains an inner portion which can be extended or retracted so that, coupled with the rotation of the whole machine, it is possible to deliver pulverised material—finally via vertical telescopic chutes—to any part of all four barges. The operator of the machine is housed in one of two control cabins located at each end of the horizontal boom of the machine itself, so that he has an unobstructed view of the delivery of waste to the barges. The operator can start and stop either or both flow lines but final delivery is by one chute only, and a barge is loaded in approximately $1\frac{1}{2}$ hours. A short fixed chute is also provided at each end of the boom so that, in emergency, pulverised waste can be delivered to road vehicles standing on the outer sides of the dock.

Pulverisers

These machines are the heart of the plant, and therefore worthy of additional mention. Three aspects which have occupied much time and thought are: (a) the hammers, (b) the grids, (c) hammer and grid reclamation and change-over.

The obvious aim is to get the maximum throughput per hammer, consequently the hammers used are double bossed and hard faced so that all four corners can be used. Much research has been, and is still being, done into hammers and hardfacing. Those now in use are to B.S. 1956 with Touca 100 facing. Reclamation by rebuilding before too much metal has been worn away is essential in the cause of economy. The aim has been to get a life of 16,000 tons per hammer before reclamation is carried out.

Grid aperture sizes are related to throughput and product size. The original grid combination was two with 6 in. x 3 in. and two with 8 in. x 5 in. apertures. Whilst this gave a satisfactory product the throughput was too low and numerous blockages occurred. The grid combination now used is one or two with 14 in. x 8 in. apertures and two or three with 8 in. x 5 in. apertures. However, this has meant a slight drop in the proportion of product below $1\frac{3}{4}$ in.

With grids, hardfacing is as important as with hammers—Armalloy 37 being used currently—but high-quality workmanship with proper heat treatment facilities are essential. It was originally intended that hammer reclamation should be done in an on-site workshop, but as experimental reclamation is not yet complete the work is being done by specialist firms.

Hammer and grid changes are facilitated by using the hydraulic rams on the mills—the work of changing a set of 40 hammers being done by two fitters in four hours.

Blockages

These are the most frustrating of troubles with pulverisers and the observation cabin incorporated at the plant represents an attempt at a solution. The feed conveyor attendants soon learn the types of material to be extracted and generally speaking such material is identified by its bulk. The most troublesome items have been textiles—in particular carpets, mattresses, textile bundles and bales, and a large tent. Obvious causes of blockages are long lengths of wood or metal, which may cause bridging across the pulveriser openings; also bundles of wire and wire ropes.

However, it is possible for damaging material to be hidden in the bed of waste on the feed plate conveyor. This happened recently, when six pieces of heavy metal entered a mill unnoticed and caused serious damage. The largest piece measured 23 in. x 19 in. x 6 in. and weighed over 5 cwt.; another measured 23 in. x 20 in. x $1\frac{1}{4}$ in., whilst the smallest was 8 in. x 5 in. x $4\frac{1}{2}$ in.

Fortunately the rotor was not damaged, although the flywheel slipped considerably due to ring feeders being used. Other damage included the bowing of the 3-ton door up to $\frac{3}{8}$ in. at each end and 1 in. at the top; the trunnions were torn off a hydraulic ram used to open the door, various bolts were broken, one 14 in. x 8 in. grid had to be scrapped, three grids needed considerable reclamation, a hammer pin was badly bent and a hammer cracked. The mill was out of action for over a week and much crack testing was required before it could restart work.

The importance of grid configurations and apertures in reducing the incidence of blockage has already been mentioned, but accidents of this nature are difficult to eliminate completely; however, the use of metal detectors is being investigated.

Fire Prevention and Explosion Suppression

In addition to the familiar fire hydrants and hoses supplied by town mains and installed at various locations, river water tanks which refill on each tide provide a capacity of 40,000 gal. for pumps to deliver to both the bunkers and the tipping bay areas in case of fire.

Experience with a slightly smaller pulveriser elsewhere in London showed the vulnerability of this type of equipment to explosive materials or mixtures being unknowingly fed into it. Considerable damage was caused, not to the pulveriser itself, but to the adjacent steelwork and conveyors when reject aerosols containing starch were delivered by a private contractor as trade refuse. They exploded when struck by the rotating hammers, after being loaded unnoticed among other waste. To provide protection at Cringle Dock, 10 bottles of 3- and 10-litre capacity of pressurised liquid chlorobromomethane are located at strategic positions. The chemical is released when the bottles are triggered by probes which sense the initial stage of an explosion pressure wave.

Electrical Protection

Various types of protective switchgear and equipment have been provided for the protection of vital machinery, particularly for the pulveriser motors against extended overloading, overheating, stalling, starting under stalled conditions, etc. Another instance is the provision of braked motors driving the input plate conveyors so that should a blockage occur in the mill itself the conveyor will stop immediately and not continue to feed waste, even by the amount delivered during the drive motor rundown.

Dust Control

In order to prevent pollution of the atmosphere through dust—and to improve working conditions—four wet-type dedusters are provided having a capacity of almost 200,000 cu. ft./min. Dust-laden air is drawn from the rear of each tipping bay as well as from above the input vibratory feeders and plate conveyors. A further set of 12 inlet fans having a total capacity of 110,000 cu. ft./min. are to be found in the bunker roof. Sludge produced by the dedusters is collected in skips which are transferred to the barges by mobile crane. The enclosure of the tipping area, pulveriser house and the dock area—except for space for ingress and egress of barges—is intended to contribute to the effectiveness of dust control measures.

Staffing and Maintenance

As operation of the plant is for 16 hours per day the staff can be divided into shift workers and day workers. Each of the two 8-hr. shifts is staffed by a senior plant operator capable of undertaking any operational work, three crude feed conveyor operators, two bunker crane drivers, one barge loader operator, and one mobile crane driver. Day work staff consists of two tipping bay attendants who assist in unloading vehicles delivering waste and keeping the bays and apron clean and tidy, one cleaner/messroom attendant and, in addition, one electrician and two fitters. It is intended to change the organisation of the duties of the electrical and mechanical fitting staff to accord with revised planned preventative maintenance schedules drawn up as a combined effort between the planning officer and plant manufacturers. These changed arrangements are the result of operational experience, the completion of all performance tests and the plant assuming greater throughput.

The smallness of the available site, mentioned earlier, imposed strict limitations on the design of the plant and in consequence on its maintenance. For instance, the need to reverse the direction of flow of waste, i.e. on the crude vibratory feeder, makes access thereto more difficult and thereby more time-consuming. In order to make this and all other maintenance work safer, a "permit to work" system is in operation. It is also considered that a standby flow line would have greatly assisted the general maintenance function. The administrative support is by two clerks and two weighbridge clerks and a storekeeper/clerk whilst overall shift responsibility rests with two shift managers. The five Thameside stations are under the control of the river's manager and his assistant, who co-ordinate the operation of barge loading and movements at all the stations, particularly as the barges are owned and manned by lighterage contractors.

An amenity floor, located at an upper level, contains offices for the river manager, shift charge managers and administrative staff as well as messroom and washing facilities, locker rooms, etc., for the staff. Additional toilet facilities are located under the weighbridge office.

Dock Facilities

The dock sides and ends are protected by Douglas fir and Greenheart timbers on rubber mountings and continued to an average of 6 ft. above the dock edge. To assist in the movement of barges—made somewhat more difficult because of the presence of a nearby sea-dredged aggregate jetty — four electrically powered capstans are employed along with the more usual sliding moorings, "cows horn" bollards, etc. As the dock is tidal quick movement of barges is essential, for they are "on the bottom" for up to five hours, although the throughput of the station is geared to the capacity of four barges.

Plant Cleanliness

The fact that solid waste transfer and treatment plants deal with "dirty" material is no excuse for failing to make every effort in both design and housekeeping to keep the place clean and tidy. Adequate dust control is an essential but so are such things as the covering of conveyors and the provision of easily cleaned surfaces. Two particularly difficult aspects are the windage produced by the pulveriser rotors and the "bounce" of pulverised waste when it lands in a barge. Considerable modifications have been necessary to permit extra expansion space below and adjacent to the pulverisers but further improvements are being investigated. It is proposed to fit a water "mist" at the lower end of the chutes which discharge into the barges, in order to confine dust and fine material and to prevent its being blown into the dock particularly when the barge is almost full. Experiments have already indicated the effectiveness of a fine spray of water even when directed from the quay side. Experience has shown that the best freights are obtained by half-filling an entire barge and then going back and complete its filling.

Recycling

Recent world events have highlighted the need for recycling our diminishing assets and in this connection the extraction of ferrous metals was included in the design brief for Cringle Dock. Extraction is effected by overband magnetic extractors placed on the product conveyor shortly after it leaves the main pulveriser building. The metal then falls down chutes to 32 cu. yd. articulated vehicle trailers located below. Analysis shows that up to 6% should be extractable, but to date the rate has averaged only 2½% although this has been influenced greatly by trailer exchange difficulties. The resolution of this problem is imminent.

Costs

The capital cost of the entire project was £2,651,000, which included £331,000 for land. Loan charges amount to £289,000 and operating costs to £263,000 per annum, giving a total of £552,000. The estimated income for a full year from the sale of metals and the receipts from trade waste are £25,000, whilst barge transport and landfill disposal cost £180,000 and £185,000 respectively; therefore the overall cost is approximately £4.6 per ton.

A set of hammers cost approximately £3000 and a set of four grids approximately £1000. Reclamation of hammers costs approximately 15p per ton of waste processed whilst grids cost approximately 4p per ton as they have about 2½ times the life of hammers. Power consumption is 7 kWh per ton.

Conclusion

The views expressed in this paper are those of the author and do not necessarily reflect the policy of the Greater London Council.

A review such as this, with all its shortcomings due to the need for brevity, must provide but a sketchy look at a station which is unique in many ways. Should further information be required the author will endeavour to provide it upon request.

PAPER 14

Pulverisation of Municipal Refuse

By G. Hitchell*

For many years the common methods of garbage disposal have been controlled tipping and incineration but during the past decade pulverisation of garbage has become increasingly popular as a reasonably economic manner of treating refuse. Most local authorities are reluctant to spend money on refuse disposal and often ask: "Why treat refuse at all when such material may be placed in the ground and covered?". This method undoubtedly is economic but in the position that the Municipality of Canterbury is finding itself, tipping facilities are diminishing at a rapid rate and Canterbury, alike with many other municipalities in the County of Cumberland, is to be faced with haulage of material to other areas. This will seriously add to the cost, depending on the mileage to be carted, or involve the Council in exorbitant cost in the installation of transfer stations.

The Municipality of Canterbury has found that in approximately two years time, no further areas for the tipping of crude refuse will be available, and it may be pointed out that during the past 20 years tipping facilities have been provided for neighbouring councils on the southern side of the harbour such as Ashfield, Leichhardt, Randwick, Waverley and Woollahra and on occasions South Sydney.

The Mayor and Aldermen of Canterbury Council, after consultation with its Senior Officers, carried out an investigation of vacant land within its area, some of which is used as open park land. It was concluded that if garbage was treated by pulverisation, such areas could be effectively up-graded by the introduction of pulverised refuse which contains sufficient humus to enable more effective plant growth on some of the heavy clay pans which occur in the municipality; and it is anticipated that these areas will give a tipping life within the municipality of approximately 25 years.

As pulverisation has not been carried out in this country, the Council resolved in November 1972 to send the Town Clerk and the author on a feasibility study in regard to pulverisation in England, Europe, Canada, America and Japan. Pulverisation of garbage was investigated first in England and Scotland. Several different types of pulverising plant is at present being used in Great Britain for the treatment of household and trade refuse. The types of pulverisers used are of the hammermill type capable of processing between 15 and 40 tons per hour, or horizontal shaft machines and in some instances vertical shaft type machines, whilst in one instance, a drum-type machine was being used. In order to breakdown the paper content of the garbage, water is added to soften the paper and cardboard.

The processing of pulverised garbage was used by small and large cities and boroughs in England, and it was claimed that the volume of garbage was being reduced by 40% to 60%. At Cringle Dock, a large twin machine has been installed for the express purpose of treating a portion of the City of London's garbage and discharging the milled refuse into open barges for transporting down the Thames to the Essex marshes for disposal (see Paper by O. G. Hughes, this Symposium).

For the Greater London City Council, which has a population of approximately 7.2 million and collects approximately 2¾ millions tons of garbage per year, it was stated that pulverisation reduced the volume of garbage by 50%.

It was apparent from the inspections carried out, that garbage in England and Scotland was reduced in size to particles of approximately 2 in. to 4 in., and in the majority of cases was tipped in open areas and no other cover was used. In certain other instances, such as London and Slough, pulverisation was carried out to reduce the transport cost by reducing the bulk to an average of about 50%, thus saving approximately half of the transport costs involved. The following advantages are claimed for milled garbage:—

(i) It is easy to spread and does not require the use of heavy equipment.

(ii) It is approximately twice as dense as raw refuse and within a period of approximately one year densifies by a factor of two; in other words, pul-

---

* Municipal Health Surveyor, Municipality of Canterbury.

verised garbage becomes four times as dense as raw refuse after approximately one year on the tip.

(iii) It is comparatively odour free.

(iv) It does not attract birds or vermin to the same extent as raw refuse.

(v) Top soil cover is not a daily requirement of freshly pulverised refuse, and in certain instances where covering is required, pulverised garbage approximately one year old is used for the cover of milled refuse.

(vi) Because the moisture is evenly distributed throughout the milled refuse and as a result of better compaction, problems of wind blown litter are minimised.

(vii) Tip fires do not occur readily on pulverised refuse tips.

(viii) Heavy vehicles may traverse freshly-milled garbage without bogging, even in wet conditions.

(ix) The leaching effect of rain water is considerably reduced in comparison with raw refuse tips, due to the fact that the milled refuse, especially the paper and the cardboard content, readily absorbs moisture and the decomposition of the organic matter in the garbage is accelerated by milling.

In Europe, pulverisation was evidently first commenced in France in the province of Champagne by the use of a Gondard horizontal hammermill and the pulverised garbage was used as a composting material in the vineyards in Champagne.

In Germany, several plants of a fixed hammermill type were inspected at various cities along the Rhine. At Giessen, the plant operated by the Municipality of Giessen, was used in conjunction with the Giessen University for which the West German Government have made a grant for the investigation and prevention of pollution of water of the Rhine. The Germans were of the opinion that where a coarser grind material (larger than 6 in.) was being used and left in winrows for a period of 6-8 weeks, aerobic decomposition was accelerated. After a period of approximately six months this pulverised garbage to a size of below 6 in., decomposed mostly to a soil-like material, with the exception of plastics; and at Giessen, experiments were being carried out with vibrating screens of triangular shape in order to obtain soil-like material for top dressing, parks and open playing fields. Experiments were also being carried out in regard to the treatment of leachates arising from garbage disposal.

As most of the larger cities are situated in close proximity to the Rhine and all sewerage is treated in sewerage plants, sewerage sludge is frequently introduced into the milling charge. Local experience indicated that the effluent arising from milled garbage to which sewerage effluent had been added, could be treated more readily and the process of decomposition was accelerated, being practically completed in about six months, whereby the decomposition of organic matter in normal garbage tips may take many years. However, it is pointed out that the areas used for the storage of winrowed material are not satisfactory as far as aesthetics are concerned, and unless a space can be found shielded from public view, heaps of winrowed garbage approximately 6 in. or more in size would not be acceptable.

The coarser the material is milled, the more economical the process becomes, and the plants with fixed hammers on a horizontal drum treating not only household but trade refuse to 6 in. sizes could effectively be driven by a motor of a horsepower of approximately 250; whereas in England, the reduction of ordinary household garbage to 4 in. sizes requires the installation of a motor of approximately 450 horsepower.

Investigation in America disclosed that pulverisers had evolved from metal and car shredding plants, using a horizontal hammermill type but are driven by larger horsepower motors. The Americans have investigated and produced a machine in which forced feeding of the material is used to improve the operation of the machine and to gain optimum efficiency.

In Chicago, approximately 70, 18 cubic yard loads of clean-up material are collected each day and passed through a pulverising plant in order to reduce the bulk; when material such as clean-up, bedsteads, bedding and the like are pulverised, the reduction rate is approximately 70%. In certain instances, metal of the ferrous type is extracted by the plant and sold to smelting works. In other cases, this material is used in tipping sites or is incinerated.

In Madison, which is a city containing a very large University, the American Government has made a grant for the investigation of pulverisation and of landfill disposal. Two types of machines are at present in use, one being the Gondard horizontal hammermill and the other a vertical Tollemache machine. The summary by the Madison University in reference to milled refuse, for which experiments have been carried out since 1966, states that during 1968 milled refuse was used for cover material on raw refuse and that this resulted in a neat, smooth working face; and also when milled refuse was used for landfill, the quality of the sanitary landfilled operations was improved.

It was the opinion of both the Project Director and the Director of Public Works in Madison, that the improved operation justified the increased cost of milling. They also stated that in any future project of solid waste management, it was apparent that new methods would involve particle size reduction for processes such as:

(1) Incineration.

(2) Salvage and re-cycling.

(3) Baling.

(4) Landfill with new concepts.

(5) Bulk grinding to create better pay loads for transferring.

(6) Composting.

(7) Advanced forms of treatment such as anaerobic digestion.

Cover is not used in Madison for the milled refuse and when observers walk on the grounded refuse the negative, often times emotional feelings regarding solid wastes, seem to change; so that it is quite likely that milled garbage may be acceptable in areas which otherwise would not be used for solid waste disposal.

The conclusions by the Madison University study are as follows:

(i) Based on 1970 costs, projections indicate that the cost of milling and landfilling 61,800 tons of refuse per year, exclusive of land charges and administration, will be $2.80 per ton with the Gondard mill. Comparable costs for the Tollemache mill handling 61,800 tons per year are $2.12 per ton.

(ii) Domestic solid wastes can be shredded in either type of hammermill without presorting and with negligible downtime due to mill stoppage.

(iii) Milled refuse has been left in a landfill without cover for up to four years without complaints having been received about odours, nuisances, offences to onlookers, blowing milled materials, insects or rodents.

(iv) Experience with the use of milled refuse without daily cover indicates that the quality of operation when compared with sanitary landfill operations at Madison is superior in respect to: travel over the fill and at the face of the fill; blowing paper and dust; tracking of mud by trucks on highways; appearance during operating hours; and maintaining a uniform high level of operations during cold and wet weather.

(v) Fully loaded trucks exceeding 72,800 pounds can drive on the milled refuse in inclement weather, facilitating continuous high quality landfill operations. Tyre problems have not been experienced when travelling on the uncovered milled refuse.

(vi) Experience and specific tests have indicated that there is less of a fire hazard with milled than non-milled refuse. Although freshly milled refuse does support a fire, it spreads slowly and is readily extinguished with water.

(vii) Density of milled refuse without cover dirt was found to be approximately 10 per cent more than the density of refuse not milled and not covered. When intermediate cover dirt was placed on the non-milled cells, the resulting effective density of the milled refuse cells was found to be from 53 to 90 per cent greater than the apparent density of the non-milled refuse which is covered.

(viii) Rats and flies are no more likely to be found on milled refuse without cover than on refuse which has not been milled but is covered as is customary in sanitary landfill practice.

(ix) Rats cannot survive on a diet of milled refuse.

(x) Under optimum weather and moisture conditions, flies can breed in freshly milled refuse; however, once the refuse has aged several months, this ability is evidently lost. Cage studies and fly counts on the landfill provided data which complements operating experience at Madison, during which no fly nuisance problems have been observed.

(xi) Tests with the Gondard Mill showed that a large percentage of fly maggots passing through the mill with refuse was killed.

(xii) Milled refuse cells which were not covered were observed to produce leachate faster and with higher contaminant concentrations than nearby cells containing refuse which was not milled but were covered. Once the period of peak productivity is passed the milled refuse cells produce leachate of ever decreasing contaminant concentrations, dropping to levels much lower than non-milled refuse cells six to twelve months after the peak. Non-milled but covered refuse cells continued to produce leachate at substantial contaminant concentrations throughout the three-year length of this study.

From investigation in Japan it was learned that four plants were in the process of being installed by various authorities. The largest of these is at Kyoto, and it is an American-designed machine of horizontal shaft with a compression feeder; it is being used by the Kyoto City Council for the purpose of milling the refuse from a city of approximately one million population.

The purpose of the milling is to conserve tipping space within the available area adjacent to the city, but the majority of the material pulverised is treated in the incinerator and the authorities have stated that this is used for the express purposes of assisting combustion in the incinerator. The plants in Japan are in the process of being erected and should be in operation within the next few months.

Canterbury Council, after consideration of the author's report and bearing in mind the possible costs of carrying material to other areas for disposal, have called tenders for the installation of a pulverising plant capable of treating all types of household and trade refuse at the rate of 40 tons per hour, at the Council's garbage depot, and it is hoped that this plant will be in operation by the end of the year. The overall cost of this machine is approximately $422,000 and the operating costs of treating 50,000 tons of garbage per year will be in the vicinity of $1.54 per ton, plus the disposal costs after treatment by the plant. This would make the total cost $3.00 per ton.

PAPER 15

High Density Baling as a Solution to Solid Waste Disposal

By F. R. Zettel*

The paper outlines the history of the solid waste baling facility constructed as a pilot project in St. Paul, Minnesota, by American Solid Waste Systems, Inc. The concept of high density baling of solid waste materials and its enonomic and environmental benefits as compared with other methods of solid waste disposal are emphasised and explained.

1. Introduction

The inherent dangers of disposing of solid waste materials in open dumps are many; such dumps attract rodents, create a breeding ground for flies and insects and, in addition, raise the possibility of groundwater pollution in addition to air pollution. The public's awareness of these dangers and the resulting emphasis on environmental controls and protection have forced industries which had never before been concerned about pollution control or waste disposal to propose, or advocate, new methods of solid waste disposal.

Composting was the new solution quickly accepted. The proponents of composting plants pointed out, that every ton of solid waste material processed and sold as compost material (claimed to fetch $74 ton on average), is a ton of waste material which would have had to be buried at a cost of $2 to $3, reducing the usable life of the landfill area. Unfortunately, only the organic constituents of the waste could be processed into compost with the balance still having to be transported to a landfill. Salvage of certain materials such as cardboard, paperstock and tin cans was strongly advocated, not only as a means to preserve natural resources, but to reduce the cost of disposal of solid waste materials through revenue received from the sale of these items. Composting plants, however, had a short life and investments of millions of dollars were lost in the following years by cities and private enterprise which had adopted this method. The primary reasons for this were high operating costs coupled with the unwillingness of John Q. Public to buy his garbage back and the limited market and low market value of the salvaged constituents.

Other methods advocated were: (1) Disposal in sanitary landfills after shredding only. (2) Shredding and baling with limited salvage efforts. (3) Baling in low, medium and high density balers. (4) Partial conversion into a usable product, such as an insulation material or iron pellets for leaching of copper ore mines, and disposal of all unusable constituents in a sanitary landfill. (5) Waste heat generation through incineration with or without inclusion of electric generators and disposal of residue in a sanitary landfill. (6) Conversion into carbon or activated carbon and disposal of non-convertible materials in a sanitary landfill.

While the inclusion or omission of salvage operations and, therefore, the demand for various pieces of equipment varies with each method, or process, it is evident that each of the aforementioned methods includes the use of a landfill area for the disposal of all or a good part of the unusable constituents of the waste accumulation. Since such land is necessarily in close proximity to the areas in which the waste materials are generated, it is also evident that such land areas are neither abundant nor cheap. The utilisation of such land areas to the greatest degree is therefore of prime importance.

This realisation stresses the importance of high volumetric reduction in order to achieve the greatest possible land utilisation. In simple words, the denser the waste material is compacted or volumetrically reduced, the more tons per acre foot can be disposed of.

2. A New Approach to Solid Waste Disposal

Volumetric reduction can be achieved by three basic methods: (1) Chemical alteration by incineration. (2) Physical alteration by shredding. (3) Physical alteration by baling.

Physical alteration of huge amounts of solid waste materials by baling into dense cohesive bales which are easily handled and transported, offers an additional bonus by changing the site where the refuse is deposited from a practically useless, valueless piece of land into a usable, revenue-producing parcel of real estate.

* American Solid Waste Systems, Inc.

After conducting tests utilising various scrap balers over a period of more than two years, American Hoist & Derrick Company's management adopted in 1968 high density baling of solid waste materials as a viable method for the disposal of these materials.

Prior to this decision the writer travelled across the United States for a period of eight months investigating all other practices of solid waste disposal. The next step, if was felt, would have to be the construction of a high-density solid-waste baling facility to ascertain the operational reliability of the equipment and the system itself. The effects of seasonal variations of solid waste composition as well as the variations in solid waste composition due to geographic location had to be determined.

In January 1971 the first high-density baling facility became a reality. For the next four months tests were conducted with a variety of materials and mixtures to determine not only the balability of them, but stresses and pressures within critical components of the baler itself. The baler was a three-compression baler similar to those which had been manufactured for the scrap processing industry for nearly 25 years.

The reliability of the hardware had already been proven in hundreds of applications in the scrap industry. The concept of high density baling of solid waste had been tested, during the time the plant was under construction, by a study conducted by Dr. Karl Wolf in connection with rail haul of baled solid waste materials. This study indicated that, in addition to elimination of problems normally associated with conventional disposal methods such as air pollution, baling offered economic advantages regarding the transportation of the solid waste materials away from the area where they had been generated.

3. High Density Baling of Solid Waste Materials

Through daily operation in the St. Paul baling plant the reliability and economy of the concept was established. Solid waste materials received from two transfer stations in addition to refuse collected by private haulers and city operated collection packers have been processed at the St. Paul baler plant since June 1971. Numerous improvements to the original baler design and infeed conveyor design were made in the following year.

The first baler was designed for a 120-second operating cycle, producing a 2800 lb bale. Current production models of the baler are designed to produce a 3 ft. x 3 ft. x 4 ft. 6 in. bale (weighing 3000 lb) every 90 sec.

When the pilot plant was designed the emphasis was on total disposal. The infeed conveyor system was accordingly designed without regard to resource recovery.

Plants currently under construction or in the planning stage reflect in their design the newly recognised need for resource recovery and recycling of valuable waste constituents such as cardboard, newsprint, miscellaneous paperstock and plastics. Such resource recovery activity can be easily incorporated into the total system.

The composition and analysis of average municipal refuse for the U.S.A. is 43.8% by weight for cardboard, newsprint and miscellaneous paperstock. The addition of commercial and industrial wastes raises this percentage to over 50% (in the St. Paul operation 54% paperstock). The recovery or reduction of the percentage of paperstock does not affect the bale stability adversely. Any reclamation of cardboard, newsprint and miscellaneous paperstock from the refuse mix for the purpose of recycling will affect the moisture content of the remaining refuse mix and, consequently, the cohesiveness and stability of the bales.

"Dry" municipal refuse has a moisture content of 20% to 25% moisture by weight. For baling, 15% to 25% moisture content has been found to be ideal, 30% to 40% acceptable, and over 45% moisture content undesirable if cohesive bales are to be made continuously. In the St. Paul operation refuse mixes as low as 30% by weight paperstock and a moisture content of 40% have been successfully baled. Stable, cohesive bales can be made with a lower paperstock content than stated above, provided the moisture content is within the "ideal" range.

Stable, cohesive bales have been made with refuse mixes containing up to 50% by weight organic materials such as grass clippings, yard and tree trimmings, food wastes, straw and cornstalks.

In order to determine the balability of different waste mixtures generated in other parts of the U.S.A. and Europe waste mixtures were blended, or synthesised, to the specifications of a particular community. Such test mixtures were made for Miami; Cleveland; Honolulu; Milano, Italy; London, England; and Berlin, West Germany.

All these tests were successful and proved a variety of waste mixtures as balable. The bales were stable and cohesive to withstand handling, transportation and stacking in the balefill. As expected, different amounts of springback were obtained as with the standard St. Paul refuse composition. (Springback is the amount of growth the bales undergo immediately upon release from the baling chamber.) The average volumetric springback for bales produced from St. Paul refuse was determined to be 35.8% with an average directional springback of 13.4% in height, 5.3% width and 14.8% in length.

Bales produced with standard St. Paul refuse averaged a density of 60.0-62.0 lb/cu ft. Tonnages as high as 670 tons/day have been processed in the St. Paul plant at a cost as low as $3.24/ton for baling, $1/ton for transportation to the balefill area and $0.82/ton for unloading and stacking at the balefill. These costs are further reduced through the use of the 90-second cycle baler. To date 221,000 tons of solid waste have been baled in the St. Paul facility and transported to the balefill 12 miles away by truck.

Bales have also been shipped to Hawaii for a study concerning disposal of baled solid waste in an ocean environment, to London (England) for a trade show and to many other cities.

4. Economic Advantages Associated with High Density Baling

The high density baling concept has some unique economic advantages over other methods of solid waste disposal. In addition to a much lower capital investment required for a baling facility v. an incinerator plant or shredding plant, the operating cost is much lower due to ease of maintenance and more efficient dollar/horsepower utilisation of the energy cost. Furthermore, balers are not as self-destructive as incinerators, especially, or as shredders.

Baled refuse is easily handled and transported. The reduction in volume results in lower cost per ton/mile and, consequently, makes longer hauling distances economically feasible. For instance, a ton of refuse can be deposited in a landfill, limited in space, for $8 on Long Island; the same ton of refuse, baled, can be shipped for slightly less to a stripmine in Ohio. This not only conserves land in a highly populated area for other uses but converts an unusable to a usable piece of land. Baled refuse can be shipped economically by rail or truck depending on the distance and tonnage involved.

When the baled refuse is stacked in the landfill it results in a neat, clean operation without all the nuisances commonly found in an ordinary landfill such as blowing papers, burning, birds and rodents. Since bales can be placed in tiers three or four high before any cover material is applied, a substantial saving in the cost of covering is realised.

And, last but not least, the balefill can be used soon after filling is completed rather than waiting for years for settlement to occur. Placing of bales can also be accomplished utilising a cell, or grid system, which will make it possible to drive piling on grid lines between cells to support heavy concentrated loads. Baled refuse can be placed forming slopes of 2:1 or even 1:1 with contouring to conform to the planned end use of the property.

5. Pollution Potential in Balefills

The baling of heterogeneous mixtures of solid wastes into high density bales contributes a number of factors which allow better environmental pollution control. Some of the more obvious are listed below:

(1) The mixing of organic and inorganic constituents in the mix prior to baling results in a more homogeneous refuse mixture, obliterating the high concentration of organic wastes in spots or cells within the landfill, thus reducing concentration of bacterial reaction and pollution potential.

(2) In the course of high density compression, excessive amounts of both air (free oxygen) and water are reduced from the waste mix, thereby limiting bacterial life and biochemical reaction.

(3) The reduction of the permeability of the waste mixture due to the bales' high density limits percolation through the refuse and, consequently, also affects biochemical reaction time and leachate generation.

(4) The high density of the waste mixture induces excellent insulation properties of the bales, thereby controlling heat generation and heat dissipation which, in turn, results in a cyclical biochemical "self-cleaning" process.

(5) Refuse mixtures which have been generated in an area-wide system incorporating transfer stations and a baling facility have a higher percentage of non-water soluble, non-decomposable inert solids and upset the carbon:nitrogen nutritional ratio.

(6) In placing bales in a balefill rather than loose refuse in a sanitary landfill, greater control over some of the factors affecting gas production such as placement and cover, infiltration, temperature and aeration, can be exercised, thereby reducing the pollution potential.

In the light of the research and studies conducted it can be concluded that environmental control measures employed in sanitary landfilling today are adequate with respect to water pollution and gas control for landfilling with high density heterogeneous solid waste mixtures. The data suggest that sanitary landfills containing high pressure compacted solid wastes reduce the potential of environmental pollution and can be considered much safer in this respect than ordinary sanitary landfills.

PAPER 16

Management and Disposal of Wastes Generated in 30-40 storey Blocks of Flats

By J. E. Williams*

The management and disposal of wastes in high rise residential buildings pose problems of relatively recent origin. Waste material has descended upon us in vast quantities and caught us unawares. Municipal Authorities, building designers and others concerned might be forgiven to some extent for not facing up quickly enough to the whole refuse disposal problem and for creating tall buildings which lack adequate facilities for handling the rapidly increasing quantities of refuse.

There is ample evidence to suggest that solid waste management has not been considered deeply enough in the early and subsequent stages of recent building design and construction, in this country and overseas. There are too many examples of Sydney residential buildings of some 25 storeys where the domestic refuse is conveyed in standard rubbish bins by way of the passenger lifts and front entrances, and where passenger lifts are monopolised for hours on end for the transport of newspapers, cartons and bottles.

Such arrangements are not likely to be tolerated in future and would certainly not have been acceptable for the transport of night soil containers with their closely related characteristics. Many existing refuse disposal facilities in the inner high density Sydney suburbs are primitive. Plastic refuse bags, for instance, fixed to wall-hung frames in the lift foyers at each floor, are by no means uncommon.

At the outset, therefore, the opportunity is taken to stress that adequate consideration be given by all designers to this essential engineering service at the conceptual stages of every proposed building development.

General Design Concepts

The refuse handling facility must be an integral part of an entire waste management system from generation to final treatment and disposal. The system must be simple, ideally maintenance free, and capable of one-man operation during strikes, mechanical breakdowns and extended holiday periods. It must be capable of efficient operation for the projected life of the building.

The system must conform with all regulations, particularly those concerned with health, fire, clean air and nuisance. It must provide adequate facilities for movement, storage and collection of refuse, combined with maximum convenience, hygiene, amenity, insulation and safety for the occupants and operators of the building. Ideally it should not be seen, heard or smelt.

Early Consultation with Authorities

At sketch plan stage it is essential that agreement be reached between building designers and the appropriate Authorities with particular reference to the following points:

1. The volume and nature of the refuse to be expected.
2. The type of service the Municipal Council or private collection contractor is likely to provide presently and in the future and the cost of such service.
3. Relevant bylaws and regulations.
4. The general method of refuse storage and collection to be used relative to layout and density.
5. The location of the proposed refuse storage areas and the access provided to them.
6. The storage capacity to be provided relative to the frequency of collection.
7. The type of equipment to be used — chutes, containers, compactors, conveyors, collection figures, etc.
8. The responsibility for cleaning and maintaining the system.
9. Air pollution control.
10. Noise control.
11. Fire control.
12. The discharge of wastes into sewers.

Likely Occupants

We are building for people and should know quite a lot about the likely occupants of every proposed building, their mental attitudes, behavioural patterns and the quantity and type of refuse which they are likely to generate.

* D. S. Thomas and Partners Pty. Ltd., Consulting Engineers.

Wide variations can be expected with each building, locality, city, socio-economic group, age, age of children, and other factors.

The following figures extracted from a survey of living unit occupants of a high rise residential tower in South Sydney reveal much about the inhabitants.

Some Tenant Statistics:

Income:		One parent families: 45%	
Under $70 week	37%	(Usually no male)	
Under $100 week	43%	Average occupancy: 2.49 persons per unit	
Over $100 week	20%		
		Average beds per unit: 2.93 persons	
Occupation:		Nationality:	
Manual worker	51%	Born in Australia	60%
Office worker	14%	Foreign	30%
Service employee	32%		
Age:		Previous Accommodation	
20-39	47%	Inner City	25%
40-60	31%	Eastern Suburbs	28%
		Inner Southwest & West	34%
Children:		Reasons for Accepting Home Unit:	
		Work in city and	
Under 4	52%	inner city	45%
4-11	27%	Lived in area	25%
12-15	21%	Early allocation	20%

Similar survey reveal significant variations which can be anticipated with respect to present and future occupants of every proposed living tower.

Quantity of Refuse

Again variation of refuse is the outstanding feature emerging from the several surveys which have been made.

The following estimated refuse yields per dwelling are given in 1972 British Code of Practice for refuse handling for residential buildings:

	1969	1975	1980
Weight	28.7 lb/wk	29.8 lb/wk	32 lb/wk
Density	9 lb/cu.ft	8 lb/cu.ft	7.5 lb/cu.ft
Volume	3.2 cu.ft/wk	3.7 cu.ft/wk	4.3 cu.ft/wk

The weight compares fairly closely with Australian surveys but U.K. density figures seem high when compared with those compiled for this country. A density of 5 lb cut.ft seems to strike a closer estimate for Australia, and the trend is for this to fall as the standard of living rises in a country.

Acceptable figures for refuse generated in Australia might be:

7 lb of domestic refuse per person per week.

12 lb of total refuse per person per week.

30 lb of total refuse per dwelling per week (2.5 persons per dwelling).

"Total refuse" includes trade, street, garden, clean-up and industrial refuse.

One Housing Authority estimates refuse volume on a base figure of 7 cu. ft per home unit per week, which correlates with several examples measured recently in Sydney. This same Authority reports tremendous escalation in refuse generation, particularly in the younger age groups.

In the more affluent towers the figures are lower but 5 to 7 cu. ft per home unit per week has been recorded where communal facilities are used for refuse disposal. People adapt to the prevailing circumstances.

Total Volume Estimates

Taking a theoretical structure of 40 storeys with eight living units per floor (although 12 living units per floor are not uncommon) generating 7 cu. ft per unit per week, we have a design figure of 2240 cu. ft or 83 cu. yd per tower per week.

Due allowance must be made for the increase in both the volume and weight of refuse in the design of the handling system, particularly with respect to the provision of adequate space and headroom.

Composition of Refuse

Composition is of particular importance with regard to segregation, volume reduction and final disposal. The following table shows an average measurement of this variable and changing material, which cross-fixes fairly well with available figures for the average type of refuse found in high rise living units.

Again, variation is an outstanding characteristic but the figures have value as a general indicator.

Components	Weight lb/person/week*	%	Estimated volume cu. ft/person/week**	%
Mixed Paper	2.47	36.6	0.49	60.0
Putrescibles	2.21	32.8	0.09	11.0
Plastics	0.11	1.6	0.02	2.3
Rags & Miscellaneous	0.18	2.7	0.02	2.4
Tins & Metals	0.47	7.0	0.05	6.0
Glass	0.43	6.4	0.02	2.3
Salvageable bottles	0.87	12.9	0.13	16.0

*Approximations from Sydney Surveys.
**Component densities based on U.K. data.

Subsequent surveys indicate the following trends in a four-year period:

Paper content — increased by 7%.

Putrescibles — decreased by 7%.

Plastics — mainly polyethylene and polypropylene used for packaging — increased by 120%.

Metal content — cans, packaging, etc. — increased about 50%.

Most other items — remained about the same.

Resource Recovery

In the natural world the term "wastes" has no meaning. Perhaps the word "residues" may be more appropriate in the consideration of this subject. With indicators suggesting that the whole world is moving from an age of relative resource abundance to an era of definite resource scarcity, due cognisance must be given to the possibility of, and provision for, increased segregation of refuse within the proposed building during the lifetime of the building, if not for the immediate present.

The major residues which could be recycled, other than compostible materials, generated in one high rise dwelling complex were paper 32% by weight, glass 16% and metal 11%, but it may be necessary to look beyond such broad categories to narrow product categories which can be recycled on a much higher quality level.

This may be possible with specialised organisation and equipment, but it does not appear to be a sound economic proposition under present-day practices. Some segregation would have to be made by the tenant, which may seem only a remote possibility in the behavioural climate of today. In fact 90% of home unit occupants do segregate refuse when required to do so and do co-operate with management and caretaking staff in such matters.

Furthermore, the new Code of Behaviour for Strata Title Living will enable management to restrain offenders effectively who will not abide by the rules laid down by the Body Corporate.

It is appropriate to mention that segregation is not always practised. All refuse is disposed of efficiently via chutes, shredder compactors and containers in some high rise living towers in Sydney.

Refuse Flow

Refuse flow will be traced through key locations and equipment with special comment upon:

Chute rooms.
Chute hoppers.
Gravity chutes.
Storage hoppers below chutes.
Shredder, compactor, container operations.
Automatic transportation, at ground and basement level.

Insufficient space is available here to discuss many of the systems used in solid waste management, including on-site incineration, many of shredding and compaction systems, and kitchen waste disposal units. Several of these have great merit, whilst others have limitations at present. No doubt we shall see rapid developments in site incineration, pyrolysis, heat recovery, garbage smelting and other refuse treatment and disposal systems, as more attention is focused upon them.

Chute Rooms

A key point in the refuse flow is that located at the entry to the chute hopper.

In the better organised buildings, newspapers, cartons and bottles are separated by occupants and placed on shelves in an air locked chute room fitted with exhaust fan, impervious floor and walls, coved corners, all cleanable by hosing down.

The segregated materials are conveyed in passenger lifts by caretakers or contractors to storage rooms at lower levels, with varying degrees of efficiency. In future, deeper consideration can be given to moving this material in multiple chutes or special lifts.

Chute Hoppers

Guidelines for the design, location and installation of hoppers can be found in British Standard Specifications and in Municipal Council Regulations.

These Codes encourage the location of hoppers in freely ventilated positions in the open air, e.g., sheltered balconies, but not within a dwelling, habitable room or place used for the preparation of food.

They should be designed and constructed to prevent emission of smell, dust, fumes, or excess noise when in the closed or open position and no part of the hopper or frame must obstruct the free passage of refuse down the chute.

Consideration could be given to developing a small device for dry shredding or grinding adjacent to the chute hopper, similar to the wet disposal units used in kitchens which discharge wet garbage direct to sewers.

Simple levered manually operated guillotines could be developed to mix, reduce, slice or smash paper, bottles, tins and wrapped garbage if tenants could be prevailed upon to operate such devices. Each chute room and each chute could be fitted with a deadlock as a safety precaution for the protection of children and others. Some Codes encourage tenants to wrap all food and vegetable refuse before placing it in the hoppers. This assists in meeting the requirements of convenience, hygiene, amenity, sound insulation and safety from fire.

There is merit in endeavours to keep refuse parcels as small as possible. This might be achieved to some extent, in the ways mentioned above, by keeping the hopper size much smaller than the chute diameter, although there is a tendency for some material to exfoliate, cartons in particular.

The Dutch have introduced a system where the chute hoppers are located inside the home unit, thus placing the responsibility of proper usage of the hopper and chute squarely with the occupants of the home unit, in this way eliminating the difficulties associated with communal hoppers.

The specially designed hopper boxes are quite small, about 10 in. x 10 in. x 10 in., so arranged as to permit garbage to fall through the bottom of the box only when the chute door is closed. More chutes are necessary, but diameters can be smaller. This is a radical departure from the concepts used in Australian practice, but it does indicate that there are many alternatives that are yet to be developed.

In my view there is need for the design, manufacture and marketing of a new range of hopper chutes, especially designed for specific purposes, in particular those for high rise living units.

Gravity Chutes

The number and siting of chutes will depend on the volume of the refuse, the layout of the building, the storage and collection system, and the means of access for collecting vehicles.

Chutes should be sited not more than 100 ft., measured horizontally from any home unit. In terms of building costs it is more economical to provide space for additional storage beneath each chute than to provide additional chutes. A minimum of two days' storage without attention by the caretaker is considered necessary.

In very high towers consideration can be given to staging the chutes for a maximum distance of about 20 floors. Thus, in a 40-floor tower two chutes would be installed instead of one, one chute would encompass 40 floors and the other 20 floors, each chute being fed from the top 20 floors. The installation of hoppers in the chute line at 20-storey intervals is another possibility.

Construction

An 18 in. or 20 in. cylindrical section is usually adequate. The chute should be vertical without offsets, constructed with durable, smooth, impervious corrosion-resistant and fire-resistant material such as aluminised steel, hardened aluminium, or stainless steel. $\frac{1}{8}$ in. plate should be considered up to the first floor to counter impingent erosion in the bottom of the chute.

The concrete chutes, which are popular in the U.K., are not favoured here because of their porous surface, but glazed stoneware linings also used appear to be effective as well as noise attenuating. The chute should be extended above the level of the roof without reduction in diameter and fitted with an approved metal mesh screen and cowl for protection against birds and weather.

Sound Control

Chutes must be insulated with soundproofing material and should not be situated near habitable rooms. The noise nuisance from descending cans and bottles can be considerable, particularly at night when the ambient noise is reduced. Notices which prohibit the disposal of cans and bottles in refuse chutes can be found in most high rise residential blocks in Sydney. These are 90% effective and likely to become more so when new legislation is introduced.

Ventilation

Authorities direct that the flow of air is from the loadside of the chute in a downward direction towards the refuse storage area, which can be maintained at a negative pressure relative to adjoining spaces.

Fire Control

Fires will occur in refuse chutes because a small percentage of people are either careless or delinquent. Approved fire prevention facilities must be built into the chute, which in the event of fire would automatically spray the chute at various levels and disconnect the chute from the compactor or refuse room when a fire occurs.

Authorities require, along with other refinements, an approved metal cut-off in the bottom of the chute which closes effectively and automatically at a temperature of not less than 10°C above normal maximum room temperature.

Chute Cleaning

An Australian designed circular weighted steel brush, which moves up and down the chute, provides an effective cleaning method combined with a satisfactory washing and disinfecting operation which is approved by several municipal councils.

Storage Hoppers Below Chutes

A method of solid waste handling much favoured by property management for its simplicity is the use of collection trucks which are loaded from uncompacted storage hoppers located at the base of vertical refuse chutes.

Refuse is released direct to the vehicle usually by single or by-parting chain-operated access gates. Daily clearances are desirable which makes the system vulnerable to industrial action. If such action is prolonged there is no alternative but to empty the hoppers into the loading bay below and then to remove the refuse manually at a later date. The monopolisation of valuable ground floor space is another disadvantage.

Refuse Rooms

Refuse rooms should be located at vehicle access level with due regard to chute location and access roads, preferably away from the main residential entrance to the building, in such a way to facilitate refuse removal without creating nuisance from noise or odours.

Dimensions will be determined by the number, size and type of containers to be used, their height above floor level, the clearance necessary for container handling and by the chute outlet details.

Construction

The floors must be constructed of solid impervious material, cement rendered to a smooth even surface, coved at the intersection with the wall and graded to an approved drainage fitment so as to permit the room to be easily cleansed.

The walls and ceilings must be constructed with similar surfaces. The room should be equipped with close-fitting doors which prevent the entry of vermin-fire rated where necessary. Where required provision must be made for artificial light, refrigeration and smell controllers.

Cleaning

An adequate supply of hot and cold water must be available for cleaning refuse rooms and the equipment in them. Hand-operated hose jets are recommended.

Shredder-Compactor-Container Operations

These can be key operations in the refuse flow lines of our 40-storey buildings. Various combinations of shredding, compacting and containerisation provide solutions most favoured today in high rise towers and the type of system selected will be determined in part by the equipment to be used for final treatment and disposal of refuse.

Improvements in this type of equipment have been quite rapid in recent years and even more spectacular developments can be expected in the near future.

Small automatic shredders with a capacity of about 1000 lb per hour and a compaction ratio of about 5:1 are giving good service in Australia.

Container compaction systems which employ a power-operated packing unit to compress refuse are suitable for chute-fed residential installations. They have considerably reduced the cost of refuse disposal, they are attractive to authorities who have collection equipment which is suited to this system.

Sack compaction systems have been introduced successfully in Australia. The system uses a power-operated ram to press refuse into disposal sacks. When used in conjunction with refuse chutes, up to 12 sacks are carried on a turntable installed in a room under each chute and a cut-off plate is provided to seal the chute from an extension tube feeding the compaction unit. The compaction cycle is started when refuse falling down the chute operates a sensing device, the cut-off plate closes and the refuse is forced into a sack by a ram. When the refuse in the sack reaches a pre-determined level the turntable rotates to bring an empty sack to the filling position.

The process is completely automatic and operative with minimum maintenance. Compaction ratios of about 4:1 result in a sack weight of about 60 lb per 2 cu. ft. sack. These sacks must be manhandled at the completion of the automatic cycle.

Machines capable of producing compaction ratios of up to 15:1 have also been introduced.

Again, the design principles to be constantly sought at this point in the refuse handling cycle are ease of handling, absolute simplicity, minimum maintenance, and pleasant conditions for the operator. (For instance, a small shredder/compactor could be installed directly below the chute and operated in conjunction with small well-sealed containers on wheels of, say, 2 cu. yds capacity, which could be moved by one man down a slightly inclined plane to an open location where they could be lifted by the compactor vehicle.) Thus full use would be made of the gravitational force along the complete line of refuse traffic.

Such an arrangement would have several advantages over the use of very large containers which are sometimes favoured. A container of, say, 30 compacted yards capacity, operated with a prime mover, could easily immobilise the system in the event of accidents, strikes or mechanical failure.

Cleansing of Refuse Receptacles

In all proposed high rise housing schemes

the Local Authorities should be consulted regarding arrangements for the regular cleansing of refuse receptacles. This is likely to be the responsibility of the owner, because receptacles are usually emptied into the collection or compactor truck on the building site. Few councils have the necessary plant and replacements needed to provide a refuse cleansing service.

Storage of Bulky Refuse

Separate enclosed space in an accessible position at ground or basement level must be provided for the storage of large unwanted articles.

Such items must be removed subsequently by the local Municipal Authority or other commercial or charitable organisations by special collection arrangements. British Codes recommend a minimum of 100 sq. ft of floor area or 1 cu. ft per person for this accommodation.

Roadways and Approaches to Buildings

Whilst automatic conveying systems could dispense with the need for the refuse collection vehicles within the residential site, road vehicles cannot be excluded entirely. Provision is necessary for the movement of furniture and maintenance equipment, segregated residues and other materials. Many high rise residential buildings are inadequately serviced in this regard.

Local Authorities should be consulted at planning stage concerning road widths, road loadings, turning circles and vehicle weights. Heavy collection vehicles are usually used for reasons of economy.

Paths between refuse storage areas and vehicle collection vehicle bays should be free from kerbs and steps, of sufficient width and hardwearing enough to withstand loadings imposed by wheeled containers. Surfaces should be level unless the slope falls away from the storage area, in which case the maximum gradient should not exceed one in twelve. Where bulk containers are used, direct unobstructed vehicular access to the refuse storage area is necessary. Special attention must be given to dimensions, particularly head room.

Automatic Collection Systems

It is interesting to examine one such system already in use overseas and being installed also in Australia. This is the Pneumatic Conveying System, extensively used in Sweden. This system requires the use of conventional refuse chutes or, alternatively, where it is acceptable that occupants carry their refuse to ground floor level, the use of chutes connected to a central loading point which can handle any refuse which can normally be put into the chute hopper.

The lower end of the chute terminates at a discharge valve where the chute is connected by lead-in tubes to horizontal steel transporter tubes of 16-24 in. diameter which are normally laid below ground level.

The tubes can be laid in tunnels or in the same trenches as water and sewerage pipes — with an inclination of 20° upwards or downwards — for a distance of one and a half miles without relay stations.

The refuse falling down the chutes rests on flat valve discs at the bottom of the chutes. The valves provide a base upon which refuse in the chute can be stored, and create a seal when vacuum is being created in the transporter tubes by the extractor turbines.

The valves in the system are opened at predetermined sequences, and the refuse is carried along the transporter tubes at high speed to a collection bin in the equipment room for subsequent shredding, compacting, baling, incineration or other treatment.

The equipment room also houses the electric motors, turbo exhausters, silencers and air filters.

The system is swift, silent and hygienic. Its function is virtually independent of the volume of the refuse. Capacity can be expanded with a longer cycle.

Compared with manual collection of refuse, capital costs per household are high, but manufacturers claim that the system will cost less than other waste handling methods in the long run because the volume of refuse can be expected to increase in the order of 5% annually, while the cost of manual collection is rising at an alarming rate.

Comparative Costs for Automatic Collection Systems

The capital cost to install the system in a concentrated high rise building area in Australia has been estimated at $A350 per dwelling unit. In the U.S.A. estimated costs of similar residential systems range between $A200 to $A540 per dwelling unit. Actual cost for residential systems in large complexes in Sweden have averaged about $A280 per apartment according to figures published during the last few months.

By comparison, the cost of a sewerage connection to a single dwelling unit is in the order of $450 for the average good quality non-luxury job.

To these figures must be added the cost of Public Authority works, made up of the neighbourhood sewerage reticulation, which ranges from a maximum of $200 per dwelling unit on a virgin site down to almost nothing on a demolition site where existing pipe capacities are usually adequate — plus the main sewage carrying system at about $200 per unit — plus the sewage treatment system which costs no less than $200 per dwelling unit also. This makes a grand total estimate ranging from about $1000 to $1500 per dwelling unit, depending upon the quality, size and location of the particular job.

The estimated cost of $350 per dwelling unit for refuse removal by pneumatic conveyor does not include final treatment and disposal by incineration and/or land fill or other means which could mount to 25% of the total of the whole operation. Nevertheless, this figure, which can vary with many factors, compares favourably with the estimates of up to $1500 per single dwelling unit for sewerage management and disposal, and in this content we cannot discount such systems on the grounds of capital cost alone.

Pneumatic conveying systems may well become the type of system best suited technically and economically to handling solid wastes in multi-storied flats. More manufacturers are becoming interested and more variations are being offered.

The Future

Current trends in high rise waste handling are towards the use of chutes, shredding, compaction, containerisation and automatic conveying.

The cost comparisons contained in the foregoing suggest that society can afford the cost of well-engineered refuse disposal systems for concentrated city living, particularly if they are conceived on a grand scale.

It is easy to forecast the extensive use of pneumatic conveyor systems which are capital intensive, largely attendant and maintenance free, with built-in flexibility likely to give satisfactory performance in all respects for the projected life of the towers. One can visualise a series of pneumatic systems silently servicing all of the inner suburban dormitory areas which will ring the great cities. The systems will discharge directly into railway rolling stock specially equipped to pulverise and compact the refuse.

The railway vehicles will be housed in large refuse buildings in strategically placed railway sidings to minimise all offence of a visual, acoustic or olfactory nature. These buildings could also be possible segregation point, for instance, for the magnetic separation and treatment of steel cans. Road vehicles could be used to bring refuse to the rail head when transporter tubes are beyond range. Compactor trucks provide the most simple and flexible removal system available at present. Swift, regular refuse trains will transport the material direct to selected tracts of barren land in country districts. Here the material will be finely ground and distributed with giant mechanical spreaders.

The economies of the operation would lie in the elimination of much of the cost associated with small piece-meal activities which depend upon many men, many contractors and many Municipal Authorities.

Such activities would be replaced with large-scale concepts utilising automatic suction systems, railway rolling stock, grinding mills and huge spreaders, seldom known to anyone except the train drivers and engineers responsible for operating the equipment, usually by remote control.

PAPER 17

Bulk Removal of Waste: Local Government Services

By H. R. Healey*

1. Introduction

The 15- to 20-gallon garbage bin is becoming less and less suitable and adequate as a storage vessel for waste arising from certain types of premises, particularly residential flat buildings, commercial buildings and industrial premises. In the industrial waste field, private contractors have already demonstrated efficient systems of waste storage and removal using containers of varying capacity to meet the demands made. The time has now arrived for local government to update its methods and plant so as to offer a variety of services in its waste removal function.

2. Changing Demands

Present-day increases in activity and population are accompanied by a need to increase community services, not the least important of which is solid waste removal. In residential flat development the 8- and 12-unit blocks of pre-war and early post-war style are being dwarfed by new buildings of up to 30 storeys, containing as many as 200 flats. The waste product generated by such a population is not such as can sensibly be stored in 2- or 3-cubic foot parcels.

The single-fronted shop and dwelling is being replaced in many shopping areas by arcades containing numerous shops and professional rooms, with several storeys of offices above. The resulting commercial and trade waste has to be measured in cubic yards instead of cubic feet and again emphasises the need to apply practical solutions to storage requirements and loading methods. In many situations, residential as well as commercial, the waste produced is of such volume as to warrant special handling systems being installed within the building. Where chuting and/or volume reduction of waste is proposed, as an in-house solid waste handling system, a bulk container is likely to become an integral part of the programme, providing both a storage vessel and a means of wheeled transportation of the accumulated waste to a collection point.

3. Scale of Buildings Requiring Container Services

3.1. *Residential*

In residential applications where conventional garbage cans are used the average family is able to contain its solid waste in one bin on a twice-weekly clearance schedule. British experience indicates that where common large containers are shared and externally stored around an apartment building, the propensity to generate waste volume can increase per family by a factor of 2.5:1. Where a central disposal system convenient to each dwelling unit is installed (such as a refuse chute in a high-rise apartment building) the waste volume generated per family is likely to increase by a factor of up to 3.75:1 over the norm.

The logical explanation of this phenomenon is that where a resident has to make do with one garbage bin for the disposal of waste, he resorts to packing and cramming in an effort to fit all of the accumulation of otherwise objectionable rubbish and putrescible matter into the available space. Where a larger common container is used the resident is not as careful or painstaking in reducing the volume of his solid waste, and is therefore likely to require more space. But as the householder has to carry his waste outdoors to the container, some volume reduction may be undertaken to reduce the number of trips. In the third case of an in-built system, conveniently located near the dwelling unit, the resident no longer exercises any restraint either on the type of waste or the reduction of bulk, with the resultant marked increase in volume per capita.

Local experience so far suggests that loose waste generated in residential buildings should be calculated at not less than 1 cubic foot per day per 2-bedroom flat in the case of communal containers and possibly a greater capacity per head where central chutes are provided. The point at which container services would therefore become logical would be, say, a 15-flat or larger building, which would be expected to generate 15 x 1 cubic foot x 7 days = 105 cubic feet per week (equivalent to two clearances of a 2-cubic yard container).

3.2. *Commercial*

In commercial applications any building generating 2 cubic yards of waste per week

* Health Surveyor, Woollahra Municipal Council, New South Wales.

would warrant introduction of container service. The only accurate gauge of waste volume for a particular building is by survey, but for practical purposes an approximation can be obtained by applying the following figures (Table 1) on use and floor area.

Table 1

ESTIMATE OF SOLID WASTE — COMMERCIAL TENANCIES SYDNEY 1971

Use	Rate (cu. ft./day)	Floor Area (sq. ft.)
Office space	1½	1000 (lettable area)
Retail store — general	2	1000 (gross area)
Department store	8	1000 (gross area)
Supermarket	16.2	1000 (gross area)
Theatres	1.1 screening/ 100 seating	
Car park — plaza, public toilets	0.25	1000 (gross area)

From the foregoing a new commercial development of, say, 4000 sq. ft. of supermarket, five lock-up shops and two floors of offices could be expected to generate:

4000 sq. ft supermarket = 4 x 16.2 = 64.8 cu. ft./day

4000 sq. ft. retail shops = 4 x 2 = 8.0 cu. ft./day

6000 sq. ft office space = 6 x 1.5 = 9.0 cu. ft./day

(no on-site parking) toilets and open space = 2 x 0.25 = 0.5 cu. ft./day

Total = 82.3 cu. ft./day of solid waste.

As this is equivalent to over 3 cubic yards of waste/day the case would appear to warrant consideration as needing on-site compaction at approximately 3:1 ratio and a container/bulk loading service on a 2- to 3-day interval.

4. Phasing-in New Services

A proposal to introduce bulk loading should be fully debated before a decision is taken, as long term commitments are likely to flow from such a change. For example, the local authority has to outlay capital both on vehicles and containers, some hundreds of which will eventually be needed to fully utilise work potential. Some containers may even by privately purchased to suit the local authority equipment, and as the service increases, more containers will be added to the old with the result that at no one point in time will all equipment reach the stage of replacement. Therefore, the first decision concerning the type and design of equipment could well be a decision in perpetuity, or one difficult to change in the future as newer and more efficient systems become available. Many buildings will be designed with service space and access only suited to the council's present system.

It would obviously be advantageous for a central authority to set standards, as in Britain, but local conditions vary tremendously and a system suitable for one area might be quite unsuitable for another. Several metropolitan councils have already committed themselves to bulk loading, and so far every system differs from the next.

Once the decision to introduce container services has been reached all new building approvals should be made conditional on compliance with on-site loading facilities for container services, and the time-lapse between approval and occupation can gainfully be used for tender and the purchase of new equipment. The council has the choice to let-out to contract the first new container services pending a time when the outright purchase of vehicles and containers is economically viable or a secondary use of a container vehicle can be found. For example, there is likely to be a variety of existing buildings housing conventional services, both residential and commercial, which could readily change-over with advantage to container services. Schools, large generators of waste, are already using container services in some locations and hospitals, and public institutions also are potential customers. Containers can be located at parks, beaches and reserves for use of ground staff. Playing flelds have large volume waste potential, particularly during the winter sport period. Containers can also be located at strategic locations to assist street sweeping staff, i.e. both the hand sweeping crews and the mechanical-broom operators. Annual or periodic "clean-up" campaigns can be programmed for container use, depending on the choice of plant.

Generally then it will be seen that upon reaching the decision to introduce bulk handling, sufficient work can be allocated to the special equipment purchased to keep it fully employed. The re-programming of many other services may be influenced by this change, but must result in more efficient handling and time and labour savings.

5. Choice of Plant

Almost every major type of waste management equipment available around the world is already being marketed in Australia. The first decision to be made is whether a specialised bulk loader should be purchased or if a garbage collection vehicle with added facilities to empty containers should be utilised.

With specialised container vehicles, high efficiency can be obtained, as the entire design is aimed at a single function. The disadvantage is that despite programming of the work schedules the container vehicle will often run parallel with normal garbage collection rosters, and two trucks and crews frequently cover the same street, each emptying containers of different types from neighbouring buildings. A further major disadvantage is the lack of backup vehicles in the event of breakdown. It is not practicable to hold costly single-function vehicles in reserve as standby, for use in emergency.

The advantages of separating the two functions, of garbage collection and bulk container collection, is that the most efficient type of vehicle body and chassis can be selected for each. Furthermore, in long-term work programming, particularly in areas where new high-density development is proceeding, or in redeveloping regions, the normal garbage rosters can remain static for longer periods without overloading. The container clearances can be increased as the need arises, greatly lessening the administrative time-consuming work studies and work reallocating which normally occur.

The other major factor in this decision is the overall labour allocation and structure of the waste collection programme. For example, authorities which collect waste only once weekly would have to consider a bulk service capable of holding 7 days' produce. A twice-weekly collection would present another situation, while daily collection services would put the emphasis on, say, speed of handling smaller containers.

Councils often have programming peculiar to their own areas, and the decision to introduce dual-function or specialised-function vehicles would be greatly influenced by the overall organisation structure.

5.1. *Choice of Container*

Almost every system requires its own specific design of container. Obviously, the container has to be of adequate capacity to fulfil the function for which it is being supplied. Two cubic yards is the minimum capacity that should be seriously considered, having in mind the bulky nature of waste today, and there is no indication that the decrease in density has reached its peak.

The container material and design should also be hygienically acceptable and have an adequate life span. By way of example, some European cities use a container manufactured of vinyl fabric with capacity up to 16 cubic feet. There is no adequate method of cleansing these bags and in practice this is not attempted, with the result that grease and putrescible material build up on the interior surfaces. As the climatic conditions are entirely different from those in Australia, as is the type of native vermin, the system operates effectively. These containers are light in weight, allowing hand carry between waste rooms and vehicles, and do not require mechanised lifting equipment on vehicles; they are also simple and effective. Nevertheless, they could not be recommended for adoption in this country. There could, in fact, be dangers in using such materials, other than for hygiene in that vinyl under fire could give off toxic gas (fire has to be associated with bulk storage of wastes).

Generally then, attention should be given to metal containers, of 2 cubic yard capacity or greater, designed for ease of emptying, smooth surfaces well jointed for ease of cleansing, enclosed to reduce odour and control vermin, weatherproof if outdoors, rugged to withstand heavy duty in lifting, possibly inverting and impact on return to ground.

If the system requires movement of the container by hand, the overall weight, manoeuverability and bulk have to be carefully looked at. Placing wheels under containers is a simple exercise, but the choice of wheel and the geometry of the container movement can be greatly improved with the correct choice. Wheels of less than 7 in. diameter x 2 in. tyre width are not recommended. The wheels should be rubber, or rubber-tyred, to avoid excessive noise. The most manoeuverable wheel pattern is of two fixed wheels and two castors, mounted to give the greatest wheelbase possible. Some systems require four castor wheels to permit the container to be moved sideways so as to engage into the loading mechanism of the truck. These containers can be difficult to control when wheeling into or out of buildings, over gutters and stormwater drains, and along corridors.

Movement of containers should be limited and controlled. In Britain, the recommended access requirement suggests that distances for crews to move containers are not to exceed 30 ft., and that there should be a 6 ft. wide corridor (for a 3 ft. wide container), which should be kept free of irregular surfaces and not contain gradients in excess of 12:1. Obviously the normal storage location for a wheeled container should be level or have a stop or recess under the parking position to arrest unintentional movement, and the loading location should be quite level.

Before selecting the system a specification of container performance (capacity, shape, manoeuverability, etc.) should be prepared. This specification could well determine system choice or at least eliminate certain systems from further consideration.

5.2.1. *Side loading garbage collection vehicles*

There are a number of vehicles of this tvpe which may be fitted with container lifting equipment. A side loading hydraulic mechanism may be fitted with a capacity to lift up to 1000 lb., and this could handle containers of 1 to 1½ cubic yards capacity. The containers would be emptied by inverting the container through the side and part of the roof of the vehicle. The normal compaction ram would be used to sweep the waste to the rear of the body and clear the loading hatch ready for the next use. The mechanism folds away when not in use so that conventional hand loading may be carried out. Coverting existing vehicles to allow the fitting of this mechanism is not recommended, but fitting this device to a new vehicle would add about $1900 to the purchase price (Fig. 1). While this system is adequate for certain functions, the container wheelbase is not very satisfactory as the bin measures approximately 33 in. wide x 36 in. deep x 66 in. high. The container cost is around $180 per bin.

Fig. 1: Side Load Attachment
Dual Purpose Vehicle

Side loaders may also be fitted with a front fork hydraulic lift mechanism. If this is designed for a 1000 lb. lift the conventional side load body bay be used; but if the 3000-lb. lift capacity is sought, to allow 2-, 3- and 4-yard container work, a heavy industrial body would have to be used. The cost of the hoist attachment is around $4000, but the total price on a heavy hoist and industrial body is around $14,000, exclusive of the chassis (see Fig. 2).

Fig. 2: Front Fork Lift
Single or Dual Purpose Vehicle

5.2.2. *Rear loaders.* Container handling can be readily undertaken by fitting retractable hydraulic arms which engage a specially shaped (tapered front) container. The container can then be lifted and partly inverted, allowing waste to slide or spill from the angled front of the container into the rear loading hopper.

Alternatively, the container can be winched from pivot points into an unloading position, in lieu of using hydraulic arms (Fig. 3).

Containers of 1-, 2- and 3-cubic yard capacity can be handled (up to 5000 lb.), but as the hopper capacity is usually only approximately 1 cubic yard difficulty can be expected in loading larger containers without some spillage. This exercise will involve several compaction sequences so as to clear the hopper while the container is being emptied.

The container attachments may be fitted to an existing vehicle; the cost of the winching attachment is approximately $975, and the hydraulic equivalent approximately $950.

5.3 *Specialised Bulk Collection Vehicles*

The cost scale of many specialised vehicles is higher than of dual-purpose vehicles, as most of the plant on the market is geared for industrial application and larger pay-loads. The chassis are also usually in the heavier class, often with tandem rear axles, also increasing capital cost considerably.

Fig. 3: Rear Load Dual Purpose Vehicle

The most widely demonstrated style of vehicle in this category is the front fork lift. It has great capability both in lifting strength and carrying capacity. It can be operated as a one-man unit, as the driver does not necessarily have to leave the cab to engage and empty containers. This vehicle is suited to industrial work; the gross vehicle weight, noise output and bulk are excessive for residential and even some commercial situations.

Spillage can occur in emptying containers by this method, although some models do have screens which can direct most of the waste into the vehicles.

Containers may be of 1-, 2-, 3-, 4 and 6-cubic yard capacity and be either wheeled or static, depending on storage location and vehicular access. The lifting mechanism is designed for heavy duty, and compaction of wastes in even the larger containers should not over-strees it. The containers are reasonably priced at around $200, are capable of holding any type of waste (including food) and can be lined with tar-paper sacks to reduce cleaning frequency. A liquids sump can be incorporated to collect and release fluids.

This vehicle, as a 1-man operated plant, and capable of up to 90 cu. yd. carry, is very efficient. For local government use, a high load demand would have to be developed to keep the unit fully engaged and to yield a reasonable return on investment. Special site requirements would be necessary in residential buildings (heavier domestic roads, clearance of 16 ft. head height for the vehicle to empty containers) and noise complaints could be expected at whatever hour the plant was used, but more particularly if work were attempted before, say, 7 a.m.

The second specialised vehicle in frequent industrial use is the Benes Marell unit, which lifts a single container on rear hoists. The container is then conveyed to the disposal point, tripped and emptied, and returned to the site. A rotary system is often used, leaving an exchange container on-site in place of the full one removed. These containers can range up to 40 cubic yards and up to 10 tons in weight.

This system, though excellent for industrial application, is not recommended for residential or small commercial work. The containers are rarely of the covered type and could only be used for garbage collection in such cases as a major residential building with a refuse chute or similar central loading point. Council with high-density housing could with advantage obviously consider this system at present, but in other cases it would be likely to form a second phase scheme.

Due to the small demand for this type of plant likely to be made on local government it would not warrant purchase unless it could be used in other functions undertaken by council, such as at transfer stations or in construction works. Similar remarks would apply to the roll-on, roll-off body units. Unless sufficient utilisation of a major nature can be regularly achieved the local authority will be well advised to leave these fields of waste removal to general contractors (Figs. 4 and 5).

There are few light-weight specialised container vehicles on the market. In the U.K., Hong Kong, and several other countries the Pallodin system is widely used. The vehicle, in this case, is not suitable for any function except lifting a circular bin of specific dimensions (1 cubic metre capacity). The bins are wheeled to the vehicle, clamped in position by hydraulic locks, lifted and inverted into the rear body (Fig. 6). Little spillage, with the exception of dust, occurs and the power take-off is able to supply hydraulic power with very low engine speed.

While the system operates effectively, the shape and size of the container are not the most suitable. The bins are constructed to British Standard 1136, have a height of 4 ft. 3 in. and are of 3 ft. internal diameter. The bins either have four castor wheels or stand on legs or a plinth and the in latter case are moved by a wheeled jack carried by the collection crew.

The containers, standing 4 ft. 3 in. high plus legs or wheels (a total of 4 ft. 9 in.) are not suitable for resident hand loading of waste unless a stair and platform are provided for the user.

Although this system is widely used overseas its local introduction is not endorsed, because of its inflexibility and because the 1 cubic metre container is too small for present demands. Furthermore, it is not very stable on its small wheelbase, and is difficult to load.

In one local handling system, demonstrated in Melbourne, the vehicle comprises a flat-top body with a centrally mounted crane. Four containers of 2 cubic yard capacity can be carried on the tray, locked down with chains. An empty container is placed on-site and exchanged for a loaded container. On reaching the disposal point the containers have to be re-engaged by the crane, lifted to rear of the vehicle and emptied singly by trapping the container body on an in-built post and lowering the base by operating the crane. This system would be suited to, say, "clean-up" work, provided the disposal point was not far removed. If this type of system were introduced in residential flat buildings for domestic garbage, the exchanging of containers would involve the

Fig. 4: Marrell Container

Fig. 5: Roll-on, Roll-off Specialised Vehicle

Fig. 6: Pallodin Bin Lift

local authority in cleansing bins after each use — a costly operation.

Another truck which has recently been locally marketed involves the use of a crane for lifting and emptying containers into a garbage compaction body. The crane is rear mounted and, on engaging the container, raises it to the loading position. The second operator releases the hinged container base and the container is returned to the ground with the base still open. The weight of the container body closes the base when in contact with the ground.

This system was initiated in Finland for 1 and 2 cubic yard containers and in many situations there the containers around residential buildings are recessed into the landscape so that storage of waste is not noticeable (Fig. 7).

This system allows for lifting containers from various levels above or below the vehicle level, and from either side of the vehicle (Figs. 8 and 9).

6. Service Charges and Hire

6.1. *Collection Charges*

Under Sections 167 and 168 of the Local Government Act of New South Wales the council may fix charges for providing services in connection with, amongst other things, the removal of garbage. There are many methods of implementing this section of the Act, and it is not intended to cover the entire aspect of normal garbage service charges. Generally, however, the fee for garbage collection is about 10 cents per cubic foot per service.

Where a container service of, say, 2 cubic yard capacity is provided (54 cubic feet) a charge of around $5.40 per clearance — with reduction for emptying less than fully loaded bins — has been normal in private-enterprise fields. This figure, although compatible with council charges, is expected to escalate due to general operating cost increases.

A council has the option of setting a special fee for bulk services and may choose to apply a cost rate that will reimburse the authority for depreciation of plant, operational collection costs, disposal charges, overheads and taxation. Close examination of all these expenses will yield an appropriate charge per unit volume which could vary from council to council. It is suggested that the end result could produce a factor which is less than 10 cents per cubic foot. The actual cost per 2 cubic yard container handling in the case of one council, where an allowance of 15 minutes' travel between collection sites was allowed, was $4 per service.

There appears to be reasonable economy in handling by bulk container services as com-

Fig. 7: In-ground Container Residential Application — Finland

Fig. 8: Bin Exchange System

Fig. 9: Local Example of Overseas Container System (Finland)

pared with the conventional small regulation bin services.

6.2. *Supply of Containers*

Again, council has the option to set charges for services rendered and may choose to provide containers to certain premises for a fee. Alternatively, under Clause 62 of Ordinance 71 (repeated in Ordinance 70, Clause 45.1. (c)) which applies to residential flat buildings, the council may require the owner of the building to provide, at his own cost, a suitable container.

If containers are leased an exercise on an appropriate charge should be attempted. The purchase price of the equiment may be found to be, say, $400 per container. The life expectancy would be up to 10 years, but it would be reasonable to repay capital within 4 years, i.e., $100 per year. The rental achieved after 4 years would offset other costs such as repainting, repair, maintenance and cost of transport in replacing containers at maintenance intervals.

7. Servicing of Buildings

Having entered into agreements covering additional services for residential, commercial and possible industrial premises, the council should prepare a code of requirements, so that with buildings either being newly constructed or substantially renovated sufficient on-site area is allowed to permit full benefit of the new services.

On-site storage and loading should be persued as an aim wherever it is possible and practical to do so. So that designers can provide for this in preparing plans, the council should have available a list of vehicle and plant specifications, including gross vehicular weight, length, height and width dimensions of vehicles, turning geometry and container storage and loading details (see for example Appendix I, prepared by the City of New York administration).

It is advisable, on safety grounds, to ask for vehicular reversing on-site so that collection trucks enter and leave the site in a forward direction. In commercial application, where all of the site is to be covered by buildings, it may be possible to set back a lay-by in the service lane so that the collection vehicle can pull in parallel to the kerb, but out of the traffic flow, to permit loading.

It will be found necessary to spend considerable time in discussion with designers when implementing new service requirements.

8. Summary

Different authorities have different local problems in waste collection, and therefore have to make a decision based on a service capable of fulfilling their particular needs.

Some inner city councils have large-scale commercial, industrial and residential development whereby containers of 10 to 20 cubic yards' capacity should be used. However, the narrow service lanes and on-street parking may restrict access of vehicles capable of handling this type of plant.

In many cases the existing industrial working conditions of staff and available capital would restrict early change, but moves should now be set in train to overcome this situation. Long-term budgets should be prepared, to allow for funding of purchases over, say, 3 to 5 years of the more expensive plant, while immediate purchase of one system may be financially viable and allow the commencement of at least one phase of container handling.

Authorities in the past have tended to allocate each item of plant to a particular department for its exclusive use. With container handling equipment, which is likely to enjoy at least some standby periods, the plant could probably be used in other Departmental functions, such as delivery and removal of sand, metal and spoil for construction works, minor transfer stations for street sweeping operations and general materials handling situations.

Authorities should also consider operating several shifts per day so as to allow either greater utilisation of special plant, or alternatively servicing of plant outside the use periods so that the greatest work potential is reached. This practice is common overseas.

Appendix I

THE CITY OF NEW YORK
ENVIRONMENTAL PROTECTION ADMINISTRATION
DEPARTMENT OF SANITATION
REFUSE COLLECTION VEHICLE DATA FOR ARCHITECTS AND ENGINEERS

Item	Maximum Length	Maximum Width or Depth	Maximum Height	Minimum Turning Radius	Max. Axle Load (lbs.)	Max. Total Weight Loaded (lbs	Minimum Operating Headroom
Bulk collection truck	27′8″	8′6″	9′1″	38′	29,000	38,000	8′6″
Collection truck (rear loading)	29′2″	8′6″	10′7″	40′	29,000	38,500(1)	11′0″
Rear loading containers:							Container weight empty
1 cu. yd.	6′6″	2′0″	3′11″	On Castors	N.A.	1,600	350 lbs.
2 cu. yd.	6′6″	4′9″	3′10″				565 lbs.
Hoist fitted chassis container truck (H.F.C.)	29′10″	8′3″	13′1″	34′	29,500	37,050	11′4″
H.F.C. containers:							
6 cu. yd.	5′0″	7′11″	5′9″	N.A.	N.A.	8,500(2)	N.A.
8½ cu. yd.	6′8″	5′10″	7′6″				
12 cu. yd.	7′2″	8′0″	7′2″				
Hoist compactor (H.C.) container truck	35′9″ w/forks extended	8′2¼″	12′9″	42′	38,000	56,000	12′10″ 20′0″(6)
H.C. containers(4)							
2 cu. yd.	72¼″	37⅝″	38″	N.A.	N.A.	7,000	N.A.
3 cu. yd.	72¼″	43⅝″	48″				
4 cu. yd.	72″	56″	48¾″				
6 cu. yd.	72″	56″	72¾″				
8 cu. yd.	72″	56″	81⅛″				
Roll-on roll-off container truck(5)	32′0″	8′0″	12′6″	47′	60,000	70,000	17′6″ 14′0″(3)
Roll-on roll-off containers 15 thru 40 cu. yd.	22′0″	8′0″	8′0″	N.A.	N.A.	50,000	N.A.

1. For snow operations a 1800 pound snow plough is added.
2. Maximum load is a function of hauling unit capacity. Hoist units capable of lifting heavier loads are available for special application in some locations.
3. Loading from a 36″ high capacity loading dock.
4. All containers must be compatible with E-Z Pack Body Model FL 70-30. Dimensions shown are approximate.
5. Preliminary data subject to revision. For further information call 566-2805, 566-5195.
6. Minimum height required to empty all containers.
7. The Department of Sanitation offers to architects, engineers and property managers, assistance and technical data on its requirements and equipment for refuse collection. It is suggested that the Department be consulted during the planning of housing developments or other structures eligible for D.S. service, so that mutually satisfactory facilities and arrangements can be made for refuse collection. Container service is available upon request and is recommended where applicable.

N.A. Not Applicable.

PAPER 18

Use of Membrane Processes in the Treatment of Aqueous Wastes

By D. C. Griffith,* A. G. Fane,† C. J. D. Fell† and R. T. Fowler†

The present state-of-the-art of both ultrafiltration and reverse osmosis as applied to waste treatment is reviewed and the advantages and disadvantages of both processes discussed. As an example of the type of development effort required in piloting either process for a specific application, experiments carried out on the ultrafiltration of starch factory effluent are described. The results are encouraging and suggest that further investigation of this method of treating starch effluent be undertaken.

The use of membrane processes (ultrafiltration and reverse osmosis) for the treatment of aqueous wastes is relatively new, but is being looked upon with increasing interest by industrialists concerned with the prevention of pollution by proper waste treatment. It is felt there are many situations where membrane processes may hold the key to economic effluent processing because of their inherent ability to handle low concentration streams at relatively low energy cost. Indeed, there are already examples of the successful application of membrane processes to specific problem effluents.

The principle by which membrane processes operate is straightforward. The waste stream to be treated is presented under pressure to a semi-permeable membrane. The solvent molecules (usually water) pass through the membrane, whilst the solute molecules or suspended solids are retained and concentrated. The distinction drawn between ultrafiltration and reverse osmosis is in the size of molecules retained, the term ultrafiltration being applied to processes in which molecules being rejected are approximately 10 times the size of the water molecule or larger.[1] Both processes require that the transmembrane pressure be sufficient to overcome osmotic pressure and to cause flow through the membrane, but only in the case of reverse osmosis do osmotic effects dominate.

In arriving at the optimum treatment method for a specific effluent, it is clear that the industrialist must give consideration to the possible use of a membrane process. The first purpose of the present paper is therefore to provide the industrialist with an indication of those situations in which each type of membrane process is likely to be successfully applied. This is done by reviewing the current state-of-the-art of both processes. Then, as indication of the subsequent development program required to establish the feasibility of a membrane process, a detailed description is given of experiments by the authors on the application of ultrafiltration to the treatment of starch factory effluent.

ULTRAFILTRATION

Industrial scale ultrafiltration is practically restricted to applications where the rejected material is a solute of molecular weight of 500 or greater, or is a finely divided suspended solid. Typically, a trans-membrane pressure in the range of 200-700 kPa (30-100 p.s.i.) is used and a membrane flux of 0.04 m^3/m^2 hr (15 gal/ft^2 day) achieved.[2]

The choice of membrane is most important. Indeed, prior to the appearance of the "skinned" membrane in 1960, ultrafiltration was not considered commercially feasible because of the low membrane flux possible with the then existing homogeneous membranes. Skinned membranes, first introduced by Souriajan and Loeb[3] for reverse osmosis, and later modified for ultrafiltration, are anisotropic and consist of a high density layer of thickness 0.1-0.5μm supported on a porous matrix of thickness 100μm.[4] The ultrafiltration characteristic of such membranes is provided by the dense layer which is sufficiently thin to offer a minimum of resistance to solvent flow. Whilst early skinned membranes were made exclusively from cellulose acetate using an evaporation/precipitation process, they can now be fabricated from a number of synthetic poly-

* Now with Spruson and Ferguson, Patent Attorneys, Kent Street, Sydney.

† Members of the Academic Staff, School of Chemical Engineering, University of New South Wales.

mers to give improved resistance to attack by chemicals and enzymes in critical applications. Skinned membranes are always used with the dense "skin" in contact with the stream being processed. In this way blinding of the micropores present is avoided.

Concentration Polarisation

Although with the advent of anisotropic membranes the maximum possible solvent flux for a given transmembrane pressure was considerably enhanced, membranes used in industrial situations will rarely field a flux greater than 5-10% of that obtainable with pure solvent. This is because of an effect known as concentration polarisation, which is the build-up of a layer of rejected solute or suspended solids at the membrane surface. The exact nature of this layer depends very much on the physical and chemical characteristics of the rejected material. If, for example, the material being rejected is proteinous, the layer may take the form of a gel. Even if gel formation does not occur, the high solute concentration will usually substantially increase the local viscosity at the membrane surface. The polarised layer therefore offers additional resistance to solvent flow and the net result is a decrease in membrane flux.

Concentration polarisation can be partially alleviated by causing the process stream to flow past the membrane surface. The shear field so established helps to minimise the thickness of the layer of high solute concentration. Indeed, the vendors of commercial ultrafiltration modules have exercised considerable ingenuity in designing their equipment to reduce the effect of concentration polarisation and a variety of membrane geometries and process side flow paths has resulted. These may be broadly classified into three types:

(i) *thin channel*: in which the process stream flows in channels of width 1-3 mm; the nature of the channel varies with module design and may be rectilinear or in a spiral;

(ii) *large diameter tubular*: in which the process stream flows inside tubes of ½-1 inch diameter; and

(iii) *hollow fibre*: in which the process stream flows within hollow fibres of diameter 0.5 mm.

The various commercially available ultrafiltration modules suitable for industrial waste treatment are classified according to the above scheme in Table 1.

From an examination of the module designs in the table it is apparent that the various

Table 1

AVAILABLE ULTRAFILTRATION MODULES

Trade Name	Local Supplier	Module Geometry	Membrane Area	Membrane Material	Typical Transmembrane Pressures (kPa)
Abcor	Mauri Bros. and Thomson Ltd.	Tubular (25 mm dia.)	From 0.2 m^2	Cellulose Acetate	280
Ajax	Dunlop/ IBC Ltd.	Thin Channel —Spiral Wound	Wide Range	Cellulose Acetate	280 to 550
Amicon	Scientific and Research Equipment Co. Pty. Ltd.	Hollow Fibre	1.0 m^2 to 3.0m^2	Co-polymer	180
		Thin Channel —Membrane covering splined rod	0.15 m^2 to 1.3 m^2	Co-polymer	550
Calgon-Havens (Osmotik)	Havens International	Tubular (12 mm dia.)	From 0.2 m^2	Cellulose Acetate	Up to 1400
Den Danske Sukker Fabrikker (D.D.S.)	Anderson Equipment Co-Op. Ltd.	Thin Channel —Plate and Frame	0.4 m^2 to 28 m^2	Cellulose Acetate	1200 to 1900
Dorr-Oliver	Dorr-Oliver Pty. Ltd.	Thin Channel —Parallel Plate	From 0.2 m^2	Co-polymer	280
Paterson Candy International	The Permutit Co. of Aust. Pty. Ltd.	Tubular (10 mm dia.)	From 0.4 m^2	Cellulose Acetate	600

manufacturers have each compromised to a greater or lesser extent in either decreasing the effect of concentration polarisation by increased process side flow (thereby promoting turbulence but incurring higher pumping costs) or in increasing the membrane area per unit volume to give a greater overall module flux (thereby increasing fixed cost).

In fundamental engineering terms is would seem that the nature of the process stream and the likelihood of gel formation may well tip the balance in favour of high turbulence, low membrane area modules in certain situations. It is for this reason that one particular module type cannot be universally recommended, and it is necessary to compare the performance of each type of membrane for each new application.

Waste Streams amenable to Treatment by Ultrafiltration

Many effluents from the chemical and related manufacturing industries can be successfully treated by ultrafiltration to yield a recyclable filtrate and a concentrate suitable either for product recovery or relatively easy disposal. Some examples are given below. [1, 2, 3]

Table 2

REPORTED APPLICATIONS OF ULTRAFILTRATION FOR EFFLUENT TREATMENT

Industry	Species Removed
Automobile	Electrophoretic paint.
Chemical	Oil, either dispersed or dissolved.
Dairy	Whey protein.
Food Processing	Protein.
Iron and Steel	Iron (as hydroxide) for pickle liquors.
Meat Processing	Blood and paunch manure.
Metal Plating	Metals (as hydroxides).
Pulp and Paper	Lignin and lignin sulphonates.
Sewage Treatment	Dissolved and suspended solids.

Okey[5] has prepared cost data on ultrafiltration for waste treatment which suggest that membrane treatment is indicated for wastes that have high value waste components suitable for recovery, high concentrations and low to moderate flows, and in situations where there is a need for space conservation. These guidelines can be used in selecting those waste streams not listed above but which are likely to be amenable to treatment by ultrafiltration.

Equipment Arrangement

Ultrafiltration units can be operated on either a batch or continuous basis. Fig. 1 shows both arrangements. Perhaps the most significant feature here is the suggested use of two pumps, one to recirculate the concentrate to obtain the necessary flows in the interstices of the ultrafilter, and the other to provide the transmembrane pressure requirement. Such an arrangement can result in considerable power savings on a large scale ultrafiltration installation.

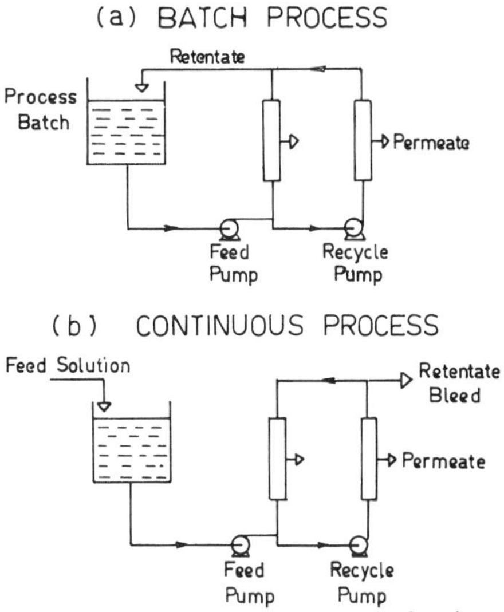

Fig. 1: Possible Arrangements of Ultrafiltration Equipment.

REVERSE OSMOSIS

Reverse osmosis is industrially applied to streams where an ionic or small molecular weight solute is to be removed. Typical transmembrane pressures are 4 to 7 MPa (600 to 1000 psi) and membrane fluxes of 0.04 m^3/m^2 hr (15 gal/ft^2 day) are possible.

The nature of the modern reverse osmosis membrane though similar to that used in ultrafiltration in being anisotropic, differs in that the active layer is denser and the membrane itself must be able to withstand much greater pressures. Currently available membranes are fabricated either from cellulose acetate or from co-polymers. Smith[6] has recently given an extensive review of those available locally.

The mechanisms occurring within the membrane are not yet fully understood. Since the size difference of the molecules or ions being separated is small, an explanation of reverse osmosis based on sieving alone does not seem credible. Indeed, various other theories to explain the functioning of the membrane have been proposed. At one extreme there are those which regard the active layer as a homogeneous continuous medium with no pores of any sort, with transfer of solute and solvent taking place by diffusion alone.[7] At the other extreme the existence of discrete pores is suggested and the separation is thought to be due to surface tension forces.[8] Between these is the more usually accepted theory that hydrogen bonding plays an important part in the transfer of solvent, with the solvent molecules moving through the membrane by jumping from one hydrogen bonding site to the next.[9]

The flux in a commercial reverse osmosis unit is dependent on both the transmembrane pressure and the concentration or viscosity of the process stream being treated. Fig. 2 shows typical membrane fluxes for three transmembrane pressures for different concentrations of a simulated wash liquor from a nickelplating bath. These results have been obtained by the present authors using a D.D.S. reverse osmosis unit fitted with 104 type 999 membranes which have a molecular weight cut-off of 200 for sucrose solution.

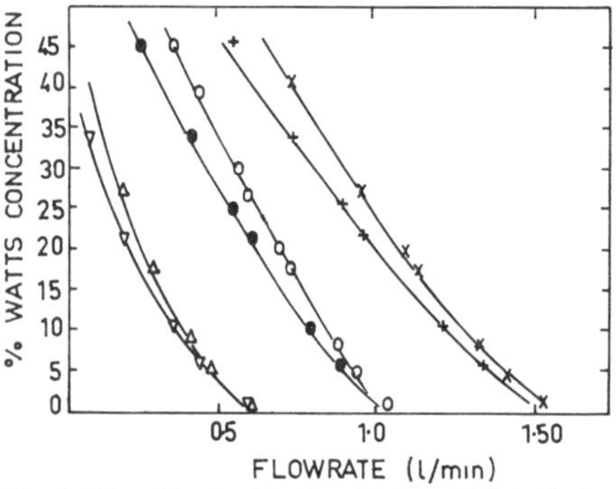

Fig. 2: Flux Plot for Simulated Watts Nickel Plating Liquor Through D.D.S. 999 Membrane.

× 5.9 MPa — No Brightener
○ 4.0 MPa — No Brightener
△ 2.1 MPa — No Brightener
+ 5.9 MPa — Brightener
● 4.0 MPa — Brightener
▽ 2.1 MPa — Brightener

Brightener referred to is organic additive to improve plating properties.

As with ultrafiltration, reverse osmosis suffers from concentration polarisation effects which necessitate a flow of liquid past the membrane surface to achieve acceptable membrane flux. Again, this has resulted in a number of different membrane geometries and process flow paths being developed. Table 3 indicates the nature of these in the reverse osmosis units presently available for industrial purposes.

Waste Streams amenable to Treatment by Reverse Osmosis

Table 4 lists process effluents which have been successfully treated, at least at the pilot plant scale, by reverse osmosis.[5 10-13]

In general it can be said that the economics of the treatment are strongly influenced by the value of the product recovered. In all of the above processes the filtrate produced is potable and suitable for immediate plant recycle.

APPLICATION OF ULTRAFILTRATION TO STARCH FACTORY EFFLUENTS

An indication of the development program necessary to establish the feasibility of a membrane process in the treatment of an industrial waste stream, is now given by discussing work carried out to date at the University of New Wales on the application of ultrafiltration to the treatment of starch factory effluent.

Table 3

AVAILABLE REVERSE OSMOSIS MODULES

Trade Name	Local Supplier	Module Geometry	Membrane Material	Transmembrane Pressure (MPa)
Abcor	Mauri Bros. and Thomson Ltd.	Tubular (12 mm dia.)	Cellulose Acetate	2.5 to 6
Ajax	Dunlop/IBC Ltd.	Thin Channel —Spiral Wound	Cellulose Acetate	4.2
Calgon-Havens (Osmotik)	Havens International	Tubular (12 mm dia.)	Cellulose Acetate	2.5 to 6
Degremont	Degremont — M.I.S.	Tubular	Cellulose Acetate	6.4
Den Danske Sukker Fabrikker (D.D.S.)	Anderson Equipment Co. Pty. Ltd.	Thin Channel —Plate and Frame	Cellulose Acetate	2.0 to 8.0
Hidraport	J. Howden & Co. Aust. Pty. Ltd.	Tubular	Cellulose Acetate	4.2 to 5.3
Paterson Candy International	The Permutit Co. of Aust. Pty. Ltd.	Tubular (10 mm dia)	Cellulose Acetate	4.2 to 8.4
Permasep	Sybron Aust. Pty. Ltd.	Hollow Fibre	Co-polymer	3.0

Table 4
REPORTED APPLICATIONS OF REVERSE OSMOSIS FOR EFFLUENT TREATMENT

Industry	Type of Effluent	Species Removed
Chemical	Catalyst washings	Cobalt salts
Dairy	Whey	Protein
Dyestuffs	Dye mother liquor	Dye
Food Processing	Pickling	Brine components
Iron and Steel	Spent pickle liquors	Iron salts
Photographic Processing	Spent processing solutions	Ferro/i-cyanides, Silver
Pulp and Paper	Mill wastes	Sugars and inorganics
Electroplating	Pickle liquors	Nickel, chromium salts
Sewage	Activated sludge effluent	Organic matter
Starch	Waste water from potato starch industry	Proteins and amino acids

Starch Production

Starch is an important raw material for many food processing operations and is produced in large quantities throughout the world. In North America it is recovered from maize.[14] In Australia and other major wheat producing countries it is extracted from wheat flour using either the Martin process or the Batter process. The key to economic production of starch from wheat flour lies in the recovery of wheat gluten, a valuable protein by-product. So important is the need to obtain this protein in a non-denatured or "vital" form that the starch recovery step must place major emphasis on this need.

In the process in excess of 10 kg of water are used for each kg of flour processed. Of this water approximately 90% is rejected from the process as liquid effluent containing about 1% of solids consisting mainly of soluble matter. The outgoing effluent contains about 10-20% of the weight of the original flour and this represents a considerable loss to the processor. In addition, the discharge of the effluent to the sewer will incur further cost penalties. There are, therefore, strong incentives for the establishment of an effluent treatment method which allows both recovery of valuable resources and of water for reuse.

The Effluent

For a medium size wheat starch processing plant, providing two grades of starch and gluten products and handling a weekly feed of 400 tonnes of flour, a total aqueous effluent of 5-6000 tonnes/week is usual. Relating this to the national scale, the 2.2×10^5 tonnes/annum of flour processed to starch and gluten by traditional methods[15] (Martin and Batter processes) produces an effluent of 3×10^6 m^3/annum (or 6.7×10^8 gal/annum) containing some 2.2×10^4 tonnes/annum of the original flour. These figures suggest a gross wastage of water and a substantial loss of potentially valuable product.

A typical analysis of the effluent from starch production is as follows:

Total solids	: Less than 1.0 wt. %
Suspended solids	: 21% (d.s.b.) of total solids
Organic matter	: 95% (d.s.b.) of total solids
Nitrogen content	: 3.2% wt. (d.s.b.)
Starch content	: 6% wt. (d.s.b.)
Grease content	: Nil
*p*H	: 3.4-4.7
Colour and odour	both present
B.O.D. 4000 mg/l.	

Although detailed information is not yet available the effluent is considered to be rich in dissolved protein, especially lysine and other important nutritional proteins in which wheat gluten is lacking.

Present Effluent Treatment Methods

Although a number of processes have been proposed for the recovery of protein solids from starch factory effluents by chemical or physical means,[16 17] none of these has yet been shown to be economic. Hence most starch factories around the world either dispose of their effluents directly to streams or municipal sewage systems, or, if required to do so, attempt some form of biological treatment beforehand to reduce the Biochemical Oxygen Demand (B.O.D.).

The biological treatment process favoured appears to be the activated sludge process. Anaerobic treatment is not considered feasible because of the large process volume requirement. Seyfried[18] has commented on methods by which the efficiency of the activated sludge process can be improved by the addition of lime and ferrous sulphate to the effluent, but despite these improvements, the cost of treating the effluent by the process still remains high, and most urban starch factories prefer to discharge their effluents directly into sewage systems and pay the necessary levy.

This is indeed the case in Australia. With the exception of the Fielders Ltd. plant at Tamworth, N.S.W., where the effluent is settled and used for spray irrigation of pastures and pig and cattle feeding,[14] all Australian starch factories dispose of their effluent by way of municipal facilities. A considerable burden is therefore placed on municipal sewage treatment plants. An economic solution to the problem of starch factory effluent disposal is therefore of considerable national importance.

Treatment of Starch Factory Effluents by Ultrafiltration

An examination of the properties of starch factory effluents shows that it is an excellent candidate for treatment by ultrafiltration both to recover the soluble protein and to provide a clean aqueous phase suitable for plant recycle. The stream to be treated contains a low concentration of potentially-valuable dissolved solids and a ready outlet for the filtrate is available.

Studies are under way in the School of Chemical Engineering at the University of New South Wales to improve the feasibility of ultrafiltration for starch factory effluent treatment. The initial experiments were carried out using an Amicon CH3 laboratory ultrafilter fitted with hollow fibre membranes having a molecular weight cut-off of 10,000 and 50,000 respectively (membrane numbers H1CP10 and H1CP50). The filtrate from the larger pore size membrane was found to be of poor quality and the experimental program was thus directed to an examination of the performance of the smaller pore size membrane.

Fig. 3 shows the effect of transmembrane pressure and temperature on membrane flux for untreated starch factory effluent and effluent treated by centrifugation at 5000 g for four minutes to remove suspended solids. For these runs the effluent composition was held constant by reconstitution of filtrate and retentate. The observed variation in flux both with transmembrane pressure and temperature is much as expected. Increasing transmembrane pressure has a decreasingly beneficial effect on membrane flux due to concentration polarisation. Increasing temperature markedly increases the flux since it reduces the viscosity of the solvent (water). The poorer membrane flux with dissolved solids present suggests pore obstruction.

Fig. 4 shows the effect of effluent concentration on membrane flux at constant temperature and transmembrane pressure. For

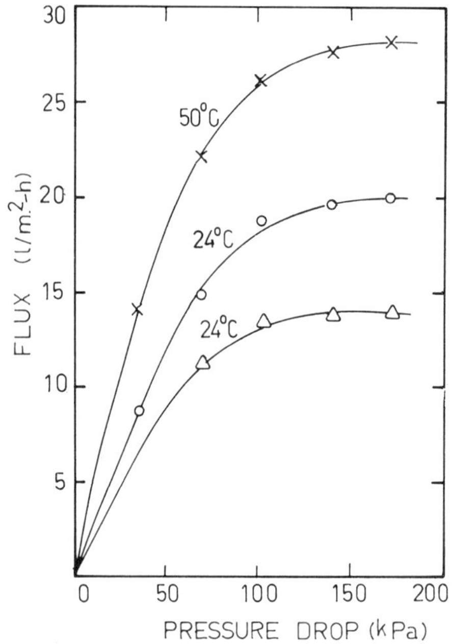

Fig. 3: Flux Versus Transmembrane Pressure and Temperature for Amicon HICP10 Hollow Fibre Membrane.
A — Suspended solids free .... 50°C
B — Suspended solids free .... 24°C
C — Suspended solids present 24°C

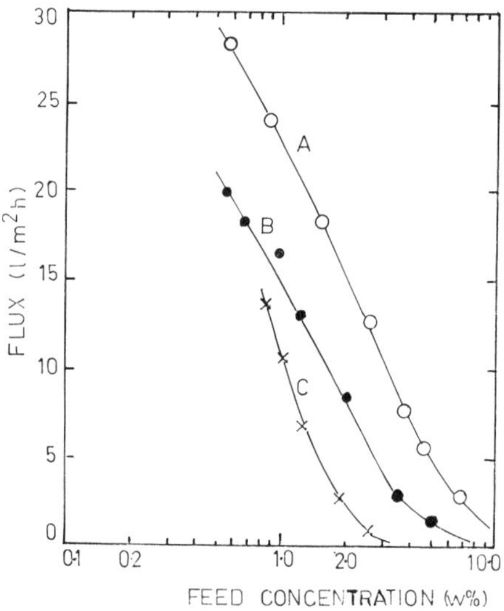

Fig. 4: Flux Versus Feed Concentration for Amicon HICP10 Hollow Fibre Membrane.
A — Suspended solids free .... 50°C
B — Suspended solids free .... 24°C
C — Suspended solids present 24°C
Transmembrane pressure for all runs 172 kPa.

these runs the retentate was recycled but the filtrate discarded. The decline in flux with increasing solids concentration is usual in ultrafiltration[19] and suggests the upper limit to which the effluent can be practically concentrated using a hollow fibre ultrafilter.

Fig. 5 shows the decline in membrane flux with time for constant effluent composition, temperature and transmembrane pressure. It would appear that the flux reaches a steady value when suspended solids-free effluent is being processed, but continues to decline if these solids are present. Hence, when a hollow fibre ultrafilter is used, pretreatment of the effluent to remove suspended solids would appear mandatory.

The fluxes obtained in the above experimental program proved to be uneconomically low for commercialisation of an ultrafiltration treatment process based on hollow fibre modules. The result is not surprising as the Reynolds number within the fibres is very low (estimated as 20), and is considered to be too low to prevent substantial concentration polarisation. For this reason it was decided to carry out experiments on a Dorr Iopor module, as the geometry of this unit lends itself to the treatment of more viscous solutions.

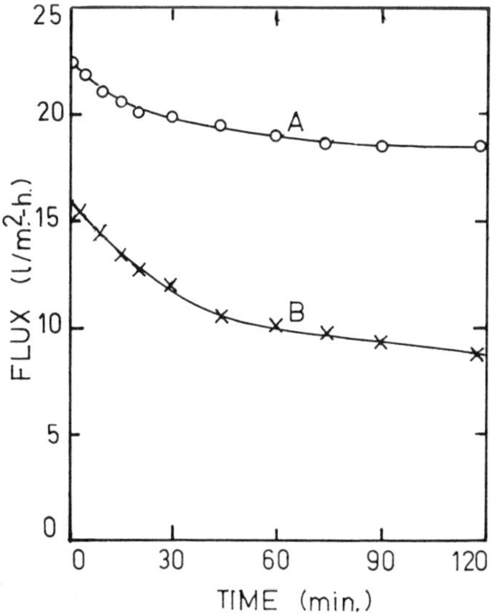

Fig. 5: Flux Versus Time at Constant Concentration, Constant Transmembrane Pressure (172 kPa) and Constant Temperature (24°C) for Amicon HlCP10 Hollow Fibre Membrane.
A — Suspended solids free
B — Suspended solids present

Figs. 6 and 7 show results obtained on the Dorr ultrafilter using both XP12 and XP24 membranes (molecular weight cut-offs 12,000 and 24,000 respectively). The results obtained, though preliminary, show that much higher membrane fluxes can be obtained than with the hollow fibre unit. These fluxes are sufficiently encouraging to warrant further detailed study of the process and this is now in progress.

Economics of the Process

Based on the flux achieved in the Dorr ultrafilter, and concentration of the retentate to 10 wt. %, it is possible to predict order of magnitude costs for the treatment of starch factory effluent by ultrafiltration.[20] The total capital investment required for a starch factory treating 400 tonnes/week of wheat flour is estimated as \$300,000. This includes the cost of effluent pretreatment and that of a spray drier to produce a dried protein concentrate. The annual operating costs amount to \$150,000. Against this can be offset the levy for effluent disposal to sewer (estimated as

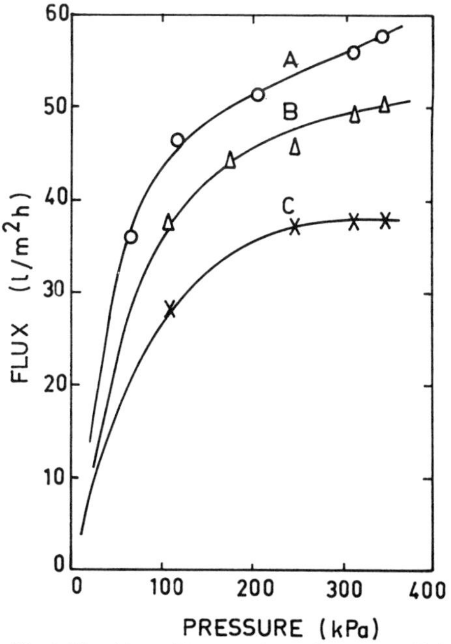

Fig. 6: Flux Versus Transmembrane Pressure at 50°C for Dorr XP12 Membrane.
A — Velocity past membrane 2.03 m/s
B — Velocity past membrane 1.47 m/s
C — Velocity past membrane 0.81 m/s

$32,000), the saving in purchase of process water ($30,000), and the revenue from sale of protein concentrate (estimated as $100,000).

Whilst, with the above costs the treatment is somewhat sub-economic, it is anticipated that by further development work the membrane area required can be substantially decreased with a large saving in capital cost. Also, if the retentate can be concentrated to greater than 10 wt. % savings in operating costs for the spray drier will occur. Clearly there is a strong case for further investigation of the ultrafiltration technique in the present instance.

Conclusion

Membrane processes are but in their infancy. With improvement in membrane characteristics and ultrafilter geometry it is to be expected that considerable improvements in capital and operating costs will be made. Those concerned with liquid effluent treatment must stay abreast of these new developments and capitalise on them when possible. Clearly in the case of starch factory effluent treatment the time when this will be possible is not far removed.

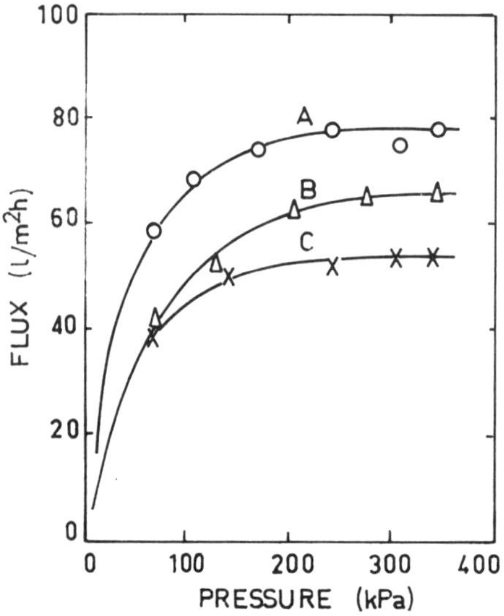

Fig. 7: Flux Versus Transmembrane Pressure for Dorr XP24 Membrane.
A — Velocity past membrane 2.43 m/s at 50°C
B — Velocity past membrane 1.63 m/s at 50°C
C — Velocity past membrane 2.43 m/s at 33°C

Acknowledgement

Grateful acknowledgement is made to the Dorr pilot scale ultrafiltration facilities provided by the CSIRO Division of Food Research and used at their Highett Laboratory, Victoria.

Acknowledgement is also made of research funds provided by the Water Research Foundation of Australia and by the N.S.W. Department of the Environment.

REFERENCES

1. Michaels, A. S., *Prog. Sep. Purif., 1,* 297 (1968).
2. McDonald, D. P., *Process Engineering,* Jan. 1973, p. 76.
3. Loeb, S., Souriajan, S. and Uuster, S. T., *Chem. Eng. News,* Nov. 4, 1960, p. 38, 64.
4. Van Oss, C. J., *Prog. Sep. Purif., 1,* 97 (1968).
5. Okey, R. W., in "Industrial Processing With Membranes", Wiley (1972), p. 249.
6. Smith, B. R., *RACT Proceedings,* September 1973, p. 257.
7. Hodgson, T. D., *Desalination, 8* (1), 909 (1970).
8. Glueckauf, E. and Sammon, D. C., *Proc. Third Int. Symp. Fresh Water from the Sea,* 1970, Vol. 2, p. 397.
9. Keilin, B., *U.S. Office of Saline Water Research Devpt. Progress Rept.* No. 117 (1964).
10. Clifton, M. J. and Fowler, R. T., *J. Instn. Engrs. Aust.,* Sept., 1972, p. 18.
11. Mahoney, J. G., Rowley, M. E. and West, L. E., in "Membrane Science and Technology" (Ed. Flinn, J. E.), Plenum, N.Y. (1970).
12. Porter, W. L., Siliciano, J., Krulick, S., and Heisler, E. C., *ibid.,* p. 220.
13. Haquoil, A. E. and Muller, L. L., *Aust. J. Dairy Technology, 27* (4), 123 (1972).
14. Knight, J. W., "The Starch Industry", Pergamon (1967).
15. Adkins, G. D., A.I.F.S.T. Cereal Chem. Group, "Forum on Gluten, Its Production, Quality and Uses", July 26 (1973).
16. Dietrich, K. R., *Starke,* 13, 11 (1961).
17. Japanese Patent, 40 184/71 (1971).
18. Segfried, C. F., Veroffentlichungen des Instituts fur Siedlungswasserwirtscharft der Technischen, *10,* 21 (1962).
19. Badenhop, C. T., Spann, A. T. and Maltzer, T. H., in "Membrane Science and Technology" (Ed. Flinn, J. E.), Plenum, N.Y. (1970).
20. Griffith, D. C., B.E. Thesis, University of N.S.W. (1973).

PAPER 19

Liquid Chemical Waste Destruction

By E. Skinner*

Industrial liquid trade wastes have become a problem. Municipal authorities are not eager to accept these pollutants into the sewage system; land disposal is becoming difficult. Incineration is now suggested as a means of disposal.

Two designs of incinerator are described, using the same principle of tangential feed admixed with the necessary combustion air. The horizontal design unit is for large quantities of waste, whereas the vertical unit can cater for factory or ship effluent.

In some cases the recovery of valuable ingredients of the waste (oil, solvents, etc.) before destruction should be considered. Inorganic waste, especially toxic or containing heavy metals, should be chemically treated before incineration. Liquid wastes containing chlorides have been successfully treated in these incinerators.

Liquid chemical residues for which the producer has no further use can be termed waste products of that process: they must be removed from the site and disposed of by some means. This does not imply that there is no ultimate use for this waste, after treatment by a recovery process. However, to put these wastes into the sewer is illegal and heavy penalties are being enforced; to dump these wastes is becoming difficult, as sites are scarce and charges high. What is to be done? Incineration is suggested in this paper; the incinerator is placed at the hub of the waste disposal plant and recommended ancillary equipment built up around it. The choice of the ancillary equipment depends on the purpose of the incinerator, viz. to be used for recovery of solvents, etc., prior to destruction or only for destruction. The type and size of incinerator to be installed will be determined by the quality and quantity of waste to be dealt with.

BASIC PRINCIPLES OF INCINERATOR

The principle of the incinerator is as follows: waste enters the combustion chamber tangentially together with air via an inlet manifold. The waste is now atomised by turbulence before being ignited by a fuel-oil fired burner. Choking by solid impurities is prevented due to the fact that there is no reduction of flow area between the feed pump and the furnace. The furnace is equipped with a pilot flame nozzle and a secondary nozzle to use additional fuel, if the calorific value of the waste is not sufficient to encourage self-burning. The waste gases are held for a short time in the combustion chamber where the majority of the combustible materials are incinerated; the rotating gases flow into an after-burner chamber where any remaining combustible matter is completely burnt off. Before being exhausted to atmosphere, the gases can be scrubbed and cooled.

The unit shown in Fig. 1 has been developed from the well-tried Celleco sulphur burner. This design of unit is used for capacities from 0.04 m^3 to 4.0 m^3 per hour liquid waste having a calorific value of about 5000 kcal/kg. Wide variations within the calorific content of the fuel constitute no objection whatsoever to incineration in the Celleco incinerator. Should the combustion heat value be low, secondary fuel is necessary to maintain complete combustion.

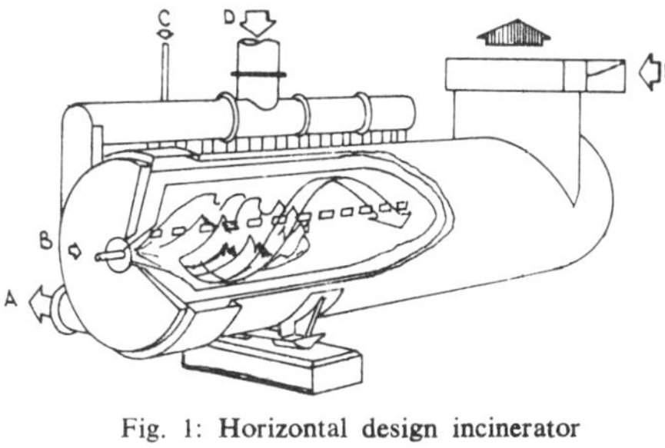

Fig. 1: Horizontal design incinerator

A. Cooling air
B. Support fuel
C. Chemical waste
D. Combustion air
E. Cooling air

The incinerator shown in Fig. 2 comprises two modules—the burner body and the control panel with blower. These can be installed either separately or adjacent as one unit. The body consists of two vertically arranged cylindrical

* Alfa-Laval Pty. Ltd., Sydney, N.S.W.

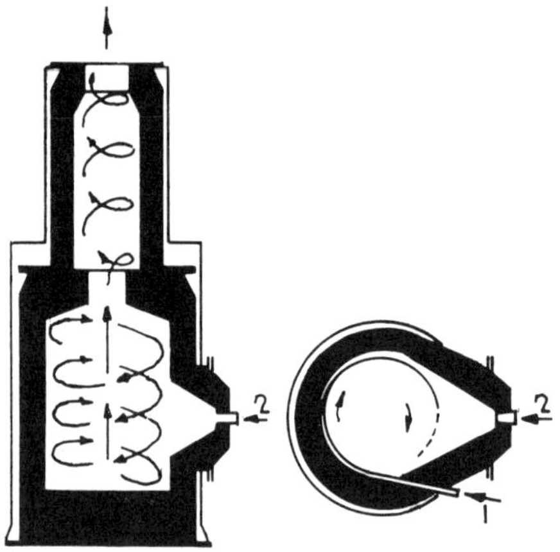

Fig. 2: Vertical design incinerator

Cutaway view of combustion chamber showing flame path.

1. Waste and combustion air.
2. Fuel oil.

sections. The lower combustion section supports the higher after-burner chamber.

This smaller design of unit is capable of a capacity of 60 to 100 kg/hr. waste with a calorific value from 0 to 7000 kcal/kg, and is used in factories and ships.

The burner body is air cooled, thus saving expensive lagging. The firebrick insulation is of standard specification to meet the thermal conditions. Combustion temperatures range from 600° to 1100°C; exhaust gas temperature varies with the pretreatment but can always be controlled to acceptable limits.

The choice of ancillary equipment to be used with the burner will depend on: (a) Nature of liquid waste. (b) Capacity of incinerator. (c) Object of incineration.

Should the liquid waste contain valuable and easily recoverable liquids, i.e., oils, solvents, etc., recovery of these liquids should be considered for use either as auxiliary fuel in the incinerator or as a by-product of the process. There are plants operating in Europe where the original purpose of the incinerator was destruction, but now there is a small chemical plant complex producing valuable solvents for market.

Central depots, based near large ports, can collect waste oil from factories, garages, workshops, and ships. This oil can be recovered to produce a light oil fraction and a fuel oil fraction. A disposal system (Fig. 3) which has been developed for this purpose will now be described.

CELLECO I.D. DISPOSAL SYSTEM — WASTE OIL

A. Receiving Department

The waste oil is delivered by trucks into a receiving tank (1) through a strainer (2) for catching of solid impurities. From the receiving tank the oil will be pumped to storage tanks (3), which are provided with steam coils for heating the oil. This makes it easy to handle. From receiving tank (1) and storage tank (3) the water will be let out from the bottom into a collecting tank for water (4). The oil separated on top of the collecting tank (4) will be pumped back into receiving tank (1). Condensate from steam coils and other equipments will be collected in a tank (21).

B. Distilling Department

From storage tank (3) the oil will be pumped through a heat exchanger (5) into an evaporator (6) provided with steam coils. The lightest gasoline vapour will be let off and cooled down in a water cooled condenser (7). Formed distillate goes into a collecting tank (8) and is then pumped further into the storage tank (delivery tank) for topped gasoline (9).

C. Separating Department

The oil not evaporated in the evaporator (6) will be transported to a self-cleaning filter (10) equipped with a circulating pump. The oil sludge reject will go to the collecting tank (11) and the oil accepted will be pumped to the separators (12). From the separators clean fuel will be pumped to storage tank (delivery tank) (13). The reject products, water and sludge, will be conveyed to respectively water collecting tank (4) and sludge collecting tanks (11).

D. Mixing Department

The sludge from collecting tanks (11) will be pumped to storage tanks (14) before destruction. These tanks work as clarifiers from which oil will be pumped from the top to a mixer and water from the bottom to collecting tank (4). Solids can be emptied at the bottom. Water from the collecting tank (4) will be pumped to mixer (15) in which the mixture, oil sludge-water, is regulated for control of the combustion temperature in the burner. In case of failure of oil, oil can be pumped directly from the storage tanks (3) into tanks (14).

E. Destruction Department

The oil sludge-water mixture will be dumped from the mixer (15) into the incinerator (16). Combustion air from fan (17) will be fed through the cooling mantles of the incinerator into the furnace. The combustion is almost complete depending on the combustion principle of the incinerator. The formed gases will then go through a waste heat boiler (18) into a scrubbing tower (19) in which the gases will be cooled down. Any remaining ashes which follow the water out in the bottom of the tower will be caught in a strainer (24). The water can be let out in public drains. The clean gases will be let out into the atmosphere through a chimney. The cooling water in the scrubber (19) can either be fresh water or water from the tank (4), depending on the water balance in the system. Fuel for the preheating of the incinerator will be taken from tank (13) through tank (23) into the incinerator. Greases and colorant wastes to be destroyed directly will be taken into a heating chamber (22). The sludge from it will be pumped into the tanks (14).

F. Power Department

Generated steam from the boiler (18) can drive a turbine (20) which is connected to the fan (17). Alternatively, a dynamo can be driven producing all power required for the whole plant, pumps, fans, light, etc. The secondary steam from the turbine (low pressure) is used for heating purposes for the heat exchanger (5), evaporator (6), storage tanks (3), and heating chamber (22).

G. Instrumentation

The instruments required for the operation of the plant are flow indicators for oil and water, pressure indicators for steam and oil, temperature indicators for oil and combustion and flow meter for combustion air as well as a CO_2-meter for the incinerator. Furthermore, the usual instruments for the boiler are required.

SPECIAL TYPES OF LIQUID WASTES

Should the liquid waste contain heavy metals, it is advisable to treat chemically and remove these metals at the source of waste before incineration. If not, the metals will be deposited with the slag and will be dumped on disposal land where the concentration will build up and could contaminate some water supply. Similarly, liquid wastes containing hazardous or toxic materials should be made less harmful or recovered before incineration. All organic liquid wastes can be treated with confidence. This includes compounds containing chlorides.

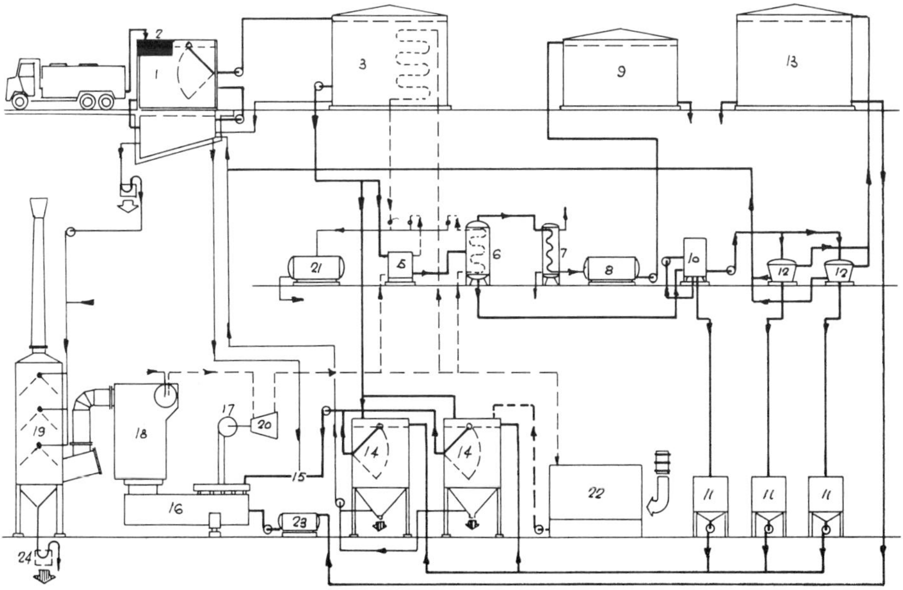

Fig. 3: Celleco I.D. disposal system for waste oil

The resultant hydrogen chloride gas can be scrubbed with lime water and the calcium chloride recovered as a sludge.

INCINERATOR WITH SCRUBBING TOWER AND STACK

A. Fig. 4 shows the slightly inclined furnace chamber in which the high temperature oxidation of the wastes takes place.

B. Below the furnace chamber are located the various feed pumps for injecting the combustible wastes.

C. The scrubbing tower is provided with large volumes of cooling water to remove the water-soluble impurities produced in the combustion process. The tower is constructed in two sections, the lower part being provided with four water sprays to facilitate the primary cleaning of the combustion gases, the upper part comprising a packed section of Intolox saddles.

D. The clean gases then pass to atmosphere through the stack.

The liquid discharge from the scrubbing tower and the stack then flows to the acid neutralisation plant.

The following is the material specification of the liquid wastes which can be handled by this incinerator.

Chemical classifications

Hydrocarbons—aliphatic and aromatic

Alcohols	Esters
Ethers	Ketones
Organic-acids	Amines

Aqueous solutions or suspensions of the above can be accepted.

Calorific value

No limitation.

Viscosity

Heated reception and blending facilities enable the most viscous wastes to be accepted (up to 3200 sec Redwood approx.).

Flash point

Intrinsically safe handling facilities permit the safe reception of the most volatile combustible materials (e.g. diethyl ether).

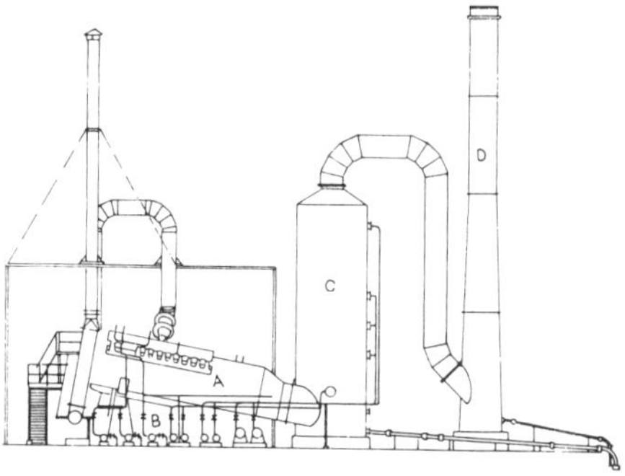

Fig. 4: Incinerator with scrubbing tower and stack

Chlorine content

Blending facilities permit acceptance of all chlorinated wastes, with no restrictions on the chlorine content of individual wastes.

Ash content

Limitation on ash content exists; the ash can be removed periodically on shutdown.

CAPACITY OF THE INCINERATOR

The required capacity of the incinerator depends on the quantity of liquid waste to be treated. Factories or ships wishing to solve their own waste problems will utilise small incinerators. Large municipal authorities, on the other hand, will no doubt form a central collecting point and blend their wastes to give a relatively constant quantity and quality of liquid to the incinerator.

Should the quantity of incoming waste to the central point be varied and contain chlorides, it is advisable to use two, or more, incinerators —one for the chloride-containing liquids, with the special scrubbing tower, and one for non-chloride, with water scrubbing.

ACKNOWLEDGEMENT

The author wishes to express grateful acknowledgement to the Celleco Company, Tumba, Sweden, for permission to present this paper.

PAPER 20

Some Australian Developments in the Use of Oxygen for Wastewater Treatment

By R. G. Shaw*

Recent development activities of most of the major industrial gas companies of the world is being directed towards finding a use for oxygen to satisfy the BOD of sewage or trade effluent. Research and Development in this field by CIG was started about twelve months ago.

The main advantages of using oxygen have been discussed extensively in literature and can be listed as follows:—

(a) Smaller area requirements due to reduced size of aeration tank.

(b) Low sludge yield.

(c) Lower power consumption/lb. of oxygen dissolved.

Sewage treatment has long been considered as a civil engineering activity, but industrial gas manufacture is basically a chemical engineering activity and, with this in mind, I wish to describe to you our process which is now well past the R and D phase.

Our first problem was how to dissolve sufficient oxygen (300 ppm) in water containing a significant level of suspended solid, dissolved solids and dissolved gases. Normal gas contacting equipment using packed columns would be clogged by the solids. The pressures required to put 300 ppm into solution also (6-7 atmospheres in pure water) seemed likely to cause pumping problems.

1. Gas Dissolver

CIG has had some success in developing gas/liquid contacting systems for the wine industry and initially these devices were tried out on water to see what levels of dissolved oxygen could be obtained. The results were disappointing but led to the development of the device shown in Figure 1. This device is one of the main features of the water treatment process described in the paper.

Waste water or mixed liquor is pumped up to the top of the vertical column and falls through an oxygen pocket.[1] At the interface[2] bubbles are formed as when one directs a garden hose into a bucket of water. The velocity of water down the column is adjusted so that the bubbles formed are entrained downwards and travel slowly to the bottom of the column.

In the phase separator,[3] the small bubbles amalgamate and large bubbles rise back up the column. The result is a column full of bubbles in turbulent motion for very little energy loss and hence an effective means of contacting oxygen with activated sludge or raw sewage.

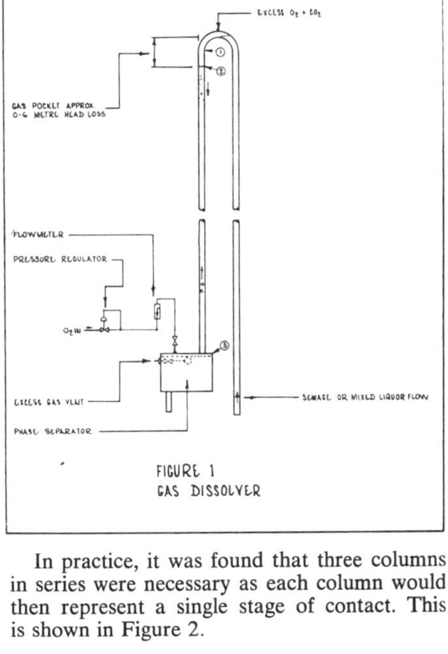

FIGURE 1
GAS DISSOLVER

In practice, it was found that three columns in series were necessary as each column would then represent a single stage of contact. This is shown in Figure 2.

2. Pressures

Some difficulty was experienced in establishing optimum pressures.

High pressure solution of oxygen at 6-7 atmospheres at first appeared attractive, but was abandoned for several reasons.

(a) High pressure (around 6-7 atmospheres) did not help the stripping of other gases

* Commonwealth Industrial Gases.

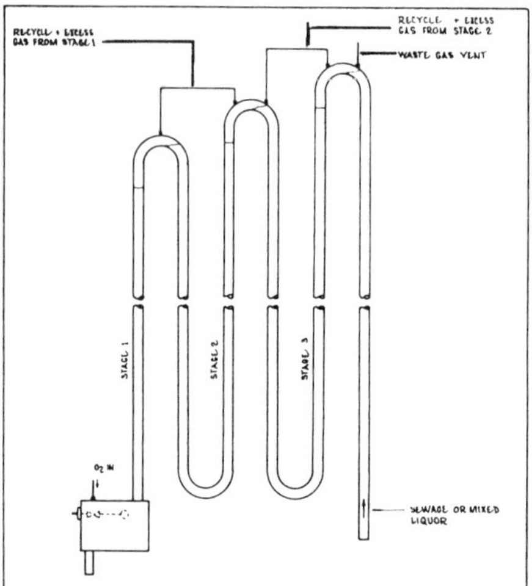

FIGURE 2
3 STAGE GAS DISSOLVER

from the liquid phase (mainly CO_2) formed by activated sludge.

(b) Pumping slurries to this pressure was not likely to be easy or mechanically reliable.

(c) Bacteria could be damaged by these pressures.

(d) Pressure vessels would be necessary to contain dissolved oxygen at this level.

(e) Energy requirements would be high.

The advantages of increased pressure to aid solution of oxygen were, however, attractive and compromise pressure of 15 to 30 ft. head of water was reached. At this level, dissolved oxygen levels of 80 to 100 ppm could be achieved if adequate CO_2 stripping could be provided. The CO_2 could be stripped with air or oxygen in a similar column to Figure 1.

The advantage of being able to feed high oxygenated sewage or mixed liquor into the activated sludge reaction vessel became apparent on literature survey.

3. Reactor/Clarifier Vessel

The combination of clarifier and reactor is not a new idea and was first proposed in 1949 by Okin of the University of North Carolina (ref. 1 and 2). This work was followed up by Budd and Lambeth in 1956 (ref. 3).

In this latter case, the need for an efficient oxygenator seems to have hampered the development work. Recoveries of oxygen were only 50%. The use of a combined reactor/clarifier assumes that sufficient oxygen can be introduced dissolved in water rather than as a dispersion of bubbles.

The CIG version of the reactor clarifier consists of a vertical open tank 5 metres high on small plants but up to 8 or 10 metres high on larger plants as shown in Figure 3.

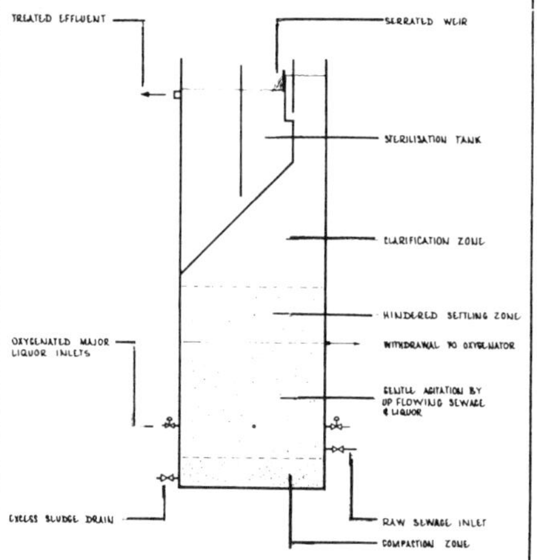

FIGURE 3.
REACTOR/CLARIFIER

The zones in the tank starting from the bottom will now be described.

The bottom zone is used for sludge compaction and results in an excess sludge withdrawal of 20,000 ppm suspended solids. Above this zone is the main reaction zone. Here activated sludge of above 5,000 ppm suspended solids is gently agitated by the injection of incoming sewage and oxygenated mixed liquor from the oxygenator. The oxygenator flow is 4 to 8 times that of the incoming sewage flow. The height of this zone is 1 to 2 metres. It extends to about half a metre above the side withdrawal to the oxygenator.

Above this zone is a hindered settling zone of activated sludge. Oxygen demand here is reduced and supply is provided by dissolved oxygen in the upward-flowing partially-treated sewage.

The uppermost zone in the reactor clarifier is a clarification zone. The treated effluent flows upward to a serrated weir where it overflows into the sterilisation tank. This is located above the clarifier in order to maintain head of water over the reaction space. Sterilisation can be by chlorine or ozone.

4. The Total System

Figure 4 represents the complete treatment system in diagrammatic form. Some additional features include the air stripper, which has a twofold function, to remove the high dissolved CO_2 concentration and as a safety device to prevent any methane coming into contact with oxygen. During normal running, dissolved oxygen levels are maintained at between 5 to 15 ppm in the reactor, but should oxygen supply be lost, anaerobic conditions would soon start generation of methane, some of which would remain in solution.

On re-starting the oxygen supply, methane could be stripped out into the oxygen stream, forming an explosive mixture. This is prevented by having the air-stripping stage before the oxygenation stages. The explosion limits of methane in air is narrower than methane in oxygen.

Capacity control is achieved hydraulically by recycling to the wet well, which can also be used as reaction space for the bacteria, provided dissolved oxygen levels are kept high.

The oxygen supply is regulated with a rotameter as shown in Figure 1. It seems likely that no major adjustments will be necessary even though hourly sewage flows vary considerably. Variations in oxygen demand will be met by allowing the dissolved oxygen levels to vary which will cause a change in driving force in the dissolver or oxygenator.

The target for complete treatment time has been set at $1\frac{1}{2}$ hours and there appears to be no reason at the present time why this should not be achieved.

REFERENCES

1. Okun, D. A., "System of Bio-Precipitation of Organic Matter from Sewage". Sewage Works Jour., *21*, 5, 763 (Sept. 1949).
2. Okun, D. A., and Lynn, W. R., "Preliminary Investigations into the Effect of Oxygen Tension on Biological Sewage Treatment". Chap. 2-4, "Biological Treatment of Sewage and Industrial Wastes". Reinhold Publishing Corp., New York, N.Y. (1956).
3. High Purity Oxygen in Biological Sewage Treatment, W. E. Budd and G. F. Lambeth. Sewage and Industrial Wastes, *29*, No. 3, 1957, p.p. 237 to 257.

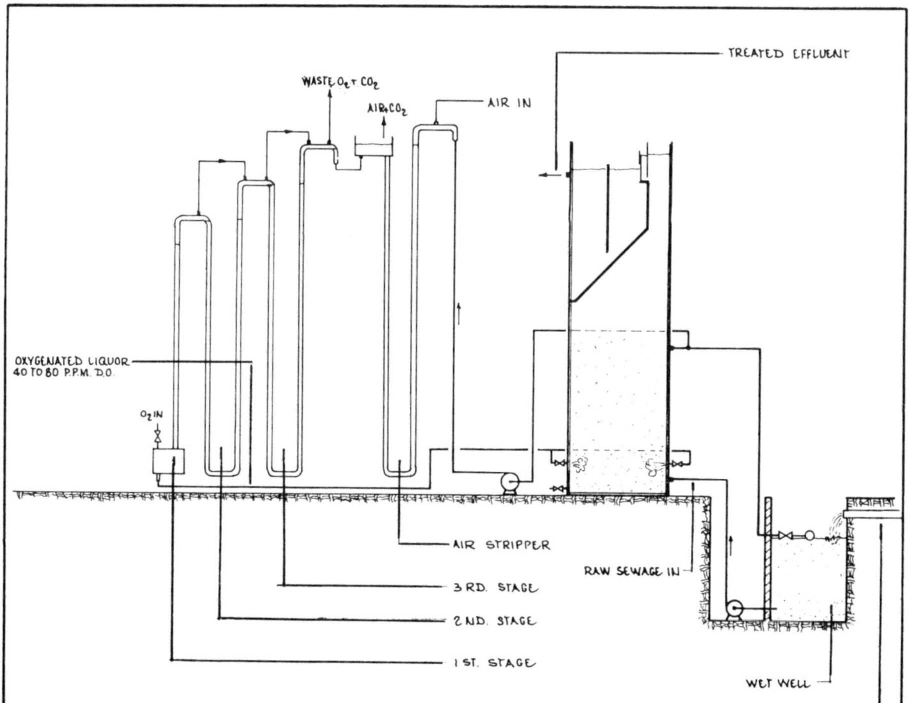

FIGURE 4

PAPER 21

Disposal of Liquid Wastes From a Large Coastal City

By B. Parkes*

1. Introduction

The very existence of life depends on the use of water. Primitive man used it only for drinking. As he became civilised the requirement increased, he needed it for cooking, washing and agriculture. In modern society water is used extensively for provision of heat and power, and it plays a big part in all industrial processes.

In a metropolis — for example, Sydney with its population of about 3,000,000 people — the water consumption averages 400,000,000 gallons daily. The law of conservation of matter states that "matter cannot be created or destroyed" and so this vast quantity of water must be disposed of when used.

Man plays a small part in the hydrological cycle, a natural process whereby water evaporates from the ocean and land, and rises into the atmosphere, where it cools adiabatically and condenses. Water is then precipitated to fall on the earth as rain, to water the ground and replenish the streams which run off into the ocean and so complete the cycle. Man collects some of the precipitated moisture in storage reservoirs or extracts it from the streams and after use discards it.

Most of the water discarded flows by way of drainage systems to the streams and waterways and ultimately to the ocean.

The discarded water is normally contaminated or polluted. In bygone times the natural biological processes in the streams purified the waste waters. Nowadays, the waste waters are so polluted that they cannot be naturally purified and so man has to treat them before they ultimately reach the receiving system.

2. The Hydraulic Systems of a City

A modern city provides complex systems for distribution, collection and removal of water.

The distribution system pipes the stored water under pressure to outlets controlled by valves.

The collection system consists of a network of open and closed channels which drain an area by gravity. Collection systems are twofold, the stormwater drainage system and the sewerage system. Some cities such as Sydney have separate systems while others, particularly the older cities, have a combined system which provides the drainage for both stormwater and sewage.

The stormwater drainage system is a series of open and closed channels which drain by gravity to discharge into natural streams and waterways. Many of the natural streams are reconstructed to form the arterial stormwater drains. A stormwater system is designed to follow the topography and the land contours.

The sewerage system though similar in principle to the stormwater systems consist entirely of closed channels or pipelines which are periodically ventilated to release foul odours. The minor sewers gravitate to pumping stations whence the sewage is pumped through rising mains to the main gravity sewers, which in turn transport it to the waste water treatment works.

Alternatively, waste waters may be discharged on site either for evaporation or to soak into the ground.

When wastes cannot be disposed of on site or discharged to a sewerage or drainage system, they may be transported by road tanker for disposal at community dumps or treatment plants.

A city the size of Sydney discards about 200 million gallons of waste waters daily to its sewerage system in which the wastes are purified at treatment works before discharge into the ocean or to inland streams.

3. Sources of Liquid Waste

Waste waters comprise natural waste waters, domestic wastes and industrial wastes.

Natural waste waters are regarded as the natural run-off after rainstorms or the drainage of agricultural waters. Mostly these waters are not contaminated, except by the dirt on the ground, and so are usually discharged to the stormwater drainage system.

Domestic wastes are the liquid wastes from

* Trade Waste Control Officer, Metropolitan Water, Sewerage and Drainage Board, Sydney.

the residential population. These wastes are the used waters from the toilet, the kitchen and the laundrics of dwellings and are collectively known as domestic sewage. Sewage is water contaminated by cooking wastes, fat, vegetable matter, soap, body wastes, faecal matter as well as paper, rags, matches, etc.

Industry and commerce use a considerable quantity of water, namely recirculated cooling water, boiler water for steam generation, wash water and process water. The former two are mostly used in closed-circuit systems and are contaminated only by inhibitory chemicals. Only a portion of these waters is bled to the drain.

By far the greatest source of trade waste is the water used for washing products, plant equipment, containers, premises and clothing. These wastes contain a variety of contaminants.

Process water can be broadly described as that used in manufacturing processes for purposes other than washing. It is used in a number of ways and except for a small proportion of product water which becomes part of the manufactured product, the wastes are usually discharged in a highly contaminated state.

Table 1 lists the types of industries and the main contaminants in their waste waters. A large proportion of the contaminants must be wholly or partially removed from the wastes before they are permitted to be discharged to municipal sewerage or drainage systems.

Table 1

SOURCES OF CONTAMINANTS IN WASTE WATERS

SOURCE OF WASTE	CONTAMINANTS
City buildings	Inhibited cooling waters, air conditioning, dust-absorbing oils
Hotels and restaurants	Kitchen grease and fat, food scraps
Factory canteens	Kitchen grease and fat, food scraps
Hospitals	Biological, pathological and kitchen wastes
Schools and universities	Laboratory wastes containing chemicals
Food processing — cereal	Starch and protein, husks, flour, cooking oils, spillage of product
Food processing — fruit and vegetable	Peel, fibrous material, sugars, juice, cooking waters, preservative chemicals
Food processing — meat	Blood, fats, small particles of meat, bone, skin, hair, paunch contents, offal
Breweries and beverages	Waste meal, sediment, yeast, sugars, hops, cork, paper, cleaners
Textiles — dyeing and spinning	Chemicals, dyes, lint
Textiles — tanneries	Fleshings, skin, hair, dirt, chemicals, lime and chrome
Textiles — wool scours	Wool, wool grease, protein, dirt, chemicals
Metal industry — machine shops	Cutting and lubricating oils
Metal industry — non-ferrous foundaries	Toxic heavy metals, acids, alkalis
Metal industry — metal finishing	Acids and alkalis, chemicals, toxic heavy metals, cyanides
Chemical and petrochemical	Oils, organic chemicals, toxic materials, acids, alkalis, heavy metals, solvents, resins, paint
Transport industry — workshops	Fuel oil, cutting and lubricating oils, solvents
Service stations	Grease and oil, dirt

4. Contaminants in Waste Water

Contaminants in waste water may be gaseous or liquid, but most contaminants are either soluble or insoluble solids. Absorbed gases may be odorous or inflammable; however, they are usually diluted in the sewer, and are not troublesome.

Liquid contaminants are either miscible or immiscible with water or are in an emulsified form. Miscible liquids rarely create problems. Immiscible liquids or oils are organic in nature and of lower specific gravity, so can be readily separated by flotation. Emulsified liquids — for example, cutting oils — are much more troublesome. Emulsions have to be cracked, usually by acidification, to separate the oil and water phases; the oil phase is separated first and the water phase then requires neutralisation before discharge.

Most of the contaminating materials in waste waters are solids. Many are water soluble and so their presence can only be determined by analysis. Among materials in this category are toxic substances such as salts of heavy metals. These, and also organic materials like starches and sugars, give the waste a high biological load or B.O.D.

Among the insoluble contaminants are a number of organic materials, which contribute to a high biological load, and inorganic materials which tend to silt up the sewers. Fats and greases in various forms are heavy contributors to this category of wastes.

A waste is normally characterised according to the effect its various contaminants have on the sewer or the receiving waters. The three main properties are described as B.O.D. (biochemical oxygen demand), suspended solids and grease. The B.O.D. is a measure of the organic loading of a waste that can be consumed by bacteria; in doing so the bacteria use oxygen and so de-oxygenate the wastes, creating septic or anaerobic conditions that cause odours. Suspended solids contribute to the colour or turbidity of the waste, as well as forming silt and sludges in sewers. Organic materials in suspension contribute to the B.O.D. Greases and fats are floating materials which, besides providing biological load, look unsightly in creeks and channels.

Other properties which describe contaminants present in waste are sulphides, which contribute towards unpleasant odours; toxic materials such as insecticides and pesticides; and heavy metals ions like copper, lead and zinc which are toxic to the biological life so necessary for the natural purification of streams and waterways and for the satisfactory operation of a sewerage treatment works.

5. Standards and Regulations

International recognition of the dangers of pollution has caused a general tightening of standards and the promulgation of numerous new regulations. Today, most countries have regulations pertaining to the discharge of wastes.

In New South Wales there are 74 different anti-pollution laws, many relating to the discharge of liquid wastes. They are administered by the many government departments or semi-governmental authorities concerned with pollution. The most important ones are the Pollution Control Commission Act (1970), the Metropolitan Waste Disposal Act (1970) and the Clean Water Act (1970). These clearly lay down the administrative responsibilities and the quality of liquid waste that may be discharged in various areas. These acts govern the effluent discharged from the various waste water treatment works and the discharges from the local authority's stormwater drainage systems.

As a result the Sydney Water Board has laid down a list of standards for liquid wastes discharged to the sewers (Table 2) and to the stormwater systems (Table 3). All wastes so discharged must comply with the Board's standards.

The regulations pertaining to the N.S.W. Clean Waters Act classify waterways and streams into different categories. Only wastes which comply with the relevant standards for the particular stream are permitted to be discharged into it.

6. Pretreatment of Wastes

A high percentage of industrial wastes are too strong to be discharged to sewerage or stormwater systems without pretreatment.

Table 2

STANDARDS FOR ACCEPTANCE OF LIQUID TRADE WASTE TO SEWERS

Temperature	Not to exceed 37°C if the waste contains grease or fats, otherwise not to exceed 50°C
*p*H	To be in the range 6.8 to 10
Grease	Not to exceed 200 mg/l
B.O.D. (5 days)	Nct to exceed 600 mg/l
Suspended solids	Not to exceed 600 mg/l
Sulphides	Not to exceed 50 mg/l for an interim period expiring on 30.6.76 and thereafter not to exceed 10 mg/l.

Toxic metal or compounds (the discharge to be subject to prior approval in each case) not to exceed:

	mg/l
Arsenic	100
Cadmium	30
Chromium	100
Cobalt	200
Copper	5
Cyanide	7
Lead	10
Nickel	100
Zinc	30
Organic herbicides	5
Organic insecticides	5

Notes

1. No discrete oil may be discharged.
2. No mercury may be discharged.
3. When two or more metals or salts of metals are discharged concurrently to the sewer in a waste, the sum of all the proportions which the concentration of each component metal or ion bears to the Board's limit for that metal or ion shall not exceed unity.
4. Cyanide baths are only accepted after detoxification. (The acceptable level of 7 mg/l in the table above refers to rinses only.)
5. Volatile solvents shall not be discharged unless miscible with water, and then only with special approval. The use of solvents in discharging the contents of grease traps to the sewer is prohibited.
6. Ferruginous pickling wastes will be accepted with *p*H not less than 5.5 in certain cases.

Table 3

STANDARDS FOR ACCEPTANCE OF LIQUID TRADE WASTE TO STORMWATER CHANNELS

Temperature	Not greater than 37°C
*p*H	Within the range 5.5 to 9
B.O.D. (5 days)	Not greater than 10 mg/l
Suspended solids	Not greater than 15 mg/l
Dissolved oxygen	Not less than 30% saturated
Colour	Not noticeable at 100 dilutions
Grease and oil	To be absent except for stable dispersions containing not more than 10 mg/l
Heavy metals	Total not to exceed 3 mg/l including not more than 0.5 mg/l of copper or zinc. The discharge of mercury to stormwater channels is forbidden

Attention is drawn to the provisions of the Clean Waters Act, which may require some variation of the above standards. Dischargers will need to make their own enquiries to ensure compliance with this Act.

If the wastes are too hot, they must be cooled by cooling towers or cooling pits. Oils, fats and floating materials are separated by flotation in suitable retaining pits. The pretreatment of emulsions has already been described (para. 3 of Section 4).

Solid material is generally allowed to settle, though sometimes flocculation by addition of chemicals is necessary to assist sedimentation.

A simple settlement arrestor pit can often be used for separation of the above contaminants. A pit generally has a volume equivalent to the maximum hourly flow, and so the wastes will be retained for at least one hour to allow adequate time for the separation of contaminating solids either by flotation or by sedimentation.

Acid and alkaline wastes require neutralisation. Solutions of the toxic heavy metals are generally present in acid solutions. Neutralisation to a *p*H of 8 to 8.5 will precipitate the metals as hydroxides for separation by settlement.

The above pretreatments are suitable for most types of wastes; however, some wastes may require specialised pretreatment.

The floating or settleable contaminants are removed as scums or sludges. The disposal of these solid or semi-solid contaminants is detailed in Section 8.

7. Methods of Disposal of Liquid Wastes

The methods available for the disposal of liquid wastes are: (i) Discharge to the stormwater system and thence to a stream or waterway. (ii) Discharge to a municipal sewer. (iii) Burial on site. (iv) Ponding for evaporation and drying. (v) Incineration. (vi) Removal from site by motor vehicle. (vii) Treatment and recirculation.

In the past, large volumes of liquid waste were directly discharged to stormwater drainage. Under current regulations this is prohibited and wastes are now diverted from stormwater drainage to municipal sewers. The sewers transport the wastes to treatment works where the contaminants are removed by biological treatment and/or physical separation. The treated water or effluent is then discharged to the ocean or to streams.

Sydney operates 15 treatment works, four of which discharge into the ocean; the remaining 11 are small inland works which discharge to the creeks of the Georges River and Hawkesbury River systems.

At the large ocean outfall works large objects, paper, rags, etc., are removed by screening and the remaining solids by either flotation or sedimentation. The sludges thus obtained are biologically digested under anaerobic conditions over long periods to render them innocuous before disposal. The liquids, after solids removal, are discharged to the sea where dilution and natural biological processes purify them.

The effluent discharged from the smaller inland works is required to be fairly pure, and this calls for secondary and sometimes tertiary treatment. Secondary treatment consists of biological degradation of the settled sewage. The microbial life feeds on the impurities in the settled sewage and so purifies it; the wastes are settled as a sludge and removed for digestion with the primary sludges.

A treatment works is virtually a factory in which the raw material is raw sewage and the main product is a purified effluent. Two important by-products are formed, namely gas and solids.

The sewage treatment process consists largely of the removal of the solid material from the liquid and the conversion of the dissolved impurities to sludges. The coarse materials (papers, plastics and rags) removed by the screens are either buried or incinerated, or, in some works, are finely chopped and returned to the flow for later treatment.

The less dense solid particles, together with oils and fats, form a scum which is scraped off from the top of the settling tanks, but most of the solids settle to the bottom and are drawn off as sludges.

The scum and sludge are transferred to the digester, a large tank-like structure in which they are stored for periods up to 30 days at a temperature of 33°-36°C. Under anaerobic conditions, bacteria convert the organic materials to organic acids which in turn are degraded to methane, carbon dixoide and water.

$$C_xH_yO_z \rightarrow CH_4 \uparrow + CO_2 \uparrow + H_2O$$

Any metallic ions present are converted to solid sulphides or carbonates, while the nitrogenous materials are broken down to ammonia.

The digester gas has a calorific value of 25,000kJ/m^3. It is collected and stored for use in dual-fuel engines to generate electric power and to provide heat for the digesters.

In Sydney, it is normal to remove 40 kg of screenings from each million litres of raw sewage treated. In addition, 10,000 litres of sludge at 5% solids are extracted from the raw sewage. In all, this amounts to 150,000 tonnes per annum of solid waste which is extracted from the liquid sewage and requires disposal.

Where the volumes are small and the ground is absorbent, it is possible to discharge the waste into the ground where the anaerobic bacteria in the soil will digest the solid materials. The soil will also filter the liquid, and the purified effluent will seep through to the water table. This principle of disposal is used in the septic tanks which are employed to treat domestic sewage in the unsewered areas of Sydney. In the Botany area, the ground is very absorbent and large quantities of industrial wastes have been discharged to the ground over a long period of time. However, the gradual contamination of the ground waters has enforced a restriction of this practice.

If much land is available, wastes are often treated by ponding or lagooning. Large areas of shallow ponds are constructed to hold the waste waters. The oxygen is absorbed into the water where aerobic bacteria degrade the wastes; the partially degraded solids settle to the bottom where they are further degraded by anaerobic bacteria. Eventually, when the ponds have filled with sludge, they are drained and dried. The sludges are ploughed into the ground and new ponds are dug.

Incineration of liquid wastes is not a common practice because it is expensive unless abundant power is available. Much heat is wasted while converting the water to steam so that the dried solids can be burnt.

The increased costs of reticulated water and sewerage and sewerage disposal (in Sydney these are 10.5c and 6.6c per kilolitre respectively) and the higher standards for the discharge of waste now laid down by municipal authorities have made the treatment and recirculation of waste waters economically attractive. However, as with sewerage treatment, one is again confronted with the problem of the disposal of the contaminated sludges and solids.

8. Disposal of Solids and Sludges

The solid phase of wastes consists of a complex mixture of grit, carbohydrates, protein, and fatty and other biological organic matter in a particulate or colloidal state; in addition there are the inorganic wastes from industry. Moisture content is 94-97%. Such sludges originate from domestic septic tanks and from industrial pretreatment plants. The solid phase of sewage sludge is said to be 70% organic and 30% mineral.

Considerable quantities of raw and digested sludges are discarded in the semi-liquid state by pumping to sea, or by transportation by special tankers for dumping far out to sea (the practice adopted at London, Manchester and some other large coastal cities).

More sludge is discharged directly to the ground than by any other method. In the United Kingdom, several works transport the sludge by tanker for discharge directly on to agricultural lands. This satisfactory method continues the cycle of fertility by returning nitrogen, phosphorus and potassium, together with organic materials, to the soil. At some works sludge is sprayed on to grass beds where biological passervation occurs.

Liquid sludges may be stored for long periods in lagoons whence the clear, settled liquid is eventually run off. When full, the lagoons are dried and filled with soil and new lagoons are dug.

Drying beds are areas of gravel overlaid by sand. The liquid sludges are spread on the sand, the liquid soaks into the ground or evaporates. The dried sludge is scraped off for disposal elsewhere and the beds are reused.

Modern practice is to dewater the sludge by mechanical means before final disposal. Dewatering may be assisted by addition of chemical coagulants such as lime or alum. Modern techniques make use of polyelectrolytes for this purpose. The dewatering process reduces the moisture content of the sludges from over 90% to about 50% and decreases the volume 7 to 8 fold.

Filtration by means of filter presses and by vacuum filtration are the most common methods of dewatering sludges. The use of centrifuges is not common because the difference in density between the sludge and the liquor is so small that only reduction to 65% moisture is possible. The method is still in the process of development.

For the disposal of sludge cake the method of landfill is the most common. Provided its toxic metal content is low, sludge may be used for agricultural purposes by ploughing into the ground. Mixing the filter cake with garbage is practised in many countries, but the relatively high cost of composting often puts this method beyond economic acceptable limits.

Incineration of sludge is quite common in Europe. At one sewerage works in Essen, Germany, the sludge has such a high calorific value (7500 kJ/kg) that up to 2000 tonnes daily are carried by conveyor half a mile to a power station where it is burnt as fuel and the combustion energy used for electricity generation. In Japan the sludge cake is compacted and used for building blocks. The sludge cake is often dried to 10% moisture content and then either incinerated or powdered and sold as a fertiliser or a soil conditioner.

All the above methods of sludge disposal may be applied to raw sludge or to digested sludge. The digestion of sludge has many advantages. The digestion process, by reason of the formation of gas and water, gives a substantial overall reduction in the amount of sludge to be disposed of. Drainability is improved because of the granular structure of the digested sludge, and it is easy to reduce the moisture content to 50%. The organic matter has a high degree of stability, so that it does not decompose rapidly; also, the odour is greatly reduced. The generation of sludge gas, which is used for producing heat and power, is an economic factor in favour of digestion.

The digestion of sludge produces 0.75 m^3 of gas for each kilogram of dry organic material digested. This has been equated to one cubic foot of gas being produced daily from the wastes of each head of population.

9. Conclusion

It is generally believed that liquid wastes are easily discharged down the drain to be later purified at a treatment plant, the effluent being discharged to a stream.

During purification of waste water, a considerable quantity of solid material is obtained from the impurities present. It is a much more complex matter to finally dispose of the separated solid matter, but it can be returned to the sea, land, or by incineration to the air.

10. Bibliography

1. State Planning Authority of N.S.W. — Sydney Region Outline Plan 1970-2000 A.D. (1968).
2. Sydney Water Board — Annual Report — 1971-72.
3. J. Nemec — Engineering Hydrology (1972) McGraw Hill.
4. Sydney Water Board — Annual Report — 1971-72 and technical records.
5. Imhoff, Muller and Thistlethwayte — Dispersion of Sewage and other Waterbourne Wastes (1971).
6. Isaacs, P. C. G. — Waste Treatment (1960) — Pergamon.
7. Buswell, A. A. — Important considerations for sludge digestion — *Sew. Wks. J., 19* (1947) 48.
8. Burgess and Wood — The properties and detection of sludge gas — *J. Proc. Inst. Sew. Purif.* (1964), Pt. 1, 3, 24.
9. Departmental records of operation of waste water treatment works. M.W.S. & D.B., Sydney.
10. Pierce and Parkes — The control and treatment of trade wastes in sewerage systems — M.W.S. & D.B. Report (1971).
11. Kershaw and Wood — Sludge treatment and disposal at Maple Lodge — *J. Proc. Inst. Sew. Purif.* (1966) Pt. 1, 75.
12. Anon — Use of poly-electrolytes for treatment of sewerage sludge — *Wat. Waste. Treat. J.*, 15 Aug. (1972) 9.
13. Simpson, G. D. — Operation of Vacuum Filters — *J. Wat. Poll. Control Fed. 36* (1964) 1460.
14. Sasano — Practical Experience of Pollution Control in Japanese Textile Mills. Soc. Dyers and Colourists Seminar — Sydney 1973.
15. Herwitz and Dundas — Wet oxidation of Sewerage Sludge — *J. Watt. Poll. Control Fed. 32* (1960) 543.
16. Porteus 1K — Sewerage Sludge Treatment at Huddesfield — *Wat. Waste Treat. J.*, 7, (1960), 543.
17. Quirk, T. D. — Economic Aspects of Incineration vs Incineration Drying — *J. Wat. Poll. Control Fed. 36* (1964) 1355.

PAPER 22

Experience With a Winery Waste Treatment Plant

By D. R. P. Jolly*

The disposal of winery wastes has been a problem of concern for many years in nearly all wine-making areas, particularly where large scale distillation is involved. Active investigation of ways to treat or otherwise dispose of the waste have been carried out from the 1930's onwards,[11] and from time to time technically feasible solutions have been presented, but acceptance of the more elegant procedures was ruled out on a cost basis. Since the disposal to rivers in most countries has been restricted by timely legislation, the industry now generally utilises lagoons to evaporate the wastes, a method which has only relative low cost to recommend it. However, the encroachment of urban areas towards distillery sites has in many cases found this practice wanting, and this, along with other problems, dictated the building of a treatment plant for an Australian winery/distillery at Griffith, New South Wales.

The decision to build a treatment plant resulted from the following factors:

(1) Six acres of now valuable industrial land under lagoons.

(2) Production potential limited by lagoon capacity and no further land available.

(3) High water table damaging winery foundations.

(4) Odour nuisance.

The choice of a treatment method followed from an appraisal of the literature and the results of two different pilot plant studies. Although the decision to build a high rate trickling filter using plastic media as the bios support was not the cheapest alternative, this method was chosen as being the best compromise between cost and practicability of operation. The characteristics desired for the plant were sufficient capacity to treat the whole volume of waste produced during the vintage (without using storage lagoons), compactness, resistance to shock loading, and ease and reliability of operation.

The plant was due for completion in February, 1970, prior to the vintage, but was not ready to receive effluent for treatment until the vintage was well under way. The slow start-up required of this type of plant necessitated holding most of the waste in lagoons for later treatment and the designed operation of the plant was not achieved until the 1971 vintage.

Design and Operating Procedures

Design: The main requirement for the plant was that it should produce a treated effluent of 5 day B.O.D. not exceeding 500 ppm suitable for discharge into the town sewerage system. An estimate of the organic load in the waste at peak production amounted to 2000 lb. of B.O.D./day, representing 50,000 gallons of waste of average B.O.D. equal to 4000 ppm (with distillery wastes accounting for some 70% of the load). In order to reduce this strength to 500 ppm, a three stage plant was required with 50% reduction achieved on each stage. The plastic chosen for tower packing was Imperial Chemical Industry's "Flocor" (a copolymer of vinyl chloride and vinyl acetate), and as the design load for this application is 10 lb./cu. yd./day, a packed volume of 200 cu. yd. was selected for the first stage, with 120 cu. yd. and 80 cu. yd. for the second and third stages respectively. The essential features of the design (Figure 1) are as follows:

(1) Wastes from the winery and still house are pumped separately to 125,000 gallon holding pits. The holding pits are roofed over to reduce odours and in the case of the still waste to slow cooling. Some anaerobic digestion takes place in the pits resulting in the lowering of the initial B.O.D. In the original design the general waste from the winery was screened by allowing the liquid to fall on to a stainless steel slotted screen, angled at 60° to the horizontal. Solids falling off the screen were removed as required. Owing to frequent blocking this has now been replaced by a vibroscreen.

(2) The separate wastes are transferred to the first stage sump of the plant via two weir boxes, the flows being adjusted to load the

---

* Company Research Officer, Penfolds Wines Pty. Ltd., South Australia.

† Adapted from a paper originally presented at the Annual Meeting of the American Society of Enologists, Cabana Hyatt House, Palo Alto, California, June 25th, 1971.

first stage at 2000 lb. of B.O.D./day. (The V notch angle of the still waste weir is narrow for measuring accuracy at low flows. This is important as the still waste averages approximately 18,000 ppm compared with the 2500 ppm of the general waste. The volume proportions to be passed are calculated from a simultaneous equation correlating flows of known B.O.D. and total load.)

(3) Each stage consists of 4 ft. x 2 ft. x 2 ft. units (made up of alternating plane and corrugated P.V.C. sheets) packed one above the other so that each layer is at right angles to the one below. The three stages are enclosed by the one tower to form a compact plant; separation of the stages resulting from the contents of each sump being distributed over only its design area of 'Flocor'.

In the first stage sump the entering waste is mixed with liquid draining off the first stage, overspill from the first settling tank, alkali added automatically in response to the pH probe, and nutrients as required. The combined feed, now of pH between 7.0 and 8.0, is lifted by a centrifugal pump to the top of the first stage at 9000 gallons per hour. (The contents of each sump are stirred by returning a small percentage of the upgoing stream. The object of this is to prevent the solids content of the sumps from settling out and going anareobic.)

The circulating rate is chosen to wet the 300 sq. ft. area of Flocor at 30 gallons/hour/sq. ft. of the packing.[2] The liquid is distributed on to the top of the Flocor via 32 nozzles designed to achieve distribution in the shape of a solid pyramid.

The liquid then percolates down through the 18 ft. packed depth where the developing microbial mass attached to the plastic absorbs and metabolises the organic load, synthesising new protoplasm and releasing carbon dioxide. The new microbial cells build up on the plastic

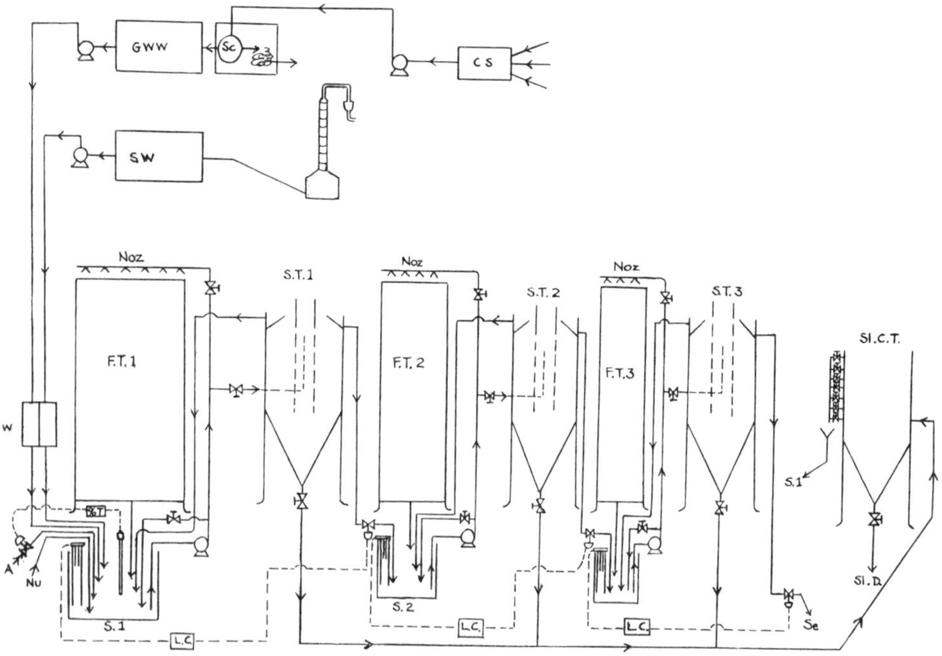

Fig. 1: Flowsheet of waste treatment plant

A — Alkali addition.
FT — Flocor tower.
GWW — General winery waste.
LC — Level control.
Noz — Nozzles distributing effluent.
Nu — Nutrients addition
S — Sump
Sc — Screen
Se — Sewer.
SlCT — Sludge consolidation tank.
SlD — Sludge disposal.
ST — Settling tank
SW — Still waste
%T — Percentage timer.
W — Weir box.

material until a certain thickness is reached when portions slough off due to the death of the inner cells attached to the plastic surface.[5] The pieces of the growth breaking away appear as a black sludge in the liqiud collecting on the raft under the tower, the raft having a 6 in. fall over a 12 ft. width to encourage return of solids to the sump.

Microbial metabolism causes a marked rise in temperature within the filter and this in turn creates a vigorous updraught of air needed for aerobic conditions. The liquids and solid collecting on the raft return to the sump for recirculation and passage to the next stage.

Fig. 2 — Plan of stage 1 of the treatment plant

A — Nozzle system for distributing effluent.
B — Roofed holding pits.
C — Pumps for transferring effluent to weir box.
D — Vibroscreen.
E — Alkali mixing and storage tank.
F — Circulating pump.
G — Side stream off circulating line for sump mixing.
H — Valve controlling Flocor irrigation rate.
I — Valve controlling side stream to settling tank.
J — Valve for desludging of settling tank.
K — "Billman" automatic valve.
L— Humus settling tank.

(4) A side stream from the first stage circulating line is passed through a settling tank and the clarified liquor becomes feed for the second stage. The design of each settling tank is conventional, incorporating a stilling compartment for incoming liquid to prevent turbulence, internal weir, and 60° angled hopper bottom (Figure 2). The movement of liquid from the settling tank to the next stage is controlled by a valve which maintains the level in the previous sump. The rate of flow of liquid from one stage to the next is thus adjusted to equal the incoming waste flow.

(5) The second and third stages are similar to the first but have lesser volumes of Flocor to maintain the same loading assuming 50% reduction on each stage. The final treated effluent leaves via the motorised valve following settling, and is piped direct to the town sewer.

(6) As up to 35% of the organic load is converted to sludge,[2] regular desludging of the settling tanks is essential. All three tanks are gravity drained (via a valve at the base of the cone) to a sludge consolidation tank. The contents of the latter are allowed to consolidate

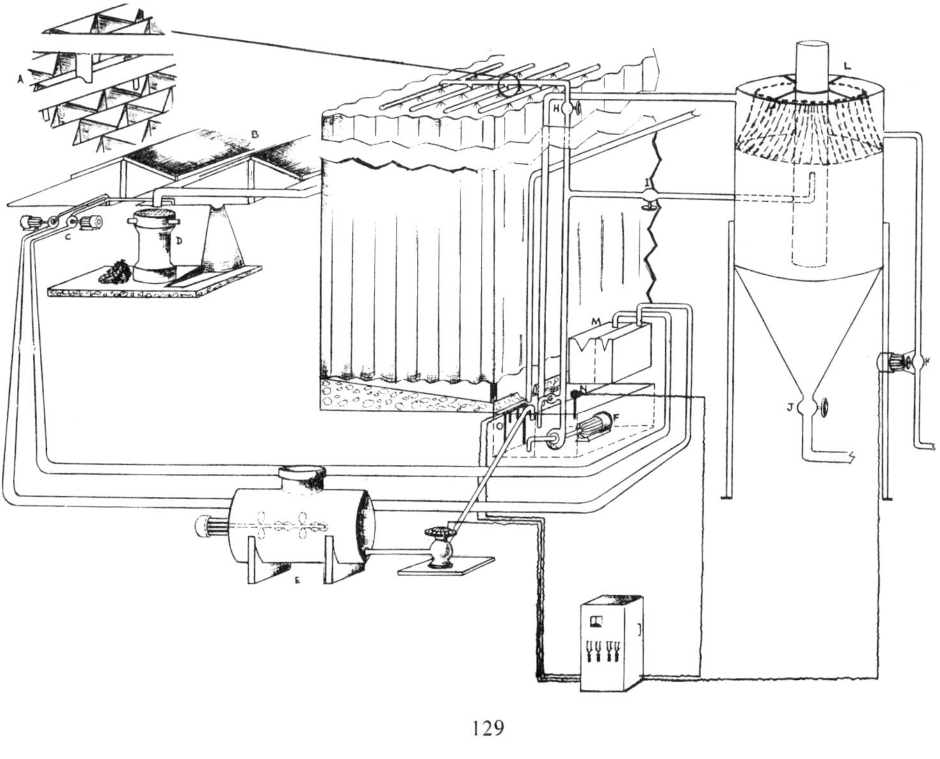

and then the thickened material is removed by a tanker for dumping. Supernatant from this tank is returned via a manifold valve system to the first stage sump as required.

Operating Procedures:

A typical sequence of operations is as follows:

(1) The organic load of each of the two holding pits (C.O.D.'s) is determined. Flows are then calculated to give a total load on the first stage of 2000 lb. B.O.D., the ratio of flows being dependent on the current situation in the holding pits.

(2) The required pH is set on the control panel. This causes an addition of alkali controlled by the percentage timer and pH probe to achieve the pH chosen.

(3) The valves (as in Figure 2) are set to achieve the required irrigation rate to the Flocor, and residence time in the settling tank.

(4) The treated effluent in the last stage is C.O.D. tested and if of acceptable quality is run to the sewer.

(5) Settling tanks are desludged once a day, and the sludge consolidation tank as required (approximately once per week).

RESULTS

Waste Control Parameters: At this stage only a simple test program has been possible and comparative results are only available in terms of C.O.D. reductions. B.O.D. testing as a check was carried out occasionally by sending samples to Melbourne. However, there are insufficient results to compare B.O.D.'s with C.O.D.'s over a period of time in order to establish a relationship between the two (as obtained by MacMillan et al[6]).

Nutrient addition before treatment depends on the composition of the waste in question. The variation of the composition of wastes can be seen from Table 1 which, with a varying waste stream, makes the calculated addition of nutrients impracticable. Generally it was considered that there was ample phosphate in the waste and hence only nitrogen in the form of urea need be added to maintain the B.O.D. :N:P ratio in the vicinity of 100:5:1 (the accepted optimum for bacterial synthesis). Aq. urea is added as a 10% solution to maintain a residual ammonia concentration in the treated effluent of not less than 15 ppm.

Start-up: The plant was not completed until April, 1970, by which time over a million gallons of waste had perforce been discharged into the lagoons. In addition to this, the pump supplying still waste to the appropriate holding pit went out of action early in the piece which negated proper separation of the wastes for weir box flow measurement. Hence both pits were used for a mixture of wastes, one being diluted by return of treated effluent from the tower.

In this way high and low strength wastes were made available. Even a "roughing" filter of this kind has to be started with low loads and under the circumstances prevailing, this proved difficult. However, by the end of May, with careful feeding the plastic media were well covered with biological growths and the feed rate was increased towards the design loading. As can be seen from Table 2 this appeared to cause overloading with an average reduction in C.O.D. of only 48%. In July/August, 1970, the ambient temperature fell so low that the general temperature of effluent in the plant dropped to 34°F with some pipe freezing

Table 1

VARIATION IN WASTE COMPOSITION

Sample Origin	Still Slop From Griffith Plant	Wine Lees From Griffith Plant	Wine Lees[1] Magill Plant
Colour	Light-Brown	Red-Blue	Yellow-Brown
pH	3.7	4.0	3.7
Suspended Matter, mg/1	684	1,958	25,424
Volatile Suspended Matter, 600°C, mg/1	652	452	23,348
Suspended Matter after 1 hour settling, mg/1	654	946	5,320
Settled Matter after 1 hour settling, mg/1	0.6	13	440
Total Solids, 103°C, mg/1	8,360	3,292	48,940
Volatile Total Solids, 600°C, mg/1	6,724	1,444	42,394
B.O.D.$_5$, raw waste, mg/1	7,300	6,670	67,400
B.O.D.$_5$, after 1 hour settling, mg/1	7,220	6,100	61,600
Ammonia-nitrogen as N, mg/1	5.6	5.6	42.0
Organic-nitrogen as N, mg/1	137	43.4	316
Lime required to neutralise to pH of 7.0 $Ca(OH)_2$, mg/1	1,380	222	4,740

1. This sample included for comparison.

taking place. The plant was shut down (to avoid ice formation on the Flocor) and started again in late August. By this time the effluent stored in the lagoons had undergone extensive methane fermentation and was of considerably lower B.O.D.

To avoid overloading the plant only a light loading of about 1½ lb. B.O.D./cu. yd. of Flocor/day was applied with the results seen in Table 3. When two samples were obtained in one day, it was found that the evening figures indicated a greater B.O.D. removal efficiency than in the morning. This was attributed to the higher effluent temperature in the plant during the day. The average removal efficiency of 72% obtained under these conditions was encouraging. For most of the first 6 months of operation, treated effluent had to be stored in a lagoon set aside for the purpose as the local council were cautious about accepting it into the sewer. Separate parcels of treated effluent found to be of low B.O.D. (in some cases well under 100 ppm) were accepted experimentally into the sewer and the performance of the town sewerage plant observed closely for any adverse effects. By mid-summer

Table 2

SOME PERFORMANCE DATA FOR THE PLANT DURING SECOND MONTH OF OPERATION

Date of[1] Sampling	Load lb./cu. yd./day	Crude Effluent C.O.D. p.p.m.	Treated Effluent C.O.D. p.p.m.	C.O.D. Removal[2] % Efficiency	Nature of[3] Samples
28/5/70	4.30	8,600	3,750	56.3	C.W./S.W.
2/6/70	4.23	8,400	4,000	52.3	C.W./S.W.
3/6/70	10.73	20,800	11,800	43.3	S.W.
4/6/70	8.28	16,400	12,000	26.8	Mostly S.W. Little C.W.
5/6/70	5.76	12,800	9,700	24.2	C.W. & S.W.
8/6/70	9.60	12,800	6,000	53.1	C.W. & S.W.
9/6/70	9.90	13,200	5,800	56.1	C.W. & S.W.
10/6/70	9.90	13,200	5,600	57.6	C.W. & S.W.
11/6/70	10.42	13,900	5,600	59.7	C.W. & S.W.
12/6/70	8.85	11,800	6,000	49.2	C.W. & S.W.
15/6/70	7.50	10,000	5,000	50.0	C.W. & S.W.
16/6/70	7.50	10,000	5,400	46.0	C.W. & S.W.
Averages	8.08	12,650	6,720	47.9	

Tower Effluent; Average minimum temperature: 8.3°C. Average maximum temperature: 12.9°C.

1. Samples are taken at the same time of day but are spot rather than composite.
2. Refers to overall C.O.D. reduction.
3. C.W. is cellar wash — general waste from winery. S.W. is spent wash — material ex still after distillation.

Table 3

FURTHER PERFORMANCE DATA OF PLANT ON LAGOON STORED WASTE

Date of[1] Sampling	Load lb./cu. yd./day	Crude Effluent C.O.D. p.p.m.	Treated Effluent C.O.D. p.p.m.	C.O.D. Removal Efficiency %	pH[2]
28/8/70	1.0	400 } COD/BOD = 10.8 BOD 37	180 } COD/BOD = 25.0 BOD 7.2	55.0 BOD 80.5	not determined
2/9/70	1.40	720	200	72.2	n.d.
3/9/70	2.33	1200	160	86.7	7.9
4/9/70	1.36	700	40	94.3	8.2
7/9/70	1.84	1840	1100	43.6	8.0
9/9/70	1.54	1540	320	79.2	8.7
10/9/70	1.09	1085	440	59.4	8.5
11/9/70	1.16	1155	360	68.8	8.5
13/9/70	1.06	1015	260	74.3	8.5
14/9/70	1.06	1015	100	90.1	8.5
15/9/70 p.m.	1.91	1905	120	93.7	7.6
15/9/70 a.m.	2.25	2355	300	87.2	7.5
16/9/70 p.m.	2.21	2205	480	78.2	7.4
16/9/70 a.m.	2.21	2205	880	60.0	7.5
21/9/70	0.19	380	160	50.0	7.0
Averages	1.47	1257	330	71.7	

1. Load consists of a mixture of effluent direct from the winery and lagoon stored effluent of B.O.D. decreased by methane fermentation.
2. pH is given for Stage 2.

most of the lagooned effluent had been evaporated which enabled the plant to be shut down for modifications and maintenance.

1971 Vintage Season: An unexpected change in the anticipated plant loading occurred prior to the vintage in that the company decided to have distillation material processed under contract. Hence the overall load on the plant was reduced by a signficant factor and this enabled lagoon storage of waste to be avoided. After the initial slow increase in load to establish ample bios the plant was adjusted to take the full volume of effluent as produced. Results for March are shown in Table 4. The two holding pits were used for flow balancing, as by returning a percentage of the treated effluent to one of the pits, the C.O.D. of the waste entering the tower could be adjusted to the optimum. The quality of the treated effluent (confirmed by B.O.D. tests performed elsewhere) enabled continuous discharge into the town sewer to be achieved for most of the vintage.

Table 5 was obtained by B.O.D. testing in the Company laboratory at Griffith once dilution water problems had been overcome. The results show the decrease in pollutional strength as the waste passes through the plant (although the overall efficiency of the plant was low compared with earlier performance, the plant having been restarted for the purpose).

Table 4

PERFORMANCE DATA FOR PLANT TREATING MIXTURES OF FRESH FEED AND ONCE TREATED WASTE, MARCH, 1971

Date of Sampling	Load lbs./cu. yd./day	Crude Effluent C.O.D. p.p.m.	Treated Effluent C.O.D. p.p.m.	C.O.D. Removal Efficiency %
2/3/71	2.16	1200	200	83.5
3/3/71	6.93	3850 COD	610 COD	84.1
	4.82	BOD 2680 — $\frac{COD}{BOD} = 1.44$	BOD 60 — $\frac{COD}{BOD} = 10.2$	97.8
11/3/71	1.44	800	40	95.0
12/3/71	2.48	1380	80	94.2
13/3/71	3.31	1840	100	94.6
15/3/71	1.37	760	10	98.7
16/3/71	4.97	2760	150	94.5
17/3/71	3.42	1900	590	69.0
17/3/71	4.50	2500	470	81.1
19/3/71	2.88	1600	280	82.6
20/3/71	6.30	3500	420	88.0
21/3/71	9.00	5000	1000	80.0
23/3/71	2.23	1240	210	83.1
24/3/71	4.68	2600	420	83.9
25/3/71	6.12	3400	500	85.3
26/3/71	3.74	2080	610	70.8
27/3/71	3.42	1900	880	53.6
27/3/71	4.32	2400	720	70.0
28/3/71	2.16	1200	440	63.3
29/3/71	2.74	1520	500	67.2
31/3/71	5.22	2900	660	77.2
Averages	3.97	2210	425	80.9

Table 5

PLANT PERFORMANCE DATA
OVERALL AND INTERSTAGE REMOVALS OF THE ORGANIC LOAD

Date	Effluent Sample	C.O.D. p.p.m.	Progressive Purification C.O.D. %	B.O.D. p.p.m.	Progressive Purification B.O.D. %	$\frac{\text{C.O.D.}}{\text{B.O.D.}}$ Ratio
27/4/71	Input	2200	—	n.d.		
	Settling tank 1 to 2nd stage	1040	52.7	n.d.		
	Settling tank 2 to 3rd stage	560	74.5	n.d.		
	Treated effluent to sewer	400	81.8	n.d.		
6/5/71	Input	2300	—	750	—	3.07
	Under-tower stage 1	1000	56.5	360	52.0	2.78
	Settling tank 1 to 2nd stage	1160	49.6	450	40.0	2.58
	Under-tower stage 2	640	72.2	220	70.7	2.91
	Settling tank 2 to 3rd stage	760	67.0	280	62.7	2.71
	Under-tower stage 3	520	77.3	160	78.7	3.25
	Treated effluent to sewer	580	74.7	215	71.4	2.70

Regression analysis of results obtained during March and April, 1971, shows that, over the C.O.D. load range for the whole plant, the relationship is of the linear form:

$$y = mx + b$$

where x = load in lb. C.O.D./cu. yd./day
y = C.O.D. removed in lb./cu. yd./day

The values obtained in this instance (for 40 samples) are:—

$m = 0.857$ (regression coefficient)
(standard error of regression coefficient $= 0.032$)
$b = -0.225$ (intercept)
Standard error of estimate $= 0.328$.

The correlation coefficient of 0.975 indicates good correlation and the plots of the data demonstrate the constancy of purification with increasing load.

A further regression analysis (polynomial of degree 3) has been attempted with C.O.D. % removal values against C.O.D. applied load. This produced a curve showing a pronounced minimum but the correlation was thought too poor to be of any particular significance. At the same time Chipperfield has shown[2] that even for wastes of widely varying compositions a plot of *B.O.D.* percentage removal versus *B.O.D.* load gives a curve tending to become asymptotic to the load axis as the applied B.O.D. load is increased.

OPERATING EXPERIENCE

For a variety of reasons the plant has never operated according to design. However, under the conditions currently prevailing it is achieving its purpose, and future design modifications should enable it to treat greater loads from the increasing grape tonnages expected to be processed. Aspects of design already modified or requiring further comment are as follows:—

Pretreatment: While still waste was being produced the contents of both roofed holding pits could be held at a temperature well above ambient caused by the influx of hot waste. This brought about a marked increase in the anaerobic digestion (commonly accepted and recently mentioned by Dague[4]) and by leading waste in at one end of the pit and removing it from the other, substantial reduction in the C.O.D. was achieved (maximum drop registered: C.O.D. 10,900 ppm down to 3500 ppm).

Provision was made to burn digestion gas vented from the roof of the pit and to remove sludge from the bottom. Neither has been found practicable. At one stage while the waste was being fed to the plant from the storage lagoons no reduction below 3000 ppm C.O.D. could be obtained even with the plant in full recirculation. Later the same waste proved amenable to treatment. It is postulated that the enzyme systems required to break down wastes of poor biodegradability are only developed slowly. Hence experimental pretreatment of the waste with certain enzymes was tried but with no significant results as yet.

Flow balancing: The weir box design for flow control has worked well but modifications to the transfer pumps were required due to the eventual non-separation of high and low strength wastes.

pH Control: The addition of alkali to adjust the pH from approximately 4 to 7 is automatically controlled by a pulse dosing system. The choice of alkali was in the first instance decided by relative cost and hence lime at $925/ton equiv. wt. was chosen. A mixer was used to slurry the lime in water in a 1000 gallon tank for transfer to the first stage sump. Excessive usage of lime prompted an investigation into the neutralising process, which is very slow, resulting in unreacted lime passing right through the plant. This both wasted lime and raised the question of Ca^{++} ion toxicity.[4] Neutralisation by soda ash at $3710/ton equiv. wt. proved essentially instantaneous and actually decreased the cost of alkali required.

Nutrient Addition: Some addition of nitrogen was thought necessary and urea was chosen as the cheapest source. A qualitative relationship between added urea and residual ammonia was found as well as pH dependence. At a pH greater than 8.5 (inadvertently obtained) the whole tower emitted a strong smell of ammonia indicating deamination of proteins, etc., as well as displacement of the NH_3 (aq.) equilibrium. Values for residual ammonia in the treated effluent varied from 0 ppm to 120 ppm, but no correlation with bios growth has yet been established.

Distribution of Effluent to Flocor: This has perhaps proved the most difficult operation to achieve successfully. The circulating pumps transfer fresh feed and recycle to nozzles suspended above the Flocor with recycle ratios for the three stages of 3.6:1, 2.2:1 and 1.4:1 respectively. The respective rates are 9000, 5400 and 3600 gallons per hour distributed through 32, 20 and 12 nozzles. The wetting pattern obtained is excellent until the 11/64 in. hole in the jet blocks. Initially this was caused by small pieces of Flocor (soon eradicated) but as time went on everything from beetles to frogs turned up necessitating increasing downtime for nozzle cleaning.

The exclusion of insects, etc., was eventually achieved only to have further blocking arising from organic matter the size of small peas. These were eventually identified as the sclerotia of the fungus "Coprinus Tuberosa".[8] Liming of the tower from the top to create a temporarily hostile environment for the fungi proved efficacious over the course of several days. (Another alternative proposed was to seed the tower with the protozoan ciliate "Childonella Cucullulus", which has been reported by Curds et al,[3] as being a potential predator for fungi in treatment plants.)

Towards the close of 1970 and again in April, 1971, the nozzles had to be replaced by splash plates in order to overcome perpetual blocking problems. Various methods have been tried to remove blocking particles from the stream (excluding expensive comminutors) such as sump screens, pump intake screens, rapifine screens, vibroscreens and hydroclones. The stationary screens failed through humus blocking, the hydroclones through the specific gravity difference between liquids and solids being too small and the vibroscreen proving effective but too costly a solution. The last reason also applies to the use of centrifuges. Currently an efficient splash bar system is being sought as an alternative to the nozzles.

Microbial growth: A heavy growth of dark slimes has formed on the Flocor in all three stages, particularly the first. The increase in weight of the modules is substantial and has justified the use of an epoxy coated steel grid as support (timber joists have been more usual). The liquid falling from the bottom of the Flocor is black in color when the plant is heavily loaded indicating extensive conversion of organic pollutants to humus. Microscopic inspection of the nature of the microbial mass was carried out (on the pilot plant) and showed the presence of large numbers of rod-shaped bacteria, colpidia and ciliates, with smaller numbers of fungal hyphae, pupae, and fibres similar in structure to microscopic grass roots.[9] According to Chipperfield 20-40% of the applied B.O.D. is converted to humus solids although the *proportion* of the applied organic matter appearing in the humus tends to decrease with increasing load.[1]

Settling tanks: On the whole these have operated quite well and on only two occasions have there been serious problems with floating sludge. Laboratory trials have shown that a few ppm of a mixture of sodium aluminate and a polymeric floculating agent (Alfloc 673) are effective for clarification. It was later observed that a longer biological adaptation period gave a sludge of satisfactory settling characteristics. (An alternative considered was the use of an internal tube settling unit which has been developed in the U.K. recently.)

Essential to the success of sludge settling is daily sludge removal from the cone of the settling tank; otherwise the sludge rapidly becomes anaerobic and floats. No problems have been found with the sludge consolidation tank and although the thickened material has not been incinerated as planned, there has been no difficulty with its disposal.

Discharge to sewer: No signs of overloading of the town works have been apparent at any stage.

Testing: Ideally, daily monitoring of B.O.D., C.O.D., pH, nitrogen, phosphate, and temperature of all three stages should be carried out. With the facilities available at the plant site only a limited testing program has been possible and this has aimed primarily at establishing whether the treated effluent was of suitable quality for sewer disposal. A secondary but important consideration has been to determine the efficiency of the plant at various loadings. Following from the work of MacMillan et al,[6] it was decided to monitor the latter by C.O.D. testing in order to have results quickly available. As the requirement of disposal to the sewer was that the treated waste should be of B.O.D. <500 ppm, a rough correlation between B.O.D. and C.O.D. was sought. C.O.D./B.O.D. ratios found varied from 10.8 to 1.4 for the crude effluent, and from 25 to 2.7 for the treated effluent. A C.O.D./B.O.D. ratio of 2.7 is typical of whisky distilling wastes treated on Flocor[7] and hence a C.O.D. of $\leqslant 500$ ppm was taken as a definite indication of a B.O.D. <500 ppm. The choice of a C.O.D. $\leqslant 500$ ppm as a practical criterion for disposal to the sewer has been vindicated by observable non-interference with the town works' operation. (Stander et al[10] have shown that 1.5% of the initial C.O.D. remains after anaerobic digestion of still wastes followed by activated sludge treatment. The same is true of still waste subjected to prolonged aerobic treatment alone. One set of results for anaerobic digestion followed by primary aerobic treatment gave a C.O.D./B.O.D. ratio of 25:1, which, followed by secondary aerobic treatment reduced the B.O.D. to less than 1 ppm showing that the organic material remaining is biologically intractable.) Further work to be carried out will be a comprehensive survey of 5 day B.O.D. reduction versus loadings, and correlation of 20 day B.O.D.'s with C.O.D.'s. (It is expected that these values will be approximately equal in that 20 days will allow full bacterial metabolism of the sample.) In addition it is hoped to establish semi-quantitative relations between plant performance, temperature, pH, and nutrient addition.

CONCLUSION

At a cost of 60,000 Australian dollars the treatment plant has proved good value in its particular location, in that the benefits accruing balance the capital expenditure.

Although there are many points of detail which can and will be improved, it has been shown that an adequate reduction in strength for sewer disposal of a highly polluting waste can be achieved on a high rate plastic media packed "roughing" filter.

Acknowledgements

The author would like to thank members of Imperial Chemical Industries (Australia) Ltd., the South Australian Engineering & Water Supply Department, the winery at Griffith, and many others whose co-operation has made this report possible.

LITERATURE CITED

1. Chipperfield, P. N. J. Experience in Great Britain by Imperial Chemical Industries Ltd. with plastic filter media. (Paper presented at the Annual Conference of the Water Pollution Control Federation, Kansas City, U.S.A. (1966).
2. Chipperfield, P. N. J. The development, use and future of plastics in biological treatment. Paper presented at Effluent and Water Treatment Convention, London, U.K. (1967).
3. Curds, C. R., Cockburn, A., and Vandyke, J. M. An experimental study of the role of the ciliated protozoa in the activated sludge process. Water Pollution Control, *67*, 312-329 (1968).
4. Dague, R. R. Application of digestion theory to digestor control. Paper presented at 50th Annual Meeting of the Iowa Water Pollution Control Association, Ames, Iowa, U.S.A. (1968).
5. McKinney, R. M. Microbiology for Sanitary Engineers, pp. 202-212. McGraw Hill (1962).
6. MacMillan, J. D., York, G. K., and Vaughn, R. H. B.O.D. vs. C.O.D. methods for determining brandy stillage purification. American Journal of Enology and Viticulture, Vol. 10, No. 4, 199-205 (1959).
7. Pearson, C. R. Research report, Brixham Research Laboratory, Imperial Chemical Industries Ltd., Devon, U.K. (1966).
8. Private communication. New South Wales Dept. of Agriculture, Science Service Laboratories, Sydney, N.S.W., Australia (1970).
9. Private communication. New South Wales Dept. of Public Health, Sydney, N.S.W., Australia (1970).
10. Stander, G. J., Cillie, G. G., Baillie, R. D., and Ross, W. R. Investigation of the fullscale purification of winery distillery wastes by the anaerobic digestion process. C.S.I.R. research report No. 270, 88-90, NIWR, Bellville, South Africa.
11. Vaughn, R. H., and Marsh, G. L. Problems in disposal of California winery wastes. American Journal of Enology, Vol. 7, No. 1, 26-34 (1956).

PAPER 23

The Bangkok Composting Plant

By W. C. Carter*

The paper deals briefly with the quantity and characteristics of the refuse, the benefits derived from fermented compost, comments on the existing plant together with outline details, and presents a flow diagram of the four plants currently being commissioned.

The Refuse

The word "refuse" in the installations under review means waste paper, waste clothing materials, food scraps, ashes, animal droppings and carcases, and includes any other materials swept or collected from the roads, markets, animal farms and other places.

The three plants presently under construction would each be capable of dealing with 320 tonnes of refuse daily, and the fourth plant with 160 tonnes of refuse daily.

The loading of the plants is based on an average feed rate of 40 tonnes per hour during an 8-hour continuous operating period, at a moisture content of 40%, and 20 tonnes per hour for the smaller plant.

The design and construction of the composting plant has been based on experience gained in this terrain and climate, and generally follows the pattern of existing refuse disposal plants installed at Bangkok and Chengmai, while the design and construction of the refuse incinerators incorporated in each plant follows a conventional pattern of units installed in many parts of the world.

Plant Arrangement

The refuse is processed through respective pulverising, ferrous metal salvaging, screening and dense particle separating equipment, to provide a fine and homogeneous mixture suitable for fermentation. At this point the items rejected during the initial processing as installed for composting, are transferred to the incinerator for disposal by burning.

The organic material for fermentation passes through a 5-floor fermentation house in which temperatures of over 60-65°C are reached, following which the fermented material is discharged to a storage area to cool and mature.

The diagrammatic arrangement of the plant is indicated on the flow diagram attached, Fig. 1.

From this it will be observed that loaded refuse collection vehicles arrive at the plant and discharge their contents into either of the two reception hoppers, which have a combined capacity suitable to receive at least half the daily input of refuse.

Each reception hopper is channelled to allow excess moisture to drain to a common bilge pump which discharges this excess moisture to an adjacent site drain. An overhead travelling grabbing crane transfers the refuse from the reception hopper into mild steel buffer hoppers.

Each of the four buffer hoppers has a floor formed by a moving steel plate conveyor which is arranged to transfer refuse at a controlled rate to pulverisers of the hammer mill type, the speed of the plate conveyor being variable, and in the event of overload of the mills by excessive volume of material the electric control system will stop the conveyor motor for a short period to allow the mill to clear itself of material before additional refuse is conveyed to it. The conveyor will automatically restart when the mill is clear.

For economical operation of the hammer mills, they are each fitted with two electric motors. For initial starting both motors are in operation, but for normal operation only one motor is in service.

After passage through the mills, refuse is discharged on to vibrating feeder trays, which transfer the pulverised refuse at an even rate on to a rotating magnetic drum which extracts the ferrous metal items such as tins, etc. These salvageable items are directed into the hoppers of a mechanical baling process. The salvaged metals are then compressed into bales of suitable size for ease of handling and transportation.

The remaining refuse is directed on to belt conveyors which receive material from the two flow lines of shuttle conveyors. Each conveyor directs material on to a rotodisc separator, the object of which is to screen the pulverised refuse and remove oversize and elongated items such as textiles, plastics, nylon stockings and

* Clarke Chapman/John Thompson Ltd., Wolverhampton, England.

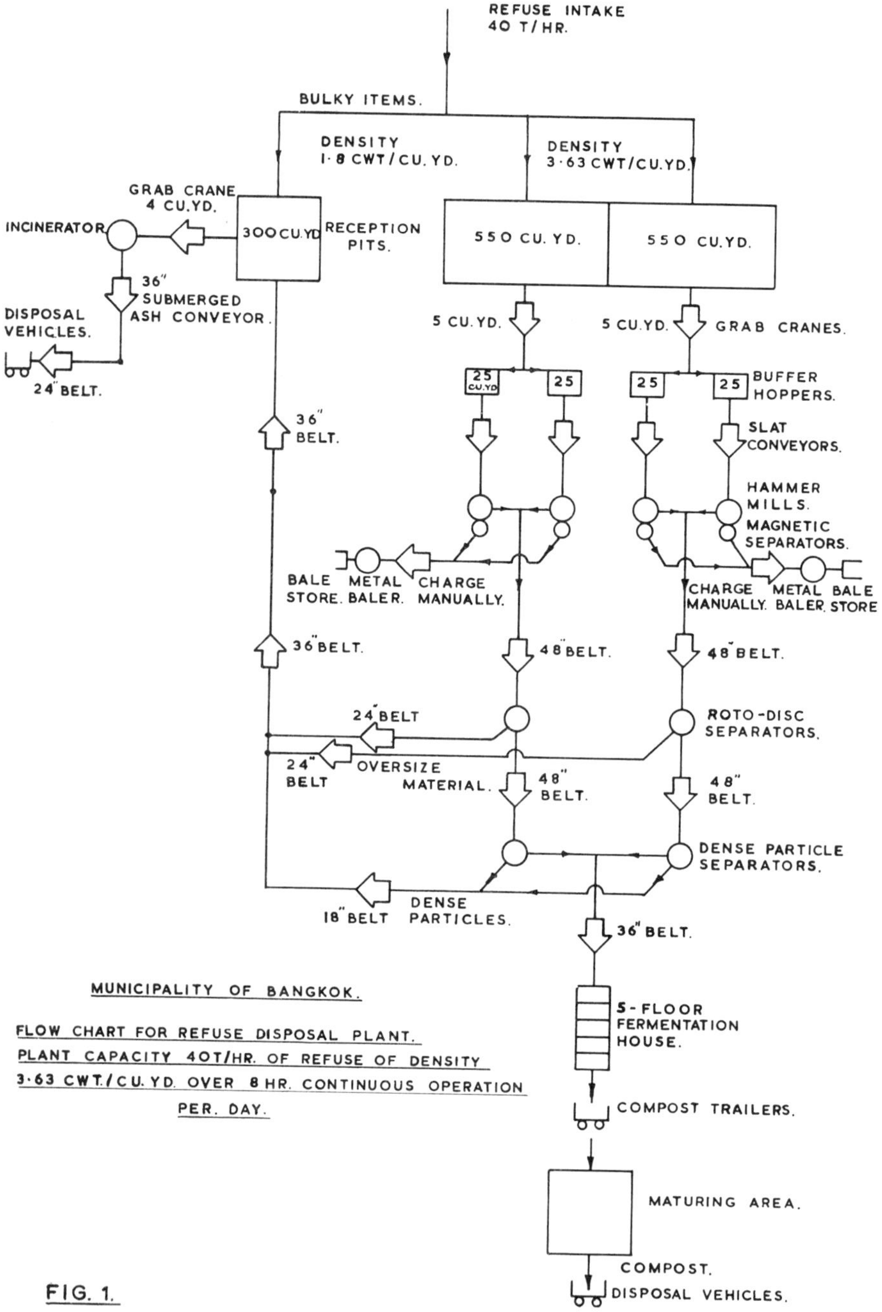

MUNICIPALITY OF BANGKOK.

FLOW CHART FOR REFUSE DISPOSAL PLANT.
PLANT CAPACITY 40T/HR. OF REFUSE OF DENSITY
3·63 CWT./CU. YD. OVER 8 HR. CONTINUOUS OPERATION
PER. DAY.

FIG. 1.

tights, etc., directing these on to a rejects belt conveyor for removal to the incinerator unit.

The separators are self-cleaning, and material passing through the screen falls on to further belt conveyors feeding the dense particles separator units. These units periodically separate denser particles such as broken glass, stones, etc., directing these on to the rejects belt conveyor for removal to the incinerator.

The main bulk of material is then directed on to a belt conveyor transporting the pulverised and separated refuse to the top floor of the fermentation house. Attached to this conveyor is a belt weigher which controls a water spray system located over the belt and the object of this is to maintain the moisture content of the material at a level suitable for fermentation.

The prepared refuse is spread to an optimum depth of a metre or so on the top floor level by means of a travelling tripper operated in conjunction with a reversible shuttle conveyor. The top floor, as are the other four floors, is formed from suitable troughs.

The material is retained in the top level of the troughs for approximately 24 hours, after which the troughs are rotated to allow their contents to fall into similar troughs on the floor below. Thus, since there are five levels of troughs and the material is retained at each level for approximately one day, during this period aerobic fermentation takes place and the discharged material is pasteurised.

For the purpose of rotating the troughs five trough turning units are installed, one at each floor level. They are mounted on wheels and travel on tracks. Each turning unit automatically travels the length of the house turning pairs of the troughs automatically successively along one side. On reaching the end of the house the engaging mechanism is automatically adjusted to allow the turning unit to return along the length of the building, turning the troughs in pairs on the opposite side to those previously turned. This automatic operation continues upwards through the building until all troughs on each side of the five levels have been rotated once each day.

After the 5-day retention period, the fermented compost is discharged from the lower level troughs to waiting vehicles for transport to a storage area to cool and mature.

During the above passage of pulverised refuse through the fermentation house, refuse is retained, as stated, in the metal troughs at each level and the fermentation must be sufficient to sustain microbic life in order to transform the pulverised refuse into material conforming to a satisfactory hygienic condition within a period of five days. The temperature of the fermented refuse within the fermentation house must therefore be maintained for a period of time above 60-65°C so as to be lethal to harmful pathogenes and weed seeds.

The material rejected as unsuitable for conversion into compost is mechanically handled to the incinerator, where it is reduced to a sterile residue.

Following fermentation as described above, the fermented refuse is transferred to a storage area where it will cool and mature before becoming suitable for application to horticultural or agricultural land. The time for maturing into organic material is that required to ensure a carbon to nitrogen ratio lower than 20:1.

Benefits from Fermented and Matured Compost

Improvement of soil structure.
Increase of moisture-retaining capacity.
Increase of the effectiveness of chemical fertiliser applications.
Prevention of leaching soluble inorganic nitrogen.
Ensuring that the nutrients are more readily available for growing crops.
Provision of additional nutrients.
Provision of food and suitable environment for soil microbiological life.
Means to bind the soil, thereby preventing erosion.

In support of the above, it should be known that the soil in Bangkok and similar areas is generally very fine and tends to pan or pack solidly. The use of compost will open up this soil and improve the structure to allow better rooting of plants, the better and more rapid absorption of moisture, and will also prevent erosion of the soil due to heavy rains.

It should also be apparent that compost is not an alternative or competitor to artificial fertilisers. Compost is humus-forming material and humus is the essential basic fuel required to support plant life. Unlike artificial fertilisers, the application of compost is less dramatic in its results, and although in the first year there is a noticeable improvement in poor soil, the best results are obtained by utilising compost regularly over 2/3 years. Results are then more lasting. The increasing rate of depletion of soil humus made by the continued, perhaps excessive, use of artificial fertilisers, makes it necessary to return organic material to the soil. Thus the use of compost is complimentary to the use of artificial fertilisers where maximum crop yields are associated with the maintenance of good soil.

PAPER 24

Design Features of the Hong Kong New Incineration Plant

By D. H. Turner*

This presentation is based upon experience gained in Hong Kong where, due to the pressure on the use of land, the most effective solution to the refuse disposal problem resulting from a population of four million people is incineration.

It deals briefly with the quantity and characteristics of the refuse, the existing plants at present in operation, together with their performance and, finally, gives an outline of the design features of the new Kwai Chung plant, shortly to be built.

The Refuse

An average of 3000 tons of refuse is collected and disposed of daily in Hong Kong, and at present approximately 60% of the disposal is by incineration.

It is mainly household and trade waste but complemented daily with about 30 tons of animal droppings from nearby abattoirs, 20 to 30 dead dogs, 600 rats and 70 to 80 tons of market waste.

All household refuse is delivered to central collection points by individual households, and consequently any items which have a saleable value (i.e., newspaper, clean cartons, tins, bottles, etc.) are extracted prior to delivery.

The refuse containers at the collection points are mainly open baskets and whilst awaiting loading can be subjected to tropical rain for as long as two hours, during which time it is possible that 25 mm of rain may fall.

Resulting from (a) the extraction of combustible material, (b) the open type of collection, and (c) the seasonal conditions, the refuse upon its arrival at the incinerator plant can vary in net calorific value from 525 to 1360 K.cals/kg., with the moisture content varying from 65% to 40%.

The diagram shown in Fig. 1 indicates the limits within which the refuse can fall and with which the incinerators must be capable of working.

The Existing Incinerator Plants

Operational at present are three plants, namely, Kennedy Town, Lai Chi Kok 'A' and Lai Chi Kok 'B'.

Each plant comprises four units, each unit having a maximum capacity of 10.5 tons per hour and is of Volund design in its complete form incorporating stepped and reciprocating grates and a rotary kiln for final burn out.

A cross-section of the incinerators at Kennedy Town and Lai Chi Kok 'A' is shown on Fig. 2,

At Kennedy Town and Lai Chi Kok 'A' gas cooling is effected as far as possible by waste heat boilers of the water tube type. Any additional cooling required to reduce the flue gas to a temperature of 400°C. is carried out by air attemperation, this method being necessary in view of the limitations upon water in Hong Kong.

The steam from the boilers is passed to turbo-alternators, the power generated being utilised for the plant auxiliaries, any surplus being supplied to local public services.

Fig. 1

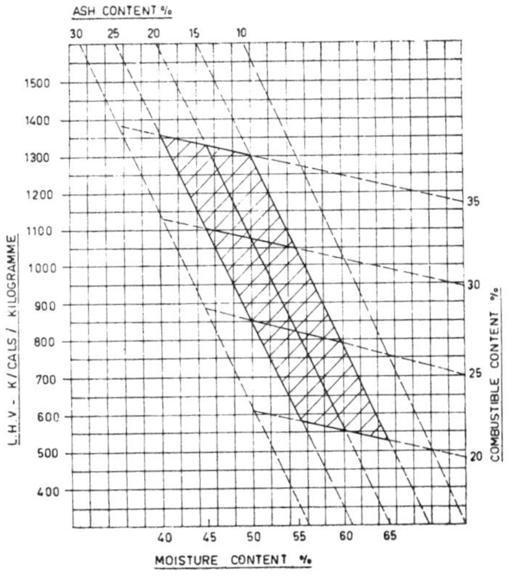

LOWER HEATING VALUES OF RAW REFUSE

---

* Clarke Chapman/John Thompson Ltd., Wolverhampton, England.

Gas cleaning is by means of multi-cyclone grit arrestors.

Operational Experience

Since the initial plant at Kennedy Town was commissioned in 1966, the ability of the design to cope with the considerable variation in refuse quality has been under survey.

It became increasingly apparent that despite the flexibility afforded by the incorporation of the rotary kilns, high moisture content refuse could not be incinerated to an acceptable degree unless load was reduced.

This problem had to be resolved on future plants required to dispose of Far Eastern or similar tropical refuse so that despite the wide variation in the refuse characteristics, full load could be maintained at all times with a constant high standard of burn out.

Attention was therefore given to the following main areas in which it was considered that re-design was necessary in order to achieve this prime requirement so essential on any incinerator plant.

(a) *Pre-drying*

The wide and frequent rapid variations in the moisture content resulted in fluctuations in the time required for the drying and combustion stages of the incinerator process, the net effect being sub-standard burn out unless the charging rate was corrected in time to compensate for such fluctuations.

It was soon evident that the concept of drying by radiation and convection was ineffective as only the top layers of the refuse bed received treatment, and under some conditions, the under layers were still high in moisture content upon the burning grate prior to the rotary kiln. This proved that even with a reciprocating grate, disturbance and turn over of the under layers was difficult when dealing with refuse containing a high moisture content.

A further noticeable effect was the retardation of drying and combustion of released volatiles above the bed when large quantities of moisture vapour were contained within the combustion chamber.

From these observations, it was considered that it was essential in the incineration of such refuse to incorporate in new plants effective and flexible means of pre-drying, such treatment taking place within a chamber independent of the combustion or burning chamber proper.

(b) *Grate Combustion Ratings*

The grates and kilns upon the existing plants were sized from a combustion rating which whilst adequate for refuse having a moisture content in the order of 50/55%, was too high for moisture contents in excess of this.

Fig. 2

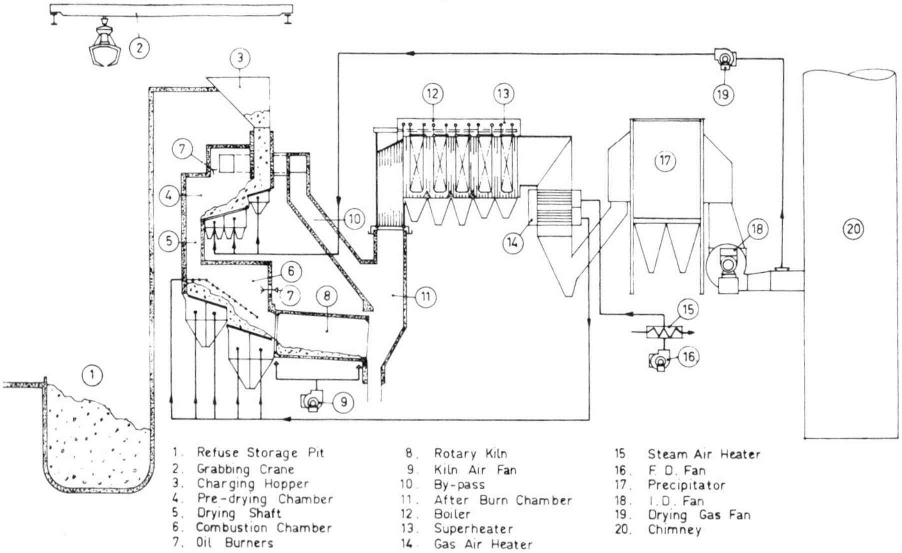

The combustion rating for grates required to deal with Far Eastern refuse therefore required review and recognisance was taken of the published experience of Mitsubishi Heavy Industries Ltd. in Japan, who had related combustion rating against the lower heating value of the refuse. Their findings showed that to deal effectively with refuse having a L.H.V. of 525 K.cals/kg. (65% moisture content) the combustion rating necessary would be about 150 kg/sq. metre/hour, with combustion air being preheated to a temperature of 300°C.

At the other end of the range in Hong Kong, the L.H.V. is approximately 1360 K.cals/kg. when the required combustion rating is in the order of 265 kg/sq. metre/hour.

A system was therefore required which, whilst providing the flexibility required to accommodate such variations, would also be economically viable. The rotary kiln had proved eminently suitable for this purpose on the existing units and it was decided that it should be incorporated on future plants whilst the overall combustion rating should be decreased. In this respect, it was considered that due to the flexibility afforded by the kiln in the latter stages of the combustion process, an overall combustion rating midway between those required for the extremes of the refuse values would be adequate.

(c) *Combustion Air*

The air supply upon the existing plants is at ambient temperature but operation clearly showed that preheated air is necessary.

Physically, the furnaces have withstood the arduous conditions extremely well and it is only after seven years' operation that a major renewal of the furnace and kiln refractory is being contemplated.

The Kwai Chung Incinerator Plant

In 1973, the opportunity arose to incorporate the revised design features in a successful tender to the Hong Kong Government for a further incinerator plant having a total capacity of 1200 tons of refuse per day. The plant was to comprise four units but under normal conditions three units only would be in operation, 24 hours per day, seven days per week, the remaining being on standby duty.

The required guaranteed performance standards were:—

(a) that the plant with one incinerator stream on standby should be capable of a throughput of 900 tons of raw refuse per day over any continuous period of seven days.

(b) The air-dried residue from the grate, free of ferrous metals, should contain by weight not more than 0.3% putrescible or fermentable substance and not more than 5% combustible matter.

(c) The maximum dust discharge with the flue gas into the chimney should not exceed 0.23 grams/cubic metre at N.T.P. (dry).

Technical Synopsis of the Plant

The refuse is transferred from the reception pit to the incinerator-charging chute by grabbing cranes and enters a separate drying chamber which consists essentially of two Volund grates of the reciprocating type surrounded by a refractory enclosure.

Within this chamber drying is effected by:—

(a) Radiation from gases emanating from the combustion chamber proper passing over the raw refuse bed and drying the top layers.

(b) The passage of recirculated flue gas at a temperature of 280/300°C. through the raw refuse bed drying the under layers.

Ventilation drying by the passage of either hot flue gas or air had been successful in Japan for some considerable time. In this instance, the choice of flue gas as the drying medium was made for the following reasons:

(a) It minimises the burning of easily combustible materials during the drying process by retardation of ignition so that the maximum heating value of the refuse is retained for the combustion process within the combustion chamber.

(b) More drying heat can be applied than with hot air as higher temperatures can be obtained.

(c) The specific heat of flue gas with its higher moisture content is higher than that of air.

Sufficient moisture will be removed in the drying process to ensure that, upon discharge from the drying gate, the refuse will be in a condition for self-supporting combustion, irrespective of its condition at entry.

The moisture vapour released during the process will enter the incinerator by-pass duct through openings in each side wall of the chamber.

The introduction of drying gas will only be necessary with refuse worse than average value, it being theoretically self supporting at average value and above without the necessity for such treatment.

From the drying chamber, the refuse is discharged into a vertical drying chamber where further drying takes place from contact with the flow of hot gas rising from the combustion chamber. In falling, there will also be a tendency for any agglomerated refuse to break up.

From the drying shaft, the treated refuse falls on to the grates of the combustion chamber, where the incineration process commences with final burn out to the condition required taking place in the kilns.

Preheated combustion air at a temperature of 230/250°C. will be introduced into the combustion chamber both undergrate through the riddlings hoppers and overgrate through high velocity nozzles.

Variations in the value of the raw refuse entering the drying chamber will be sensed by temperature change within the chamber and modulation of the supply of undergrate drying gas to match the change of refuse value will be under the control of such a temperature signal. The admission of drying gas will, however, have a tempering effect upon the drying chamber temperature and to compensate for this, additional heat will be provided by oil from the oil burners located within the chamber. The input will be modulated from the same temperature signal utilised for modulating the drying gas.

Theoretically, the supply of drying gas and oil should only be necessary for refuse values between 525 and 860 K.cals/kg. and through this range a set point temperature of 550°C. is maintained within the chamber. Above the mean refuse value, although gas and oil should not be required, a small quantity of drying gas will continue to flow to maintain the duct work at temperature and also allow the controls to react immediately should the refuse value worsen. Similarly, the oil burners will not be shut off, but maintained on "low fire" or "on/off" control. At these conditions, the drying chamber temperature will be allowed to find its own level.

Decrease in refuse value below the average will also result in an increase in gas flow from the drying chamber and variable resistance in the form of a damper is provided within the by-pass duct.

With a constant input of 12.5 tons of raw refuse per hour, there will be a variation in the quantity of moisture removed in the drying chamber with each variation in refuse quality, and consequently the weight of refuse entering the combustion chamber will also vary.

To maintain a constant set point temperature within the combustion chamber, the combustion air would also have to be modulated. To eliminate this added control loop which could create operational difficulties, the combustion air quantity will be set at a constant value irrespective of any variation in refuse quality between worst and average, this quantity being based upon that required for average conditions.

Such a procedure would affect temperature as the refuse value worsens, and to compensate for this, additional heat would be provided by means of an oil burner located within the combustion chamber, this being controlled from a temperature signal at a set point of 850°C.

With an increase above the average value of the incoming refuse and drying gas and oil at their minimum flow rates, the combustion chamber temperature would rise and would be allowed to do so until the maximum temperature of 950°C. is reached. At this point, the combustion chamber temperature control would change over and modulate the air quantity for maintenance of the 950°C. set point.

To enable combustion to take place in the rotary kiln and also limit the temperature of the combustion products, further air is required which is introduced through the transition box between the burning grate and kiln, side ports and seals. For physical and thermal reasons, this air is at ambient temperature and supplied by a separate fan. Control of the air supply is determined by the kiln outlet gas temperature and this is maintained at a set point of 950°C.

Flue gas from both by-pass and kiln then enter the "after burn" chamber where mixing takes place prior to entry into the radiant section of the waste heat boilers.

The waste heat boiler is of the forced circulation type, chosen to enable the best arrangement of the convection tube bank to be used and to ensure a safe flow of water through the tubes under the varying conditions of heat input. The steam and water mixture return to the steam drum where separation is carried out by a series of cyclone separators.

The boiler enclosure is formed from welded water walls with fins welded between, and provides a fully water-cooled enclosure from the vertical entrance to the radiant inlet chamber to the outlet from all of the convection surfaces.

The convection heating surface is arranged in the form of pendant elements with vertical tubes, the pitching of the elements being such that the gas velocity is maintained at no more than 8 metres/second. Such a configuration minimises build up and erosion of the tubes by dust and grit carryover, and also enables any dust fall out within the boiler to fall clear of the heating surfaces and into the hoppers below the boilers.

A superheater has been incorporated and this, once again in the form of pendant elements, is located at the rear of the boiler where it will be subject to the lowest gas temperature. Coupled with the low degree of superheat required, the possibility of corrosion will be reduced to a minimum.

The steam conditions required at the superheater outlet are 260°C. at 17.9 kg/cm^2. In the event of low steam flow under adverse conditions, a higher steam temperature could be experienced and therefore a design figure of 370°C. has been chosen.

A gas air heater is mounted in a vertical gas pass after the superheater, the air entering the heater having been initially preheated by a steam air heater in order to maintain acceptable metal temperatures within the gas air heater and minimise corrosion.

The flue gas temperature required at the outlet of the air heater is 300°C. irrespective of the incinerator throughput and quality of refuse. Control of this is accomplished by a system of dampers arranged in by-pass ducts around part of the convection heating surface, the effectiveness of this then being modulated to elevate or depress the outlet gas temperature to the required value.

In the design of the boiler, due cognisance has been taken of the four main types of corrosion which may be experienced, namely:

(a) High temperature liquid phase corrosion.

(b) Corrosion due to non-uniform furnace atmosphere.

(c) Corrosion by HCl.

(d) Low temperature or dewpoint corrosion.

The design parameters incorporated, both physical and thermal, should reduce the possibilities of attack by any of these.

The generated steam feeds a back-pressure turbine coupled to a brushless, salient pole alternator having an output of 2000 kW. Exhaust steam is condensed to a low pressure air/steam condenser and live steam surplus to the turbine requirements is condensed in a separate high pressure air/steam condenser, the condensate being returned to the de-aerator.

A separate saturated steam main supplies steam to the steam air heater associated with each incinerator unit, the condensate also being returned to the de-aerator.

The flue gas discharged from the air heater at a temperature of 300°C. is directed into an electrostatic precipitator having an efficiency of 96.7% and providing a dust outlet concentration within the gas of 0.23 grams per cubic metre at N.T.P. (dry). This concentration is equivalent to a dust emission by weight of 11 kgs/hour/incinerator or 0.88 kg/ton of refuse burned.

A single multi-flue chimney is provided, each incinerator line having its own flue in order to maintain high efflux velocities over the range of operating conditions. The actual chimney height is 122 metres and the efflux velocity from each flue at design condition is 23 metres/second.

ACKNOWLEDGEMENTS

(1) Matsumoto, K., Asukata, R., Kawashina, T. (Mitsubishi Heavy Industries Ltd., Japan) — Proceedings A.S.M.E. 1968 National Incinerator Conference.

(2) Public Works Dept. — Government of Hong Kong.

PAPER 25

Woodwaste Disposal in Urban and Country Areas —The Future for Incineration

By J. D. Court,* J. R. Harry* and J. H. Scheltema*

A.—WOODWASTE DISPOSAL IN URBAN AREAS

1. The Problem

Liversidge[1] estimated that during 1969/70 the equivalent of 70 million superfeet of woodwastes were generated in processing plants, as distinct from sawmills in N.S.W. Most of this (assumed 60 million super) would have been in metropolitan Sydney, and would have comprised seasoned softwood (typical density 550 kg/m^3). A conservative projection[2] puts growth in the Australian consumption of sawn timber between the years 1970 and 2000, as rising from 1767 to 2530 million superfeet per annum. Apply this growth to the estimate for waste, the following amounts of woodwaste for disposal can be compared with total municipal refuse[3]:

			1970	2000
Woodwaste	Sydney	tonnes/y	80,000	115,000
Municipal Refuse	Sydney	tonnes/y	760,000	3,600,000
Woodwaste as % of Municipal			10%	3%

Thus wastes from urban joineries constitute a relatively small fraction of total municipal waste. They are also only a small part of the total timber industry waste. Yet, their disposal has caused significant local air pollution, i.e., the nuisance of burnt and unburnt wood particle fallout and acrid smoke.

The Regulations under the N.S.W. Clean Air Act limiting emissions from incinerators and boilers have been tightened since their introduction in 1964. A joiner operating such equipment must now comply with particulate emission limit of 0.4 g/m^3 at s.t.p. and corrected to 12% CO_2 in the flue gas. He must also submit the design of any new unit to the Health Commission for review if there will be equipment on his premises capable of consuming more than 300 kg/h, formerly 2240 lb/h, of fuel or waste. The emission limit for this new plant is 0.25 g/m^3 at s.t.p. and 12% CO_2 with a smoke limit of Ringelmann No. 1. The practical effect of this is that it is now considerably more expensive than several years ago to install and operate a woodwaste boiler or incinerator and comply with the law.

At present, many joineries and small timber yards in the inner suburbs are relocating in newer industrial districts. Thus the tightening of air pollution requirements comes at a time when the industry is seeking to establish new woodwaste incinerators and boilers. Consequently, because of the increased cost of this traditional method of disposal, other means are being closely considered and in some cases adopted.

2. Woodwaste Burning

Controlled incineration is potentially a very satisfactory method of disposing of woodwaste. Combustion is self-sustaining except for a few species such as turpentine and white mahogany. Quality, including moisture content and calorific value, is consistent. Ash and sulphur contents are low.

Leaving aside incineration practice at sawmills (none in Sydney) woodwaste is incinerated either in a multiple chamber refractory unit (on a stationary fuel bed) or a McCashney cylindrical unit (where it is burnt in suspension). The multiple chamber refractory unit predominates in Sydney.

2.1. *Multiple Chamber Refractory Incinerators*

The theoretical concept behind this incinerator is that if it is carefully designed and fed at a controlled rate, the waste should be completely combusted and emissions should comply with the Clean Air Act without the use of auxiliary burners or gas cleaning equipment, except for a simple settling chamber.

Many tests on two and three chamber incinerators have been conducted in N.S.W. and the results reported by Murphy.[4, 5, 6] These results indicate that properly designed three chamber (i.e., primary combustion chamber, mixing chamber and secondary combustion chamber) incinerators are generally capable of meeting the emission standard of 0.4 g/m^3 at 12% CO_2 for existing plant and possibly the 0.25 limit for new plant. This is provided the units are fed at a controlled rate and at or

* Air Pollution Control Branch, Health Commission of N.S.W.

near design capacity. A grey smoke is usually associated with this type of operation presumably due to unburned volatiles, ash and condensed steam. The intensity is low, if combustion is efficient and it only becomes black smoke exceeding Ringelmann 2 when one of the essential elements of combustion is lacking. The species of timber, condition (i.e., wet, air dried, kiln dried) and ash content have not been recorded for these tests, although most of these incinerators would have been burning untreated, seasoned softwood.

On the basis of this work and design data established in U.S.A., the Air Pollution Control Branch, N.S.W., included the design of incinerators for woodwaste in its first issue of Incinerator Standards.[7] These called for three chamber incinerators, and specified maximum heat release rates, and gas velocities in the mixing and secondary combustion settling chambers. The Standards were subsequently modified to achieve the intent of the original standards expressed in the more easily, manipulated terms of geometric ratios between primary combustion, mixing and secondary combustion chambers. At this revision, woodwaste incinerators were excluded until the need for a secondary burner was resolved. A design guide has been appended to the Standards, however, to assist in the assessment of woodwaste incinerators. It extends the principles of the Standard to wood burning with relaxation of chamber dimensions in some areas and calls for careful control of the feed.

According to this guide the burning rate for small units (100 kg/h) could range from 90 to 130 kg/h/m^2 of grate area up to 120 and 180 kg/h/m^2 for large (1 tonne/h) units. The corresponding heat release rates for seasoned Radiata Pine would be, for small units 170 to 370 kW/m^3 of primary and secondary combustion chambers volume, and 80 to 200 kW/m^3 for large units. There is a high percentage of volatiles in woodwaste (approx. 80) and hence most of the combustion will take place above the fuel bed. The U.S. design[9] recommends 200% excess air apportioned as follows:

Primary Combustion Chamber	
Overfire	60%
Underfire	6%
Mixing Chamber	17%
Secondary Combustion Chamber	17%
	100% = 200% excess air (= 7% CO_2)

Incinerators tested in N.S.W. range from 2-8% CO_2 indicating typical operation from 150 to 1000% excess air.

To meet the new emission standard of 0.25 g/m^3 at 12% CO_2 consistently under normal operating conditions without a grit arrestor probably requires an auxiliary burner and careful assessment of the ash content. However, the Health Commission will assess each application on its merits until more test data are available. Figure 1 shows the percent of ash content which must be retained in the incinerator to meet the two current emission standards, assuming complete combustion. Allowing a settling chamber with recirculated water floor to be 50% efficient (liberal), it is apparent that the upper limit for ash in the timber for new plant is 0.4% and for existing plant 0.65%.

Figure 1

THEORETICAL RELATIONSHIP BETWEEN EMISSION LEVEL AND ASH CONTENT FOR SOFTWOODS

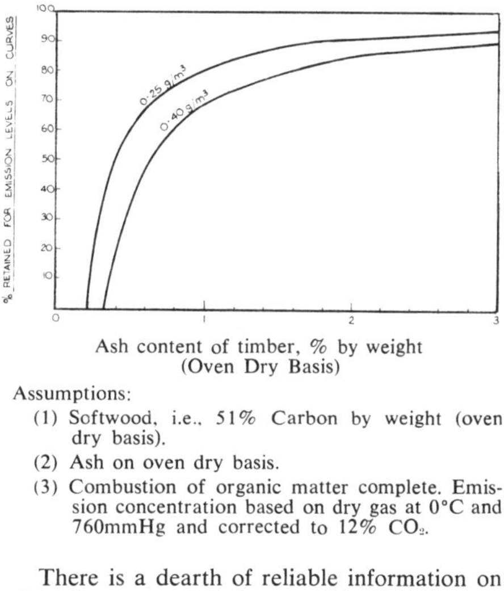

Ash content of timber, % by weight (Oven Dry Basis)

Assumptions:

(1) Softwood, i.e., 51% Carbon by weight (oven dry basis).

(2) Ash on oven dry basis.

(3) Combustion of organic matter complete. Emission concentration based on dry gas at 0°C and 760mmHg and corrected to 12% CO_2.

There is a dearth of reliable information on the ash content of timbers used in N.S.W. Most of the literature simply states that the ash content of timbers grown outside the tropical zones is less than 1% dry (dry basis). But this does not dispose of the matter when it is proposed to operate an incinerator without gas cleaning equipment against a tight emission limit. More analyses are needed, although one of the problems is variability within a species.

Some generalised cost estimates have been

made and presented in Figure 2 for indicative purposes only. The basis is 2300 h/year operation, 80% utilisation, 10% amortisation, maintenance at 8% capital and the operator at $6000 p.a. The step in the curve represents the change to auxiliary firing once the schedule limit (300 kg/h) is passed. The comparison with tipping as an alternative depends heavily on the attitude of the tip operator towards accepting the waste. (The figure may well be substantially higher in view of the rapid exhaustion of tipping space in some areas of Sydney.)

Figure 2

ESTIMATED COST OF WOODWASTE DISPOSAL BY INCINERATION AND SANITARY LANDFILL IN 1973

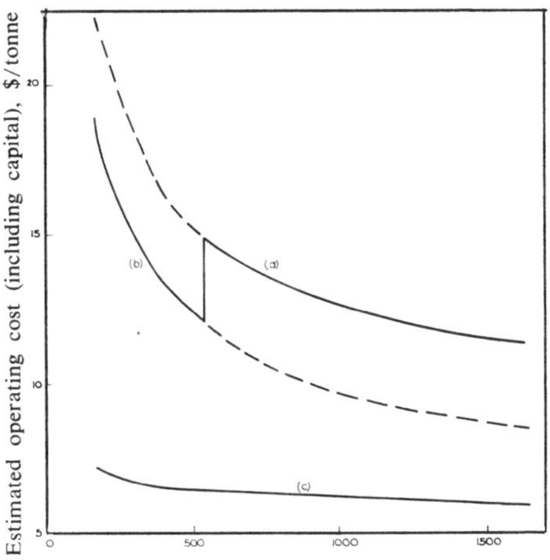

(a) With auxiliary firing (assumed over 540† schedule limit).
(b) Without auxiliary firing (assumed under 540† schedule limit).
(c) By cartage and sanitary landfill.

2.2. *Woodwaste Boilers*

A small and decreasing proportion of woodwaste is being used for steam-raising in Metropolitan Sydney, mostly in establishments where there are drying kilns. Because the waste produced is not usually in balance with the steam required, a method of disposing of the surplus must be available even when the woodwaste is used as fuel. There are a few isolated examples of woodwaste firing outside the timber and pulp industries, notably at a hospital in Perth and for public power generation at Mount Gambier.

Woodwaste is a low ash fuel and easy to burn, although bulky and awkward to handle. Thermal efficiency, or ratio of steam to fuel, is low because of both the low calorific value compared to conventional fuels and the low thermal efficiency due to the relatively large amounts of water vapour (and hence latent heat) carried out in the flue gases.

Combustion is usually achieved in a Dutch Oven on a stepped grate, with the flue gases passing forward to a fire-tube or water-tube boiler. In practice, thermal efficiences of from 50 to 70% are common in well-controlled plants, with CO_2 levels of from 8 to 14%. High grate loadings (300-400 kg/h/m^2) have been quoted in the literature as applicable to stepped grates with assisted draft; however, experience in Sydney has suggested that spasmodic heavy smoke emissions can occur at much lower loadings (120 to 170 kg/h/m^2). Overfire air can be of assistance,[4] but the root of the problem seems to be to achieve an even flow of the burning fuel bed down the stepped grate. Characteristically, slippages seem to occur, suddenly releasing excessive volatiles and causing short but very dense black smoke emissions. The following would be ideally necessary to avoid this situation:

(1) Even feed to the grate at a controlled and adjustable rate.
(2) Even sizing of material fed to the grate. (Offcuts invariably disturb the bed and should be eliminated by hogging or fed in very limited amounts to a flat grate at the base of the stepped grate.)
(3) Adjustable grate angle.
(4) Controlled air.
(5) Overfire air.

High efficiency mechanical grit arrestors are invariably required to meet the emission limits. Wickham[10] has given details of fitting high efficiency multicyclones to a wood waste boiler with close attention to the control of fires in the fliers. Murphy has reported typical emissions from boilers burning woodwaste.[4, 5]

3. Conclusion

Disposal of woodwaste in urban areas by burning has become considerably more expensive in recent years. Air pollution control requirements have been a significant factor in this. With careful design woodwaste incinerators and boilers can operate cleanly; it remains to be seen whether cheaper and more attractive methods of disposal than the traditional one of burning will gain ground.

B.— WOODWASTE DISPOSAL IN COUNTRY AREAS

1. Present Practice

Whilst woodwaste in city areas is derived largely from the further sawing of large pieces followed by dressing and moulding, woodwaste in country areas is mostly associated with the sawing of logs into sizes suitable for house framing. In the city, the disposal of this woodwaste by incineration has imposed restrictions on the procedures used because burning is done close to neighbours and requires clean techniques, calling for rather expensive refractory incinerators. In the country, the mills are often at a considerable distance from neighbours who might be affected by any pollution and the large quantities and size of the waste to be burnt would impose a very considerable expense if refractory incinerators were required to be used.

Hence, the refractory incinerator, with its capability of clean burning if properly constructed, maintained and operated, has so far seen little application in country areas. The answer to the country disposal problem has commonly been open burning or the use of the elementary tepee incinerator. The disposal method used must be capable of handling fairly large tonnages of woodwaste. Aside from the wastage of branches, leaves, roots and stumps in the forest, the sawing of the logs themselves can involve a loss of 30 to 50% discarded in the form of hearts, offcuts, edgings and sawdust.

In the most primitive form of incineration, these waste materials are conveyed to a suitable portion of the site where they are simply burnt on the ground. Sometimes a pit is dug and sometimes the area is surrounded by a sheet metal fence or the woodwaste may simply be tipped on the bank of a nearby river and set alight, but in any event the uncontrolled burning which occurs is often of a smouldering nature and always results in large volumes of whitish fumy smoke with a definite objectionable odour. Under windy conditions this may be accompanied by ash, charcoal particles and unburnt sawdust.

The tepee is an improvement on open burning but nevertheless is basically an open fire with a more effective fence around it. The name tepee immediately suggests its shape, i.e., a truncated cone, fabricated from sheet steel. It is fitted with tangential airports around the circumference near ground level and a screen across the outlet at the top to prevent the larger woodwaste particles and sparks being discharged. The incinerator is usually fed by a conveyor belt discharging through a hole part way up the tepee shell, so that the woodwaste falls to the centre of the floor below. The offcuts and edgings are conveyed on this belt sometimes together with the sawdust, but the sawdust may be pneumatically conveyed to a low efficiency cyclone, which discharges directly from its base into the tepee.

Much of the failure of the simple tepee to provide clean burning is associated with the fact that its shell is fabricated from sheet steel of limited resistance to high temperatures. In order to avoid high temperature failure, a large amount of excess air is introduced so that the outgoing flue gases have a temperature of about 120°C and a carbon dioxide concentration around 2%. The resultant low temperature combustion is partly quenched and incomplete, resulting in a substantial discharge of odorous visible organic material as well as charred particles. Fallout of these particles in the area can be severe. Normal operation of such tepee units results in the discharge of a white fumy smoke, much like that derived from open burning, and is accompanied by ash, charcoal and unburnt woodwaste particles. Woodwaste particles are often blown out of the top of any cyclone fitted and also from the feed conveyor belt. Even when a mill has a tepee installed, the hearts from the logs are often separately transferred to an open spot on the site and burnt with their own separate contribution to pollution.

Open burning and the simple tepee certainly fail to meet the requirements of the Clean Air Act by a wide margin and the poor performance of the tepee has been previously reported, the particulate emission concentrations from one series of tests being presented in Table 1.[5] In the past these excessive emissions have been tolerated very largely because the mills were erected away from neighbours, but in more recent times townships have grown towards the mills. In any case, the white fumy smoke can travel considerable distances whilst still being apparent both by appearance and smell, with the result that a mill on a town's outskirts can contribute significantly to the haze within the township proper and be the most significant polluting source noticeable in the district. This situation is common, and a recent limited survey on the North and South Coast of 50 timber mills placed close to neighbours showed that waste disposal was being conducted by 26 open fires, 25 simple tepees and seven refractory-lined incinerators.

Table 1

PARTICULATE EMISSIONS FROM TEPEE BURNERS

Test Series	Test Number	Flue Gas Flow Rate m^3/min @ stp	Average Flue Gas Temp °C	Average CO_2 % V/V	Particulate Emission g/m^3 stp 12% CO_2	Calculated Burning Rate kg/h	Smoke
1	1	243	126	1.4	0.46	215	Voluminous
19 ft. dia.	2	330	192	1.5	0.99	312	White
Simple	3	302	239	2.5	0.99	475	Fume
Tepee	4	378	265	3.1	1.04	738	
2	1	15.3	465	9.6	0.76	92.5	
28 ft. dia.	2	30.5	444	8.1	0.43	156	
Improved	3	57.8	572	14.4	0.21	525	
Tepee	4	75.3	547	17.7	0.07	840	
	5	87.7	569	14.5	0.32	802	
3	1	196	396	7.0	0.26	865	Slight
28 ft. dia.	2	196	411	7.3	0.23	906	Fume
Improved	3	177	400	5.6	0.36	626	
Tepee	4	149	355	6.3	0.32	591	
	5	156	396	5.9	0.30	580	
	6	184	403	6.9	0.18	800	
	7	179	417	7.2	0.23	812	
	8	224	419	6.8	0.44	961	
4	1	667	436	6.5	0.29	2740	Clear
45 ft.	2	545	439	3.5	0.48	1200	Flue
Improved	3	562	511	4.7	0.21	1670	Gases
Tepee	4	568	506	5.2	0.34	1860	
	5	634	497	4.2	0.39	1680	

The situation is actually worse than this because the survey is by no means complete and the vast majority of tepees examined were found to have been improperly maintained. The steel shells of these tepees are frequently buckled, with sprung circumferential and vertical joints; large gaps are evident at the bottom of the shell near the ground; ill-fitting access doors are often left wide open; excessively large openings are cut into the side of the tepee for discharge of the conveyor belt from which clouds of smoke and woodwaste are emitted; and tangential airports are left in all stages of disrepair and incorrect positioning. The result is that few tepees display the internal cyclonic flame action and turbulence which the bottom tangential airports are supposed to promote, and combustion is maintained in a smouldering manner.

In order to improve the rate of combustion, fan-assisted underfire air can be provided which results in an intense flame in the vicinity of the air entry and a shower of sparks being lifted off the fire. With open burning, the ash collects, giving a reservoir of objectionable material to be windswept subsequently. As far as the refractory incinerators are concerned, their performance is often similarly poorly controlled and maintained, and it is a fairly rare situation to find a unit working to its design capability.

2. Development of Improved Tepee

This situation has plagued mill operators in the United States in much the same way as in this State. As a result there has been some development of the simple tepee incinerator in an endeavour to improve its performance. Since a large part of the poor air pollution control characteristics of the simple tepee is associated with the fact that the combustion zone is at too low a temperature, because of the very large amount of excess air admitted intentionally and unintentionally, development has been focused on control of the air quantity. Inadequate turbulence in the combustion zone is another aspect of the problem. The position has now been improved considerably by design intended to exercise greater control of these factors. Over the last couple of years a tepee incinerator has become available in this State in which both underfire and overfire air streams are introduced under controlled conditions, so that the amounts can be regulated in conjunction with the positioning of a damper in the outlet to give the desired temperature. Unfortunately, the light steel shell of the tepee still imposes considerable limitations on radiation from the combustion zone and the temperature of the gases inside the unit.

For this reason the amount of air introduced is still considerable, giving carbon dioxide contents in the flue gas of about 7%. However, this air control has raised the temperature of the combustion zone, resulting in the temperature of the outgoing gases rising from about 120°C to about 450°C, and this has resulted in the elimination of the usual large plume of white smelly smoke. The provision of fan-assisted overfired air through tangential ports around the incinerator base has given much greater control of swirl conditions inside. The tendency to lift incompletely-burnt particles out of the top of the incinerator is much reduced, and the provision of automatic control and a degree of instrumentation improve the standard of the units above that of those previously employed.

Three series of tests have been conducted for particulate emission on these improved units with results as shown in Table 1. The first series was conducted on a prototype and the sampling conditions at the top of the unit were bad. However, the emission concentrations obtained showed quite clearly the improvement between the last three runs at a high outlet gas temperature and the first two runs at a comparatively low temperature. These trials also showed that the discharge of unburnt sawdust particles characteristic of simple tepees had been obviated and that the collected particles were confined to char and ash. The unit operated smoke-free.

The second series of the same prototype unit, but with improved sampling facilities in the form of a chimney extension on the top and an anti-swirl device before the sampling point, substantiated the improvement found over the earlier results on normal tepees. On the basis of this improved performance, a second unit with a 45 ft. base and design capacity of 4.3 tons per hour was then erected. This unit burnt offcuts, edgings and sawdust from hardwood timbers, and the test results demonstrated that it has a capability of meeting the earlier standards of the Clean Air Act, but it has not yet demonstrated its ability to maintain consistently the 0.25 gram per cubic metre limit of Regulation 20 for new equipment.

3. Conclusions

There is no doubt that a very substantial improvement has been made in the type of units used for woodwaste incineration. It has been found that even when the units need maintenance and are not performing as well as they should, they are still capable of giving smoke-free operation with no obvious fallout problems in the vicinity, at least in less sensitive areas.

The cost in comparison with the refractory incinerators used in city areas, is very much reduced and it appears an economic solution for those mills situated in areas where their incinerator emissions cannot result in a nuisance to neighbours.

Under these circumstances, by the authority given by the Clean Air Act, the Air Pollution Control Branch is no longer approving the old type tepee incinerator and is requiring these newly-developed units. No doubt further improvements will be obtained but at the present time, in order to meet the 0.25 g/m^3 limit required by Regulation 20, their approval is subject to the provision of a grit collector.

Acknowledgement:

The authors wish to thank the Health Commission of N.S.W. for permission to publish this paper.

REFERENCES

1. Liversidge, R. M., "Utilisation of Sawmill Residue", Australian Forest Industries Journal, *39* (1) 42-47 (Jan., 1973).
2. Economic Study Group Report — Part 7. *The Australian Timber Journal, 37* (4) 23-27 (April, 1971).
3. Kirov, N. Y., "The Age of Pollution", *The 1971 Aust. Waste Disposal Conference,* Paper 1.
4. Murphy, R. P., and J. F. Pottinger, "Air Pollution from Wood Waste Incineration", *The 1971 Aust. Waste Disposal Conference,* Paper 16.
5. Murphy, R. P., "The Measurement and Control of Emissions from Wood Waste Fired Boilers and Incinerators". D.S.I.R. Conference, 1966, on Burning of Wood Waste, Auckland, 1966.
6. Murphy, R. P., "Domestic and Industrial Incinerators and the Clean Air Act". First Australian Refuse Disposal Conference, 1967, No. 7.
7. Pottinger, J. F., "New Standards for Incinerators in New South Wales", *Clean Air,* March 1969, pp. 29-34.
8. "Minimum Recommended Standards for Commercial and Industrial Incinerators Designed to Burn less than 300 Kilograms of Waste per hour." Division of Occupational Health and Pollution Control, November 1972.
9. Netzley, A. B., and Williamson, J. E., "Multiple Chamber Incinerators for Burning Wood Waste" in *Air Pollution Engineering Manual,* Editor J. A. Danielson, U.S. Dept. Health Education and Welfare, Cincinnati, 1967, pp. 436-447.
10. Wickham, F., "The Prevention of Fly Ash Emission from the Sawdust Fired Boilers at Hollywood Repatriation Hospital" D.S.I.R. Conference, 1966.

PAPER 26

The Problem of Wood Waste Disposal in an Urban Environment with Special Reference to Singapore

By F. H. C. Kelly*

1. Introduction

The primary processing of wood is commonly a rural industry in Australia and in countries which have their own timber resources. The industry as a simple sawmill or a complex including such units as plywood production or paper manufacture may be associated with a small town, as are most towns in Australia, by comparison with more densely populated countries. More advanced processing such as furniture manufacture will occur in cities as well as larger towns.

The situation in Singapore is unique as a city state with industrial development. Logs are imported from surrounding countries with up to 1000 miles of sea transport and probably from longer distances as development progresses. At some stage no doubt the bulk importation of chips may be introduced as in Japan, but that time is not yet.

The largest proportion of the waste disposal problem is related to the primary processing of logs, the total waste from which may be as much as 50 to 60% of the raw material. Plywood factories seem to produce a larger proportion of waste wood than sawmills.

The total annual importation of timber to Singapore is approaching a million tons which represents a substantial waste disposal problem in what is essentially an urban environment, in whatever part of the 580 km^2 it may be produced. The telephone directory lists nearly one thousand users of wood, approximately 12% (by number) calling themselves sawmills, 7% termed plywood factories and the remainder are furniture makers of one type or another all of whom produce sawdust.

2. Classification of Wood Waste

The simplest primary classification of waste produced from wood is:

(i) Bark.
(ii) Sawdust.
(iii) Mixed shavings and sawdust.

A secondary classification separates these into different varieties of wood for which some 20 trade names are used and which is the general classification for the purpose of government statistical records. In fact two or more trade names may be used for one particular botanical species.

A further classification which has some justification from the point of view of application is to refer to the materials as heavy, medium or light hardwoods. Softwoods as such are only imported in finished articles or paper pulp and currently do not present a waste disposal problem. On the other hand, separation of the hardwoods into the three groups is not always a simple exercise for borderline types. Of the 20 trade-recognised varieties, two are commonly referred to as heavy hardwoods, five as medium and nine as light, with four of variable classification.

The light hardwood known as Meranti (Shorea multiflora) has been the most popular of all timbers and accounts for 37% of total imports, 63% of this being used by sawmills.

At any one site it may be expected that there could be a rough separation of waste products according to the initial classification. Generally speaking, not more than two or three different types of wood might be expected at any one site.

3. Problems of Disposal

The problem is taken essentially as a waste disposal problem and discussion will be confined mainly to its understanding and defining as a necessary antecedant to obtaining a solution. It is unlikely that any single solution will be suited to the complexity of the problem and careful costing becomes necessary in each situation.

The disposal problems can be divided into two groups:

(a) On-site disposal.
(b) Centralised disposal.

Then we might adopt a further disposal classification according to the means adopted, viz.:

(a) Incineration.
(b) Physical and/or chemical processing.
(c) Landfill procedures.

Whatever the disposal system one is confronted with a variety of raw materials.

On-site disposal requires that the quantity

* Professor of Applied Chemistry, University of Singapore.

be large enough to justify the cost of a technologically-sound and operative system of disposal. Small operators have often been of the the opinion that if they are small enough then on-site incineration is unlikely to trouble others or attract attention. Either there develop too many such small operators for this to be socially permissible or small operators become large operators and correspondingly socially undesirable.

If on-site treatment is to be considered, the question must also be faced of whether to treat all of the waste product or to treat the different fractions separately.

This question must also be answered if centralised processing is adopted.

Centralised processing requires transportation which may be provided by individual wood processors or by a central collecting agency — either government or contract operated. Wood processors usually consider themselves to be in business to process wood and do not take kindly to transporting their own waste material to a disposal centre. A more effective procedure seems to lie in the establishment of a collecting agency. In an urban situation distances are moderate and a central disposal system should then have maximum feasibility benefits.

4. Properties of Wood Waste Fractions

Since the different properties of wood waste fractions need to be considered irrespective of the disposal location, some attention will be given to this. The introduction will have given a preliminary indication of the complexity of this aspect of the problem, becoming even more complex as attention is focused on each section of the arbitrarily-defined classification.

4.1. *Bark*

This component varies widely with type of timber and constitutes approximately half the weight of timber waste materials with strongly-barked trees. Debarking operations commonly involve the use of substantial quantities of water with techniques varying from very high pressure jets to soaking pits or river soaks. One result is for the moisture content of the waste material to be high, values of 30 to 45% being not uncommon. This makes direct incineration difficult.

Bark may have some chemical value either as a source of tannin or occasionally as a source of alkaloid. The latter is unusual as trees, which are most suitable for their timber component, are generally deficient in alkaloid in the bark. Useful concentrations of tannin in the bark are more common but variable. Tannin extraction involves firstly a milling of the bark and ultimately disposal of the water insoluble components which constitute 90 to 98% of the original dry material — occasionally it may go as low as 80 or even 75%. The water soluble extract is never wholly "tannin", the proportion of "tannin" to "non-tannin" is very variable and careful analysis of any situation is of paramount importance to successful development.

The insoluble material requires dewatering and may be dried to 50% water content by suitable pressing, at which point it is burnable in suitably designed furnaces and steam generated can be used in the tannin production process. It is unlikely, however, that the process would be thermally self-supporting because of the large volumes of water requiring evaporation. A favourable price and quantity of tannin can however make the process economically interesting.

More commonly bark wastes must be disposed of by some form of incineration technique which for its effective operation requires homogenising prior to incineration in properly designed furnaces.

4.2. *Sawdust*

This is the term used to describe dustlike material coming away from the saws used in the wood processing industry. Not only do different varieties of wood result in different varieties of dust, but we must also recognise different types of product from different types of saws. For example, a chain saw which is very often used for the first sawing operation tends to produce a long fibred dust. Laboratory experiments have indicated the possibility of this being suitable for paper pulp production, whereas the fine dust from circular saws seems to be much less suitable.

Unfortunately, not more than about 10% of the dust produced in all sawing operations is of the more-favourable longer-fibred type. The relationship between this differentiation and the variety of timber is not quite clear as all timber processors do not necessarily employ the chain saw technique.

Sawdust is also usually not as homogeneous as might be hoped, being mixed with an uncertain amount of chips of irregular size. Any disposal procedure which intends to take advantage of the apparently homogeneous character of sawdust will need to take account of its real heterogeneity and provide either for a separation technique or direct homogenising. Pneumatic separation seems to offer one reasonable means for dealing with this.

Where a central disposal system is operated, or even with on-site disposal where more than one type of timber is processed, there will be the problem of differing properties even with sawdust from the same type of saw, and the material must be suitably homogenised. The first and most obvious difference is the moisture content which may vary from 10 to 25%. The conditions of processing will also play a part depending upon whether water transport or water storage may have been used for the logs, or whether water is used at some other stage before the sawing operation. In which case even higher moisture contents must not be ruled out.

The carbon, hydrogen, oxygen proportions in the dry material of the sawdust do not convey very much in the way of useful information in regard to variability in character. This information is, however, necessary if any form of incineration is being investigated. Other more specialised tests will be needed to define the properties of concern.

Sawdust is a recalcitrant material to incinerate, but many of the problems concerned with its incineration appear to be closely related to those experienced with the incineration of bagasse — the waste fibrous material from sugar cane milling. After more than 100 years of development, bagasse is being burned in suspension very efficiently at high rates of thermal release in large water-walled furnaces to generate steam at whatever pressures the end uses may require. Sawdust seems to be an easier problem. The biggest difference between the two problems seems to be in siting. There is usually no insoluble problem in on-site incineration of bagasse.

Sawdust commonly has a lower moisture content — maximum of 30% compared with a minimum of 45% for bagasse. Each has inhomogeneity problems of comparable magnitude. In the case of bagasse, a primary fibrating process aimed at maximising small particle homogeneity is practised because of its advantages from the point of view of the sugar juice extraction processes. Sawdust and associated wastes from wood processing must be accepted as they come and subjected to homogenising prior to incineration as may be required.

4.3. *Mixed Materials*

The differentiation between bark and sawdust is seldom clean-cut and invariably there are other waste components in the nature of shavings, chips and even blocks of wood. The amounts of mixed materials vary according to the arrangements of any individual operator. This accentuates the problem of inhomogeneity.

5. Conclusion

It is evident that the two main cost factors involved in the effective disposal of wood waste products are collection and homogenisation. For small scale processes to develop specialised uses such as tannin extraction, wood flour production or some other specific product, suitable separation rather than homogenising techniques must first be implemented followed usually by homogenising of the separated fraction.

Processing for pulp production is seldom a viable proposition. Pulp processors usually prefer a suitable variety of wood and do not countenance the mixing of wood waste materials in existing pulping plants. Where wood pulp is produced from timber, there is still a waste disposal problem which is commonly solved by incineration. Where wood chips are processed for pulp, the main waste disposal problem is merely transferred to the chipping site where again the problem is solved by incineration.

Incineration requires its own design attention, involving ultimate disposal of clean chimney effluent to the atmosphere, the ashes usually being transported for land reclamation, or occasionally processed for their own intrinsic silicate value.

Incineration may or may not involve associated steam generation. The economic viability for this must be determined according to local conditions, and is largely related to the quantity of material to be incinerated, and in its turn this relates to the magnitude of the scale of primary wood processing or the effective and economic solution of transport and collection problems. The main problems in wood waste disposal therefore are:

(1) Material distribution requiring collection and transport.
(2) Inhomogeneity requiring separation or homogenising or both.
(3) Incineration without air pollution.
(4) Economic steam generation and utilisation.

Wood waste is a re-generating source of energy and is worthy of consideration from this point of view where large scale production is involved. Where production is on a small scale, then a specialised use with a high value-added component may be of greater significance.

PAPER 27

Performance Guarantees and Acceptance Testing of the Waverley-Woollahra Incinerator

By K. F. Napier* and N. Y. Kirov†

1. Introduction

The planning, building and commissioning of Sydney's first modern municipal refuse incinerator in Botany Road, Waterloo, marks an important milestone in Australia's waste management program. The plant was constructed by Von Roll-Process Plant Construction Joint Venture for the Joint Councils of the Municipalities of Waverley and Woollahra and serves in addition the needs of three other Municipalities — Marrickville, North Sydney and Randwick, thus providing immediate waste disposal facilities for a combined population of over 400,000 people (refer Table 1).

The plant consists of the initial installation of two furnaces with continuously-fed, self-stoking moving grates, which are operated on a continuous basis to incinerate a combined daily total of 520 tons of raw refuse. Provisions exist for the installation of a third furnace, at a later date, with an additional capacity of some 300 tons per day. The detailed planning of this project was previously described by Stait and Toner (ref. to Paper 19 of 1971 Waste Disposal Conference). A comprehensive survey to characterise the physical and chemical composition of refuse generated within the Municipalities and its likely variability was also described by van den Broek and Kirov at the 1971 Waste Disposal Conference (see Papers 5 and 6).

The process equipment is designed to provide for flexibility of operation and for a variability in the calorific value of the refuse within a range of 2800 and 4600 B.t.u./lb. The incinerator capacity, and the quality of refuse burnout and gas cleaning are guaranteed for a wide range in refuse mixture composition (Fig. 1) within the hatched area of the diagram shown in Fig. 2.

The dotted line in Fig. 2 indicates the maximum design overload capacity for operation of the furnace; this is outside the performance guarantees.

2. Contractual Performance Guarantees

The plant performance was guaranteed in the contractual agreement for each refuse destructor unit together with all its associated auxiliary and ancillary equipment when operating under specified conditions, separately or collectively with respect to the following.

* Project Manager, Crooks Michell Peacock Stewart Pty. Limited.

† University of New South Wales, Process and Combustion Engineering Consultant for this Project.

Table 1

ESTIMATE OF POPULATION GROWTH AND REFUSE GENERATION IN THE FIVE MUNICIPALITIES FOR THE PERIOD 1968 to 1981

MUNICIPALITY	POPULATION			Refuse Generated, lb./person/day			Total Refuse tons/week**		
YEAR	1968	1971	1981	1968	1971*	1981*	1968	1971	1981
WAVERLEY	65,200	65,800	67,300	1.65	1.85	2.60	335	389	547
WOOLLAHRA (incl. Paddington)	60,200	61,000	63,500	2.07	2.32	3.25	390	442	645
RANDWICK	115,000	119,000	137,000	1.53	1.71	2.39	550	636	1,024
MARRICKVILLE (incl. new city area)	92,000	93,000	99,000	1.76	1.97	2.76	505	572	854
NORTH SYDNEY	55,000	60,000	78,000	1.66	1.86	2.60	285	349	634
GRAND TOTAL	387,400	393,800	444,800	Average for 5 Councils 1.70	1.90	2.66	2,065	2,379	3,704

* Assumed to increase at 4% p.a.
** Averaged over a four-weekly period during September, 1968.

RANGE OF REFUSE COMPOSITION VALID FOR GUARANTEES

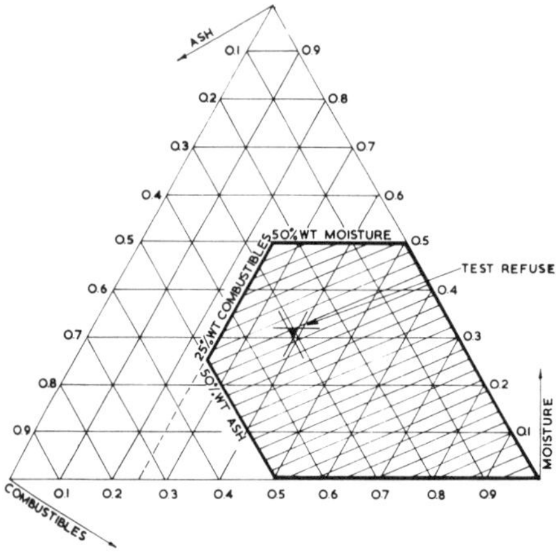

FIG. 1

2.1. *Incinerator Capacity*

Each furnace, together with all its ancillary equipment when operating under the specified conditions, would be capable of handling and fully incinerating per twenty-four hours, 260 long tons of refuse with the specified composition and density and a gross calorific value of from 2800 B.t.u./lb. to 4600 B.t.u./lb.

Each furnace would be capable of maintaining within the performance limits, a "steady state" uninterrupted operation at a capacity of 11 tons/hour.

2.2. *Independent Operation of each Furnace*

Each furnace would be capable of being operated independently of the other at the capacity of 11 tons/hour and within the performance limits.

2.3. *Control and Completeness of Combustion*

Full incineration would be achieved within the performance limits to produce a discharge of sterile and odour-free clinker. The maximum

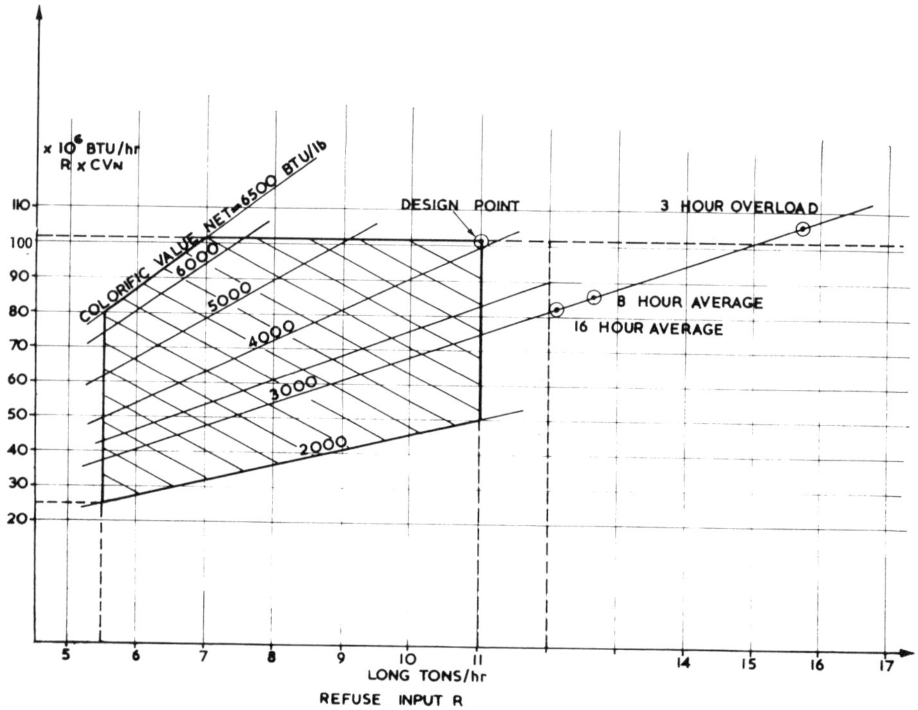

FIG. 2

COMBUSTION CAPACITY OF INCINERATOR RELATIVE TO THE C.V. OF THE REFUSE combustible matter in the residue shall not exceed 5 per cent by weight of the dry residue and the content of putrescible matter shall be less than 0.25 per cent by weight of the dry residue, determined by the T.U.V. (Düsseldorf) method.

2.4. *Gas Cleaning*

Over the entire range of guaranteed operating conditions, the total mass of soot, dust, fly ash, or other solid particles in the flue gases discharged from the top of the chimney would not exceed 0.1 grain per cubic foot of dry gases at N.T.P., when corrected to a basis of 12 per cent carbon dioxide.

The escape and deposition of large charred paper flakes shall not constitute a nuisance in the immediate neighbourhood, and not more than 10 per cent by weight of the maximum permissible emission (as above) shall be of a size greater than 50 microns.

2.5. *Control of Furnace Temperature and Gas Cooling*

Automatic and completely effective control arrangements would be provided to maintain the furnace temperature at the acceptable working levels of 900°C. Efficient spray cooling would be provided to ensure that the combustion chamber temperature would not exceed the design limit of 950°C on a continuous basis or 1050°C for short durations (½ hour).

Gas cooling would be provided in a spray tower under thermostatic control and emergency water supply under automatic control to limit the temperature of the gases entering the precipitators to a maximum of 350°C.

2.6. *Noise Level Limits*

The maximum noise levels from the incinerator plant or any of its auxiliaries, measured at the boundary of the site, would comply with the acoustical limits stipulated by the Sydney City Council, namely:

	Day	Night	Weekend
Portman Lane	69dB(A)	60dB(A)	60dB(A)
Portman Street	70dB(A)	54dB(A)	62dB(A)
Portman Street (adjacent to South Sydney Hospital)	62dB(A)	54dB(A)	60dB(A)

2.7. *Maximum Water and Power Consumption of the Plant*

When the plant is operated under the full load conditions stipulated above, the total water consumption and the total electrical consumption of the plant used for the process would not exceed a specified maximum.

Electrical power consumption	27 kWh/ton
Water consumption	14,700 g.p.h.

2.8. *Plant Extension*

Adequate provisions of refuse and ash and clinker handling capacity, plant layout, and general services would be made for the installation at a later date of a third furnace of similar capacity to that of each of the other two furnaces.

3. Conditions for Commissioning and Acceptance Tests

3.1. *Preliminary Tests*

The object of these tests was essentially to demonstrate that mechanically the plant complied with the requirements of the Contract. All preliminary testing of individual items of plant, motors, control gear, etc., including the calibration of testing instruments and temperature pressure draught and gas analysis equipment for controlling the plant performance, and the adjustment of control relays on site, were carried out.

These tests were carried out progressively under the supervision of the Engineer during construction of the plant. On completion of sections of the plant, each section or sections were tested as a whole.

3.2. *Incinerator Acceptance Tests*

Efficiency and acceptance tests were carried out to prove that the plant met all the operational performance guarantees. These tests were carried out toward the end of the commissioning period over a controlled period of twenty-four hours during which the plant was operated continuously at the specified loading.

All measuring instruments and other apparatus used in connection with these tests were checked and calibrated prior to the tests.

Refuse consists of large-size bulky heterogenous material, and hence a major contributing factor to inaccuracies in its analysis is often inadequate sampling and sample reduction methods. For the purpose of the acceptance tests, samples of the raw refuse and ash and clinker were taken at regular intervals during the test and were analysed in detail.

3.3. *Deferred Precipitator Acceptance Tests*

The final precipitator tests is to be carried out after 2000 hours of continuous and satisfactory commercial operation. During this 2000 hours of operation preceding the acceptance testing, the Contractor may carry out his own tests and make such adjustments, modifications and additions to the plant as may be approved by the Engineer but under no circumstances is he permitted to clean the collecting or the discharge electrodes other than by normal operation of the rapping gear provided.

The precipitator plant would not be accepted until this deferred acceptance test has been satisfactorily carried out and the relevant guarantees fulfilled.

The performance of each precipitator plant would be determined in terms of total weight of dust leaving the precipitator plant. Multipoint sampling would take place at the outlet of the precipitator plant. The samples of dust would be examined for particle grading and percentage combustible content.

During the test the incinerator and precipitator plant would be operated by the Client's staff under normal operating conditions.

4. Preparations for Testing

Individual plant items were tested as they became available. Defect lists were made, defects corrected and the item retested until accepted by the Engineer.

As sections of the plant were completed they were tested as a whole to ensure the correct functioning of the interconnecting plant items.

From the beginning of the dryout period the plant was manned by the permanent staff of the Client, 24 hours a day, 7 days a week and operated under the direction of the Contractor's commissioning staff. After completion of the furnace dryout the full plant stayed in continuous operation and was gradually adjusted to reach optimum performance. This period also served as a training period for the plant operators.

Preparation for the acceptance test commenced as soon as the Contractor was satisfied that the plant was performing satisfactorily. A temporary weighing device was built to weigh each grab load of refuse before charging it into the furnaces, for the purpose of measuring the exact throughput. This was calibrated, then tested in actual use for a period of 8 hours and the calibration checked.

A laboratory was set-up on site and was prepared to handle the refuse and residue sampling during the acceptance test. The sampling, sorting, crushing, drying and grinding was done on site while the chemical analysis and the calorific value determinations were done by a registered N.A.T.A. laboratory.

Other preparations for the acceptance test included installation of temporary meters to measure the electric power and process water consumed, and various temporary measuring instruments throughout the plant.

All these temporary items installed specifically for the acceptance test were checked and tested before the commencement of the acceptance test.

5. Acceptance Test

The refuse for the acceptance test was selected from various Councils to approximate as close as possible the mixture on which the initial plant design criteria were based. This was thoroughly mixed and left standing for 24 hours.

The acceptance test commenced when all operating parameters of the plant were at a steady state to the satisfaction of the Engineer. The total test period of 24 hours consisted of 8 hours stabilisation period during which time all measurements were recorded, an 8 hours' operation with two furnaces at varying capacities to prove the plant capability against the final quality of residue burnout, and a final 8 hours' operation of one unit to prove the degree of residue burnout at the nominal design load.

All measurements were recorded at one-hourly intervals and were used to calculate the performance of the plant. Samples of refuse and residue were taken at two-hourly intervals.

5.1. *Test Data*

The following final data was obtained from the measurements and samples taken during the acceptance test:

(i) *Incinerator Capacity*

The furnaces were operated at the average throughput of 12.6 tph each during the first 8 hours which is equivalent to 300 tons per 24 hour. The average throughput for the full 16 hour test was 12.1 tph equivalent to 290 tons/24 hour.

During the first 8 hours of the test one of the furnaces was operated at an overload of 40% for a period of 3 hours giving an average throughput of 15.7 tph.

Taking the average calorific value of the refuse as 3000 Btu/lb net, the average combustion chamber heat release rate was 10,000 Btu/hr/cu.ft for the 16-hour test period with an average max. of 13,000 Btu/hr/cu.ft during the overload test.

Even though the test refuse analysis gave only 3600 Btu/lb gross calorific value, the furnaces were found to have ample reserves to be capable of incinerating at least 11 tph of 4600 Btu/lb refuse.

(ii) *Refuse Analysis*

Refuse samples of approximately 300 lb. each were taken at two-hourly intervals and weighed. Successively three lots of samples were bulked together to give a total of three sample groups at the end of the test.

Each group was manually separated into paper, plastics, textiles, wood, putrescibles, metals and glass, and each component for each group weighed.

The samples of metals and glass were analysed for moisture content and discarded. The samples of paper, plastics, textiles, wood and putrescibles were shredded and split into three samples. One sample was analysed for moisture, ash, combustibles and calorific value plus C, H_2, N_2, O_2 chemical analysis.

The chemical analysis was carried out by a N.A.T.A. registered laboratory and gave the following final result:

COMPOSITION OF REFUSE
(percentages by weight)

Component	Composition % of Refuse	Moisture % of Component	Gross Calorific Value btu/lb dry basis	Gross Calorific Value btu/lb wet basis
Paper	37.7	33.8	8,000	5,300
Plastics	2.8	15.7	14,140	11,920
Textiles	3.6	18.8	8,730	7,090
Wood	0.7	16.4	8,000	6,700
Putrescibles	33.8	48.6	5,560	2,860
Metals	8.5	6.3	0	0
Glass	12.9	3.6	0	0

Proximate Analysis

Moisture	31.4%	(by weight)
Volatile Matter	32.9%	,, ,,
Fixed Carbon	5.4%	,, ,,
Residue	30.3%	,, ,,
	100.0%	

Ultimate Analysis

Moisture	31.4%	(by weight)
Carbon	19.5%	,, ,,
Hydrogen	2.8%	,, ,,
Nitrogen	0.1%	,, ,,
Oxygen	15.9%	,, ,,
Residue	30.3%	,, ,,
	100.0%	

Calorific Value

3600 Btu/lb gross
3000 Btu/lb net

These results compare well with the values stipulated in the original feasibility study.

(iii) *Residue Analysis*

Clinker samples of approximately 200 lb. each were taken every two hours from each furnace. Successively three lots of samples were grouped together to give a total of three sample groups at the end of the test.

Each group was mixed and the metal manually removed. The moisture content of the metal for each group was determined and the metal discarded.

Each clinker group was split into three separate samples, and one sample of each group was analysed by a N.A.T.A. registered laboratory for ash, combustible matter and putrescible matter and gave the following results (dry basis):

Metal	40.0%	(by weight)
Ash	59.6%	,, ,,
Combustibles	0.4%	,, ,,
Putrescibles	0.0013%	,, ,,
	100.0%	

The 0.4% combustibles and 0.0013% putrescibles were substantially better than the guaranteed values of 5.0% and 0.25%, respectively.

(iv) *Combustion Gas Analysis*

From the recorded data the calculated analysis of the combustion gas before the addition of any cooling water was:

	Wet basis	Dry basis
Nitrogen	67% (by weight)	80.0% (by volume)
Carbon dioxide	11% (by weight)	8.4% (by volume)
Oxygen	11% (by weight)	11.6% (by volume)
Moisture	11% (by weight)	

The oxygen content of 11.6% (by volume, dry basis) is within the specified maximum 121% excess air.

(v) *Electrical Power Consumption*

The average process power consumption was 476 kWh or 18.9 kWh/ton at the average test throughput of 12.6 tph.

The equivalent consumption at the specified design criteria was extrapolated from this and its value was 24 kWh/ton as against the guaranteed maximum of 27 kWh/ton.

(vi) *Water Consumption*

The total process water consumption was approximately 9000 gal/hour or 360 gal/ton of refuse.

This also was within the guaranteed maximum value of 14,700 g.p.h., even after adjusting to the specified design criteria and allowing for other process variations.

(vii) *Noise Level*

This test was performed at a later date and the requirements of Section 2.6 were adequately met.

6. Summary

Generally the acceptance test was carried out in accordance with the Contract and was successful in proving its performance to be within guaranteed limits, viz.:

	Test Results Actual	Test Results Extrapolated	Guarantee Limits ± 5%
6.1. *Incinerator Capacity*—			
ton/hr. av.	12.1	11	11 min.
ton/24 hr.	290	260	260 min.
ton/hr overload	15.7	12.1	(12.1)

It was thus established that the plant has a substantial overload capacity.

6.2. *Independent Operation of Each Furnace*—proved.

6.3. *Control and Completeness of Combustion*—

combustibles, % of dry residue	0.4	0.4	5.0 max.
putrescibles % of dry residue	0.002	0.002	0.25 max.

6.4. *Gas Cleaning*—subject to a deferred acceptance test.

6.5. *Control of Furnace Temperature and Gas Cooling*—deferred.

6.6. *Noise Level Limits*—

Noise level produced by the plant falls well below the criteria detailed in Section 2.6.

6.7. *Water Consumption,*

g.p.h.	11,240	12,500	14,700 max.
Power consumption			
kWh/ton	18.9	24.0	27.0 max.

6.8. The acceptance test proved that the plant performed well within the design criteria and satisfied other general operational requirements.

Finally, the results obtained from these tests clearly demonstrate the importance of detailing the required performance plant guarantees and of establishing the basic testing requirements and actual test procedures at the pre-contract stage, thus producing a clear result of plant capabilities in the terms of the Contract.

PAPER 28

The Packaging Industry and Solid Waste Management

By J. D. Honeysett*

In presenting this paper, I want to cover the following specific areas:—

(a) What is the packaging industry?
(b) What is its role?
(c) Why the Packaging Industry Environment Council?
(d) Packaging in Waste.
(e) What is the solution?

The Packaging Industry

The packaging industry has enjoyed a remarkable growth since World War II, and today ranks as one of Australia's major industries.

It represents some $700m turnover per year in relation to packaging alone (i.e., excluding transport), which is some 2.8% of all Australian manufacturing.

The industry can be considered as a service industry, providing a service to its customers, be they producers of food, beverage, cosmetics, tobacco, confectionary, chemicals and paints, or Portland cement.

As well as having its customers spread over this diverse section of manufactured and processed articles, the industry itself uses many types of raw materials, either singularly or together, and a wide range of technology and processes in meeting consumer demands.

Table 1 gives an indication of this, and of the growth in packaging, which has been necessary to keep pace with the growth in manufactured and processed goods.

The Role of Packaging

A minority of people today look upon packaging as an unnecessary evil, an excessive addition to our garbage can, and hence to the solid waste system, or something which the manufacturer puts around his article, so that the unsuspecting consumer thinks he is getting more for his money than he actually is. Some people look on a package as a potential piece of litter, and others, fortunately a minority, tend to confirm this observation by littering.

In defence of both the packaging industry, and its customers, and for the benefit of this minority, I would like to discuss each of these aspects in turn.

Is it evil or worthless? Packaging carries out a number of functions which contribute to the well being of the community and these are not intended to damage the environment. These functions are necessary for the following reasons:—

(i) *Hygiene*

All food deteriorates with time, particularly if exposed to the air. It was in 1806 that Nicholas Appent was awarded 12,00 francs by the French Government for a method of preserving food suitable for the armed services. Without a suitable package, in this case the can, this method would never have evolved.

* Executive Director, Packaging Industry Environment Council.

Table 1

VALUE OF CONTAINER OUTPUT — BY TYPE OF MATERIAL

	1962-63		1965-66		1967-68		1974-75(a)	
Material	$'000	%	$'000	%	$'000	%	$'000	%
Containers made of—								
Paper and paperboard	114,502	40.0	148,845	42.7	177,189	42.5	312,000	45.1
Metals	83,659	29.5	97,787	28.1	120,174	28.6	193,000	27.9
Glass	27,696	10.0	33,076(a)	9.5	36,745(a)	8.6	53,000	7.7
Plastics (flexible)	22,134	8.0	26,415	7.6	30,846	7.3	48,000	6.9
Plastics (rigid and semi-rigid)	7,868	3.0	12,881	3.6	23,322	5.5	45,000	6.5
Timber	21,228	7.5	22,223	6.4	23,571	5.5	27,000	3.9
Textiles	5,991	2.0	7,195	2.1	8,648	2.0	14,000	2.0
Total	283,078	100.0	348,422	100.0	420,495	100.0	692,000	100.0

Source: Commonwealth Statistician.
(a) Estimated by Department of Trade and Industry.

Packaging gives preservation, retains the flavour and the nutritious value, protects from dirt and disease and allows a quality standard to be maintained.

(ii) *Safety*

As well as protecting the contents from the environment, packaging also protects the environment from the contents. Typical examples are pharmaceuticals and drugs, poisons and other hazardous products, particularly used in the household.

(iii) *Convenience*

The characteristics that are designed into containers render them suitable as a form of packaging for the manufacturer, distribution, retailer and ultimate user. For example, cans are designed to withstand the thermal processing to which their contents are subject during manufacture. Bottles are designed to stand the internal pressures that arise from carbonation. Flexible packages can be designed to be resistant to light, air and moisture. Packaging permits articles and foodstuffs to be purchased in convenient usable portions so that there is no fear of deterioration or waste, if all the contents are not consumed initially.

(iv) *Distribution*

It is essential that products harvested in one season, or manufactured articles from one production run, should be conserved at the one time. Packaging makes this possible and facilitates transportation so that consumers all over the world can enjoy commodities indigenous to certain countries and geographical areas often thousands of miles distant from their own homes.

Packaging is designed to protect its contents during distribution between the primary producer and the consumer. Modern packaging techniques facilitate multi packing, sorting, counting and final dissemination.

(v) *Time Saving*

Modern packaging techniques permit a quicker sub-division of the product on a mass production line than at the point of sale or consumption. Meats in a supermarket are typical examples and, for this reason, are prepacked in consumable quantities. This also saves the shopper considerable time and does give him or her some choice of product.

Packaged foods which have been processed such as meat, fish, fruit, and puddings may be eaten cold without further preparation whereas foods like soups, vegetables and some meats normally eaten hot, require only a short period of heating. Housewives particularly are seeking more leisure time and the convenience of shorter food preparation time, which packaging allows, has become an important factor.

(vi) *Less Wastage*

It is more efficient to handle solid waste disposal in bulk at the point of manufacture, than at the point of consumption. Semi-prepared foods, such as frozen fish or vegetables or orange juice, create less domestic garbage than the original unprepared foodstuffs. In most cases, therefore, the entire contents of the package are consumable and putrescible waste is reduced to a minimum.

(vii) *Price Stability*

Packaging allows a mass purchase by manufacturers of seasonal foodstuffs at the time when they are cheapest and then canning or freezing them. This technique has contributed to the holding down of prices and has no doubt been a factor which has led to an actual reduction in the percentage of disposable income which the average family spends on foodstuffs. In 1961-62, the average family spent 23.8% of its income on food, while in 1971-72 this figure had dropped to 19.6%, and it is forecast that by 1976-77, the figure will be 17%. (The Commonwealth Bureau of Census and Statistics, Items 7.1 and 7.5.)

(viii) *Tamper and Pilfer-proof Aspects*

Perhaps one of the most important aspects of food and beverage packaging is their tamper-proof characteristics. This aspect is even now receiving further consideration in Victoria where all containers by regulation may soon have this mandatory requirement.

Industry is sometimes accused of the over packaging of small items, in particular. This apparently excessive packaging is one of the major means which a modern self service store uses to reduce stealing.

In Australia we currently have a paucity of statistics on the percentage of packaging which appears in our solid waste. It is estimated that it is between 13% to 15%, but this will, it is assumed, vary between urban, suburban, and country areas. The percentage of materials from packaging within this 13 or so per cent, will also vary but basically these are understood to be as shown in Table 2.

While packaging has increased in recent years, it has been off-set in our solid waste system by a reduction in putrescibles. The fish heads, offal, peels, stalks and bones seldom enter the market place now, let alone our homes. They are more often processed at their point of origin for the betterment of the environment.

It is not the role of the packaging industry to produce or encourage deceptive or excessive packaging. Where this has apparently occurred, there is usually a good reason for it, such as, to prevent pilfering, or as a result of settling after filling. The industry is competitive and wishes to limit costs and minimise the use of raw materials.

Litter, as distinct from packaging, receives more than its share of publicity, particularly from the ecologist, the media and the politician. It is that visible and highly decorative part of the solid waste stream which is misplaced or uncontrolled. While aesthetically offensive and often dangerous to health and human well being, litter is only a very small portion (approximately 1%) of the solid waste which society generates.

Why P.I.E.C.

Late in 1971, five main companies in Australia decided to form the Packaging Industry Environment Council. They did this because they were aware that:—

- packaging does contribute to the solid waste stream;
- the industry had few facts about itself;
- there was no prepared case to present on its economic importance to the nation;
- there was no plan to co-operate with governments to reduce the environment problems;
- and there was no cohesive body to represent the industry's important role.

This body was formally established on 8th May, 1972, with the appointment of a full time Executive Director. The Council has since initiated action to become incorporated in the A.C.T.

Table 2

PER CENT MUNICIPAL WASTE BY MATERIAL

	Australia	U.S.A.
Paper (all types) and Putrescibles	62%	68%
Brush and Garden Waste	10%	10%
Rags, Plastics, Rubber, Oils, Paints, etc.	4%	6%
Bottles, Glass and Ceramics	10%	8.5%
Metals	7%	7.5%
Ashes and Dust	7%	

Source: U.S.A. — U.S. Department of Health, Education and Welfare.
Australia—Municipality of Waverley, N.S.W.

The Council seeks to establish the implications of packaging on the environment and to ensure that the overall benefits of packaging are evaluated whenever discussions on the environment are taking place.

It is the Council's intention to be active in the prevention and control of litter; to develop expertise in the art of solid waste management including resource recovery; to demonstrate to the community the responsible attitude of the packaging industry as a whole to the environment; to promulgate information relating to solid waste management; to acquire knowledge of proposed or enacted overseas legislation related to the packaging industry and environmental issues; to be aware of proposed Australian legislation, and, where necessary, to represent the industry in offering advice and co-operation to governments, and to create more public awareness of the vital role packaging plays in our highly urbanised society in the distribution and consumption of goods.

The current membership of the Council is 23, and this represents manufacturers of the raw materials, converters and fillers. It is hoped to include retailing interests in the near future.

A current list of membership includes the following:—

A.C.I. LIMITED
ALCAN AUSTRALIA LIMITED
ALCOA OF AUSTRALIA
AMATIL
A.P.M. LIMITED
AUSTRALIAN COUNCIL OF SOFT DRINK MANUFACTURERS
B.H.P. COMPANY LIMITED
CANMAKERS INSTITUTE OF AUSTRALIA
COMALCO LIMITED
CONTAINERS LIMITED
J. GADSDEN AUSTRALIA LTD.
GLASS CONTAINERS LIMITED
W. R. GRACE AUSTRALIA LTD.
NATIONAL CAN COMPANY
NATIONAL PACKAGING ASSOCIATION
PLASTICS INSTITUTE OF AUSTRALIA
RHEEM AUSTRALIA PTY. LTD.
SANDS & MACDOUGALL
TETRA PAK (AUST.) PTY. LTD.
UNITED PACKAGES
HYGIENIC-LILY LIMITED

Since its formation P.I.E.C. has been involved in a number of discussions with Governments, and in particular New South Wales, Tasmania, South Australia, and the Federal House of Representatives Committee on the Environment. In each of these cases, the question of litter has been seen to be the main environment issue. However, while we shall continue to contribute where we can on this issue with governments, we see our prime objective in the future being conservation of resources through better solid waste management.

Packaging in Waste

While it is fairly reliably predicted that some 13% to 15% of solid waste in Australia is derived from packaging and while a percentage of the various components are known in specific areas (See Table 2), no definitive figures are available for the whole of Australia.

If we are to manage properly our solid waste, and if we are to recover from it resources which can be recycled or reused, it is necessary that accurate information on the composition of the raw material be available before an economic and practical method of solid waste management can be decided. P.I.E.C. proposes that such a study, to obtain this information, be carried out under its aegis, as a priority task.

It is of course apparent that the composition of solid waste includes some 85% of resources, which are derived from other than the packaging stream. An accurate survey of their composition is also necessary before the heat and energy value, the residual volume, and the overall value of recoverable materials can be determined.

Stated another way, the Environmental Protection Agency in the United States of America found that:—

"52.7 per cent of the solid waste consisted of volatile matter, 7.3 per cent was fixed carbon, 20 per cent was ash and metals, and the other 20 per cent was moisture. The combustible portion of the waste (volatile matter and fixed carbon) which made up 60.0 per cent of the total weight had a heat content of 8766 Btu per pound. This would mean that the overall heat content of a typical pound of refuse, which would include the moisture, the ash and the metal fractions as well as the combustible material, would be 5260 Btu per pound or slightly over 10.5 million Btu per ton. This heat content is most nearly comparable to that of low rank lignites which have a heat content of slightly less than 12 million Btu per ton. The refuse, however, has a lower moisture level (20 per cent) compared with the average of 50 per cent found in a low-rank lignite."

Many considerations must be examined before a decision on the economics of costs, energy and environment can show the most advantageous way to treat our solid waste stream, to meet each particular situation. This is one of the areas where P.I.E.C., working with industry and government, sees its future role.

Almost in parallel with this step to determine the composition of the raw material, must be a critical examination of the various methods of solid waste management from sanitary landfill through compositing, incineration and pyrolysis. Such an examination must ascertain such costs as transportation, capital amortisation, and operating costs of the various schemes. Only then can the value of the raw materials, costs of treatment, and the value of resulting products be assessed to determine which is the most suitable scheme for a particular area, with regard to both recycling and environment costs. Even then, it must be decided—and this is the difficult decision—what weighting would be given to each of these cost figures.

The Solution

P.I.E.C. considers that until the answers to the questions I have postulated have been provided, we are in no position to make firm decisions on the most suitable method of solid waste management. We must first obtain the answers on a national and co-ordinated basis, and not leave it to local governments to do their best with the limited resources they have. In this field, P.I.E.C. is prepared to play its part and would welcome the co-operation of other industries from which solid waste orgininates, together with cooperative governments, be they local, state or federal.

PAPER 29

Waste Management in the Lead-Zinc Industry

By B. G. Ratter*

Introduction

The metals industries generate enormous amounts of waste material; due partly to the large scale of the operations, partly to the amount of material that often has to be handled to gain valuable metals, and partly to the variety of products obtained. Because of this the pollution control problems cover a wide range.

A major problem of the lead-zinc industry is the toxicity of lead itself. Lead accumulates in the body and damages the central nervous system, including the brain, and it has been suggested by some workers that the lead blood levels of our urban population is rising, and in some locations exceeds the danger level. While not yet proven, this is a matter of some concern.[1, 2]

The problem of sulphur dioxide in lead-zinc processing, although not considered in this paper, is also a major one. Discussion of the sulphur dioxide problem and acceptable standards are available in various references (see for example Smith[3]).

Brief Description of the Production of Lead and Zinc

Three processes will be considered as they account for nearly all of the lead and zinc production.

(i) **The lead blast furnace:** Lead ore, coke and a flux are charged into a furnace; air is blown through the charge and the lead oxide is reduced to lead, which is tapped periodically. The flux combines with the remaining material to form a slag which is also tapped periodically; the lead then goes through a number of refining processes to remove copper, silver, gold, cadmium, iron and zinc. The slag varies considerably from place to place but is mainly composed of varying amounts of silica, iron oxide, calcium oxide, and small residual amounts of lead and zinc.

(ii) **Zinc production by the electrolytic process:** The zinc sulphide ore is roasted to produce zinc oxide, giving off large quantities of sulphur dioxide. The oxide is then dissolved in sulphuric acid to produce a zinc sulphate solution, which, after purification, is used as the electrolyte in a cell with a zinc cathode and an inert anode. Zinc is deposited at the cathode and oxygen is given off at the anode, leaving sulphuric acid in the bath. The latter is bled-off and reused as more zinc sulphate solution is added.

(iii) **A combined process, using the Imperial Smelting Furnace.** A combined lead-zinc ore is fed into a blast furnace which operates at a temperature above the boiling point of zinc, but below that of lead. Lead metal collects at the bottom and is tapped off as usual. The zinc vapour comes off at the top of the furnace where it is shock-cooled to a liquid by molten lead droplets. The resulting alloy is cooled so that it separates into two phases — a lead-rich phase which is recirculated to cool more zinc, and a zinc-rich phase which is further refined, often electrolytically.

The pollution problem from these processes is fourfold:

(a) Burning sulphides inevitably produce large quantities of sulphur dioxide, which can be used to produce sulphuric acid.

(b) Large amounts of water are used for cooling purposes, and for washing or cleaning. This water is used in two ways, either as indirect cooling, where it does not actually touch the products and so remains clean, or as direct cooling or washing where the water becomes heavily polluted and has to be treated before discharge.

(c) Blast furnaces are notorious for the production of dusts — a serious problem in the lead industry — which must be controlled for health reasons.

(d) Slag disposal, probably the most difficult of all the problems, because of its varied composition.

In what follows, the last three pollution problems will be considered.

---

* Student in the M.App.Sc. Course in Environmental Pollution Control, School of Chem. Eng., University of New South Wales.

(i) *Water Pollution Control*

Water is used in several places in lead and zinc extraction, including the blast furnace, cooling of the sinter, and in the acid plant if one is connected to the works. This water can contain many pollutants including lead, cadmium, arsenic and cyanide, so that treatment before discharge is essential. However, the treatment of waste waters is fairly standardised for different types of industrial plants. Hence, as a typical system, the treatment plant at the No. 4 Imperial Smelting Furnace complex at Avonmouth, U.K., is described.[7]

In this treatment, water, usually in the form of a sludge, is treated from three places, the blast furnace, the sinter plant and the acid plant. Sludge, from the acid plant, is neutralised by milk of lime, and is centrifuged into two streams. The 70% solids slurry goes to the cooling drum of the sinter plant and the decanted water — to a reactivator. Sludge from the blast furnace is thickened, further treated and sent to the cooling drum. The sinter plant sludge goes straight to the reactivator, is thickened, and some water is discarded, the rest being reused. The circuit is shown in the diagram, Fig. 1.

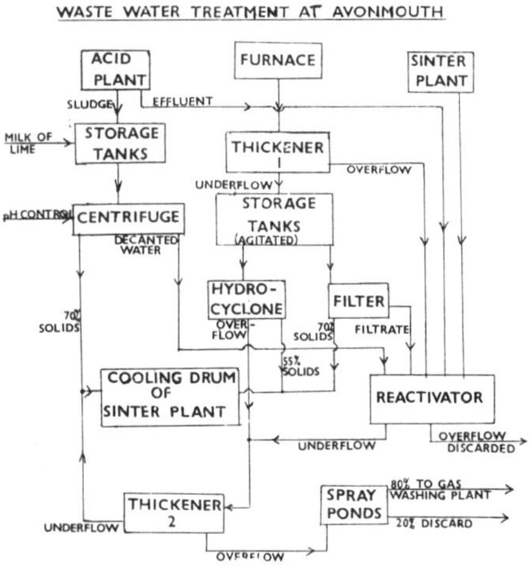

Fig. 1

The final discharge has the following analysis:

Total suspended solids	not	more	than	40mgm/l
Zn	,,	,,	,,	5mgm/l
Cd	,,	,,	,,	7mgm/l
Pb	,,	,,	,,	2mgm/l
As	,,	,,	,,	7mgm/l
CN	,,	,,	,,	1mgm/l

While not particularly spectacular, this treatment may be sufficient for most cases.

(ii) *Dust Control*

This is not a problem specific to the lead-zinc industry, but is highly important because of the toxic nature of lead. Here again the control techniques are fairly standardised. The Hurculean Lead Smelter at St. Joe Minerals in U.S.A. can be used as a model.[8] Dust collection hoods collect the dust by vacuum from over the sinter machine, furnaces and other equipment. The dust is then collected in a baghouse and reused. The storage areas are adequately ventilated to prevent build-up of hazardous dust. Apart from the dusts collected in the manner mentioned above, there is also a lot of particulate matter in the exit gases. This obviously has to be cleaned before discharge to the atmosphere.

(iii) *Utilisation and Disposal of Slag Residues*

This is probably the most difficult problem as each slag is different and so has to have an individual solution. It is only fairly recently that the idea of environmentally-planned residue utilisation and disposal has gained any ground. Previously, and still to a large extent, residues from metallurgical industries have been dumped into huge piles. For instance, the dump at Port Pirie in South Australia contains some six million tons of slag.[9] Similarly, at Risdon in Hobart the Electrolytic Zinc Company have been stockpiling slags from the refinery since 1940.[10] These are by no means unusual and, if anything, are minor compared to the dumps from similar operations overseas. These dumps are environmentally detrimental for several reasons:

(a) They are usually open so any dust associated with them will be blown around causing dust pollution.

(b) Rain falling on them can leach out certain substances, including toxic ones, which can be carried into water supplies.

(c) They are unsightly to look at and take up valuable land, usually poisoning the land underneath and directly around them.

(d) There is a danger of collapse.

(e) They frequently contain valuable components.

For these reasons it is important that the residue is treated in an environmentally consistent manner, and many processes have been, and are being, devised to do this. A few of these are discussed below as general examples.

The Balen plant of S.A. Vieille-Montagne at Balen in Belgium[11] has a very complex slag treatment. The plant produces lead by smelting and zinc by the electrolytic process. Slags from the lead are reduced to give copper-silver mattes, iron slags, and zinc and lead oxides. These are all used in some way, giving valuable by-products, such as germanium for use in the semi-conductor industry. The final waste is disposed of as a liquid containing As, Sb, In, Ga, and F. This plant then obtains all possible valuable components from the residue before discard, thus lessening considerably the amount of waste which has to be disposed of.

In Russia, Tosheva et al[12] have successfully used slags from a lead manufacturing plant as a source of raw material in the production of Portland cement clinker. The slag analysis was:

FeO 38-47%	Al_2O_3 4-6%	S 2%
SiO_2 22-26%	MgO up to 2.5%	MnO 1%
CaO 11%	Zn 2.5-3.5%	Na_2O 1%

and small amounts of Pb, Cu, As, and K_2O.

Up to 20% of this slag was added to the normal raw materials with satisfactory results.

Abdeev et al[13] have been studying the use of Altai Zinc slags (Russian) for extraction of valuable components. A two-stage system was tried in both laboratory and experimental industrial tests. These showed a high degree of copper extraction, recovery of zinc and lead in the slag, and extraction of iron for use as cast iron. The final slag from this process was used for slag pumice. This method results in a total slag usage, and shows good promise of being economic.

One of the most successful processes for the recovery of valuable constituents is the slag fuming process, such as applied at the Cominco Smelter in Trail, British Columbia.[14] The process involves mixing finely-divided slag with coal and air to maintain strongly-reducing conditions. The volatile metals, such as zinc and lead, are "fumed" off at the high temperatures involved. These are carried off in the gases, from which they can be re-oxidised, collected, and returned to the process. The Trail plant is particularly interesting as it is set in a beautiful natural environment, and so was designed to keep pollution to a minimum. The remaining slag, after all valuable metals have been removed, is returned to the mine as filling.

The slag fuming process is gaining widespread acceptance. The lead plant of the Broken Hill Associated Smelters[15] at Port Pirie is currently treating its blast furnace slag for recovery of zinc by slag fuming. The company has a stockpile of six million tons of slag, and this is being slowly reused, together with the slag currently being produced.

The Waelz process for treating leach residues is being used at the Nisso Smelting Company's works at Aizu in Japan.[16] The process is a combination of slag fuming and a complex processing operation. Leach residues are first sent to a slag fuming furnace where the volatile metals — zinc, lead and cadmium — are fumed off and recovered for further processing. Nearly all of these three metals are removed in this process, being in part replaced by carbon. This slag is then crushed and magnetically separated to remove iron. The remainder is sent to a furnace where a copper matte (also containing silver and gold) is tapped off and the metals recovered. The remaining slag is water granulated and sold.

The important thing to notice from these processes is that the residue can be treated to obtain valuable products which can be sold economically. Frequently, a company is making considerable profit from what was once considered waste. Clearly situations like the one at Port Pirie, where slag was just dumped, even though it contained considerable percentages of zinc, should not be tolerated.

It is not always possible to use all products from a large, complex manufacturing operation; there will always be a certain amount of waste to discard. After processing, the remaining residue often contains high percentages of SiO_2, CaO, Al_2O_3, and Fe oxides, which are all non-toxic and fairly inert, and so they can be disposed of easily in an environmentally consistent manner, without producing large dumps. Such material, being inert, can be used as covering material in landfill operations, and as fill in mines. Some people are experimenting with slags of high CaO content and traces of phosphorous for use as fertilizers when combined with a nitrogenous fertilizer, particularly in acid soils.[17]

What is often needed is a little imagination and a great deal of work in order to come up with acceptable methods for disposing of many of the unwanted materials of the metal processing industries.

REFERENCES

1. Mills, A. L. "Lead in the Environment." Chem. Brit. *7* (4), p. 160 (1971).
2. Semrau, K. T. "Sulfur Oxide control and metallurgical technology." J. Metals *23* (3), p. 41 (1971).

3. Smith, R. G. "Air Quality Standards for Pb." Air Quality Monogram, 1969, No. 69-11.
4. U.S. Dept. of Health, Education and Welfare. "Air Quality Criteria for Sulphur Oxides." Public Health Service Publication No. 1619, Washington D.C. Govt. Printer 1969.
5. Bailleul, M. R., et al. "Sulfur and Sulfur Oxide Developments 1971." Am. Inst. Chem. Eng., CEP Tech Manual, New York, 1971.
6. Ross, F. F. Combustion. *43* (3) p. 6.
7. AIME. "World Symposium on Mining and Metallurgy of Lead and Zinc." St. Louis 1970, Volume II Extractive, Chapter 21.
8. Ibid. Chapter 24.
9. Butcher, A. J. "Recovery of zinc from Pb blast furnace slag at Port Pirie." Mech. Chem. Eng. Trans. *4* (2), p. 187 (1968).
10. AIME. Op Cit, chapter 16.
11. Ibid, chapter 6.
12. Tosheva, D., et al. "Use of fuming of lead slag in production of Portland cement clinker." Stroit. Mater. Silikat. Prom. *10* (12), p. 34 (1970).
13. Abdeev et al. "Complex Use of Altai Zinc Slags." Sb. Nauch. Tr., Vses Nauch-Issled. Gomonet Inst. Tseret. Metal. (1967) No. 10, p. 142.
14. AIME. Op Cit, chapter 12.
15. Ibid, chapter 27.
16. Ibid, chapter 15.
17. Popova, O. A., et al. "Use of waste waters from the metallurgical industry as agricultural fertilizer." Sb. Stud. Nauch. Rab., Mosk. Seb' Shokhoz Abad (1968), No. 16, p. 212.

PAPER 30

Emplacement, Disposal and Utilisation of Colliery Waste

By C. T. McElroy*

1. Introduction

The disposal of colliery waste is a world wide problem of increasing magnitude. The present report deals with the source, quantity and nature of the waste and the methods of disposal, emplacement or utilisation of this material.

The management and technology of colliery waste disposal have been subject to detailed study in recent years, especially since 1966 when the collapse of a spoil heap at Aberfan in Wales resulted in the death of 144 people, including 128 children.

No completely satisfactory solution to all the problems of colliery waste disposal is to hand, but it seems that emplacement as planned environment-compatible landfill offers the best solution in the light of present technology and conditions generally.

2. Source of Waste

Colliery waste includes coal and stone, timber and metal scrap, rejected brattice and belting, sludges, human waste, water and a diverse assortment of disused drums, boxes, frameworks, machinery parts and other fabrications.

For many uses, coal must be beneficiated by washing before it is used. This is particularly important if the coal is to be used for iron and steel production or for export. The only significant use of unwashed coal is in power generation.

In the present context, we are concerned with that which overwhelmingly dominates the composition of colliery waste, that is, the stony coal and coaly stone, together with the fine fractions of these in sludges, which are produced either directly in the mining process or as discard from coal washeries. The simple term "stone" will be used here to describe all those variations of inferior coal, shale, sandstone and other rock types, including igneous material and inorganic mineral matter generally occurring in colliery waste.

In the working face of a coal seam may be seen some examples of the type of material constituting the main source of colliery waste. For example, these may be light-coloured bands such as occur in the Wongawilli Seam, Southern Coalfield, New South Wales, which consist of flint kaolin (or claystone) beds which are essentially continuous over a large part of the coalfield. Bands such as these, together with thin lenses and wisps, of various compositions, finely divided particles of stone and calcite, all dispersed throughout the coal, constitute the great mass of washery discard.

In some cases, irregularities, such as "stone rolls" in the floor, must be mined in order to maintain grades and working height. The term "pit-stone" is commonly applied to more or less chunky, handleable stone derived from roof and floor, thick bands or lenses or irregular bodies of stone which have been separated from the working section of the seam or igneous rock from dyke and sill with accompanying cindered coal and hybrid rock types.

3. Quantities of Waste

In the United Kingdom there are some 2000 spoil heaps owned by the National Coal Board, containing a total of 3000 million tons of waste covering almost 50 square miles. More than 50 million tons of discard from washeries is accumulating annually. The estimated cost of the production, handling and disposal of spoil may be up to £500 million annually (Round, 1968).

In France, colliery shale disposal is only a problem in the Nord and Pas-de-Calais area; here there are 200 tips totalling 500 million tons, to which is added 14 million tons annually. In Germany, colliery waste is accumulating at 65 million tons per annum (Gibson, 1970).

Australian coal washeries have been producing waste in significant quantities since 1953. Statistical data is given in the Annual Reports of the Joint Coal Board, with special detail provided for New South Wales. From a present total Australian production of approximately 60 million tons of black coal, approximately 38 million tons was produced in New South Wales. There were two washeries in New South

* Clifford McElroy and Associates Pty. Limited, Geological and Environmental Consultants.

Wales in 1953 and 37 in 1973. Over this period of 20 years the percentage of discard has risen from 10% to the present figure of around 24%. Washery input has increased from 1.1 million tons to 27.8 million tons. Thus, in New South Wales, the present rate of accumulation of washery refuse is in excess of 6 million tons per annum.

It should be noted that proportion of washery discard varies from seam to seam and from area to area. In 1972-1973, in New South Wales, proportion of rejects in coal industry washeries, ranged from 15.4% in the South Maitland Coalfield to 29.3% in the Newcastle Coalfield. The demand of the current market is the principal factor determining coal quality and this strongly influences the proportion of rejects. The other major influence is mechanisation. It is, at times, impossible for the continuous miner to stay accurately on the floor, so increasing the stone content of the seam mined. Selectivity has inevitably suffered with mechanisation. The increased rate of advance in the mechanised mine has lowered the standard of packing in gate roads in longwall mining and stone waste-packs to assist roof control at the face find little place in mechanised methods; increased contamination follows.

4. The Nature of the Waste

Although all existing spoil heaps are not composed of washery waste, this is nevertheless the most important component being added today and will be increasingly important in the future. In Australia, it is estimated that by the year 2000 A.D., with due allowance for unwashed coal for power generation, the projected production of 200 million tons per annum will yield 25 million tons of washery refuse (Edwards, 1972). It is on this basis that emphasis is given, in the following sections of this report, to washery waste; disposal problems are essentially similar, whatever the source of the waste.

Typically, the medium-sized reject (3 cm x ½ mm, with ash content up to 70%) dominates the reject fraction. Coarse reject (+3 cm with ash up to 90%) and fine reject (—½ mm with ash up to 35%) may together comprise, say, one quarter of the total reject.

5. Disposal of Waste

Disposal, which is not a really suitable term for the process, might best be considered from two viewpoints *emplacement* and *utilisation.*

5.1 *The Problems*

The following summarises the deleterious effects of the most commonly recognised methods of disposal of waste.

(a) *Safety*

The sliding of the Aberfan spoil heap, with resultant great loss of life, brought to light records of numerous other such slides.

(b) *Visual Pollution*

Some people regard the sight of a spoil heap as distasteful, even though this may be orderly and well controlled.

(c) *Air Pollution*

Burning spoil heaps constitute a very real element of air pollution in the limited areas where this applies. Dust from spoil heaps under high-wind conditions is also objectionable.

(d) *Water Pollution*

Uncontrolled discharge of waste, especially fine waste, into watercourses, lakes and the sea and accidental discharge by collapse of retaining banks of lagoons, can result in serious water pollution.

(e) *Land Pollution*

Spoil heaps or lagoons physically occupy land space which might be considered as fit for a better use. The bulk of a spoil heap also inhibits utilisation of the topographic element of which it is a part.

(f) *Ecological Impact*

The existing faunal and floral occupation of a site is virtually terminated when used as a disposal area.

In the following discussion of the methods of emplacement and utilisation of waste, reference will be made as appropriate to the above problems, particularly in relation to their solution.

5.2 *Emplacement of Waste*

The term "emplacement" is employed to cover the transfer of waste from the mine or washery to close by (on-site) or more distant locations; the prime objective here is to deposit the material elsewhere, rather than necessarily use it to some good purpose. The waste is not "disposed of" in the permanent sense; perhaps disposal at sea, unacceptable as this is, or underground stowage, more closely approach the true spirit of the term. Although there is an obvious overlap between emplacement and utilisation, for convenience the aspects considered in section 5.3 Utilisation, imply some form of physical treatment of the waste.

(a) *Spoil Heaps*

The term "spoil" is synonymous with waste, refuse, reject, discard or chitter; "heap" is synonymous with tip, dump, bank or pile.

In the United Kingdom, legislation and codes have been laid down in great detail in a number of publications in recent years (National Coal Board, 1968, 1970, 1971). This was consequent upon the Aberfan tragedy, which was judged to be due to "ignorance on the part of those charged at all levels with the siting, control and daily management of tips; bungling ineptitude on the part of those who had the duty of supervising and directing them; and failure on the part of those having knowledge municate that knowledge and to see that it was of the factors which affect tip safety to comapplied". (Aberfan Report, 1967.)

Irrespective of the justice of this judgement, the physical condition precipitating the slide of the Aberfan tip was a build-up of water in the combined tailings and coarse waste comprising the tip. The point was reached where virtual liquefaction occurred and a large section rapidly flowed downslope to the village below, burying the school house and several homes. The tip was 111 feet high, containing 297,000 cubic yards of refuse including 30,000 cubic yards of tailings. The basic physical factor contributing to the conditions was the geological situation of the site. The tip was located on glacial drift overlying the Brithdir Sandstone. This sandstone is an aquifer, which discharged into the drift and hence into the base of the tip.

The codes of the National Coal Board provide for—

(i) Safety measures in terms of stability analysis including reference to shear strength, most critical cross-section and influence of geology and groundwater.
(ii) Heights of heaps.
(iii) Slope of ground and angle of repose.
(iv) Surface drainage around and on the heap; relief wells where appropriate; filter drains.
(v) Clear space around heaps.
(vi) Construction and maintenance of terraces on heaps.
(vii) Compacted banks at toe of heaps on slopes.

The foregoing are all quantified with drawing details, dimensions and proportions spelled out.

Analogies to the Aberfan tip were recognised in the setting of the spoil heap at Coalcliff, New South Wales, where the village is sited below the mine and its accumulated waste.

A detailed investigation (Longworth, 1968, 1969) resulted in the setting up of adequate safeguards by hydraulic monitoring in bores, thorough drainage below and around the spoil heap, terracing and establishment of kikuyu grass on the terraces.

Great publicity has attended the proposal to emplace the spoil from Coalcliff in a small valley in the plateau surface above the mine. Turnbull (1972) has conducted an Environmental Impact Study into the proposal and concluded that such a site, subject to stringent conditions of maintenance and management generally, would result in least impairment of the quality of the environment. Turnbull's findings have recently been supported by an exhaustive Public Inquiry conducted by the State Pollution Control Commission.

Campbell (1971) has considered the special problems of burning spoil heaps. For example, in 1964 Pennsylvania had 117 burning heaps, some of which had been burning for more than 20 years. Some of these heaps are up to 300 feet high and hundreds of acres in area.

Air pollution results from slow combustion of coal refuse and produces CO_2 and toxic, odorous and irritant gases—H_2S, SO_2, etc. Discomfort to persons follows, especially those with respiratory ailments. Paintwork may be blackened and silverware discoloured. Additionally, fogs are induced near smoking dumps.

A residual problem is that after relinquishing the lease, fires in dumps are not the legal responsibility of the mining company.

(b) *Landfill*

Reclamation of lowlands or irregular lands by coarser spoil landfill is a relatively major, but limited, mode of emplacement. Included are—

(i) The establishment of industrial building sites, parking lots and recreation areas including playing fields.
(ii) Filling of extracted areas, such as open-cut coal mines and industrial sand deposits.
(iii) Sanitary landfill operations, such as those in progress at Port Kembla, where municipial refuse is being covered by spoil.
(iv) Embankments in road and rail constructions and, to a less extent, road base.
(vi) Agricultural use, of limited application to date, merits special mention.

Nicholson (1970) has recorded the filling to a depth of 7 or 8 feet of a marsh in Lanarkshire with minus 2 inch washery refuse of low coal content. One month after levelling, fertiliser was added and corn sowed direct onto the waste. No soil was added. The crop was harvested 6 months later; 5 years later the area was indistinguishable from good agricultural land.

In the Wollongong district of New South Wales, eucalypts have been planted in spoil heaps with a surrounding soil ball, and a 6 inch surface layer of soil. These trees are growing well after about 12 months, but have not yet sustained a severe climatic test. In a prolonged period of low rainfall, it is possible that available moisture in the spoil may be insufficient to meet the needs of large trees. Swamp oaks have been similarly planted in low-lying areas in washery refuse, and these are also thriving to date.

No research appears to have been done on the relative nutrient status of the various spoil materials, which may differ depending upon source of the refuse. It would seem that the establishing trial plots of trees in spoil of known nutrient values is a logical step in rehabilitation research. Reference has already been made to the successful use of grasses in the stabilisation of heaps.

(c) *Underground Stowage*

An apparently obvious solution to waste disposal is to return it or leave it underground. The practice of underground stowage has been extensive in the past in thin, non-mechanised seams. Additionally, packwalls of waste, bridged from roof to floor, are used to hold up roadways through goaf. In Germany, France, India and the United Kingdom it is still used to varying degrees. It has not been practised in Australia, except for an unsuccessful experiment in the Cessnock area in the early 1950's, when sand was hydraulically pumped into extracted areas.

Underground stowage of colliery waste is technically feasible, although a costly operation in itself. However, the cost of stowage is only one of many factors involved; decrease in production rate to accommodate the stowage activity must necessarily follow; the airborne dust hazard is increased; clogging of stowing pipes conveying tailings has proved to be a great impediment; the underground stowage of tailings is considered technically infeasible.

In the United Kingdom, underground stowage is still employed in exceptional cases to improve the conditions of faces and underground roadways and to minimise surface subsidence where necessary.

In Germany, stowage problems have decreased the application of the method. In 1958, 50% of all coal was obtained in conjunction with solid stowage; in 1967 this was 16%. It has been concluded that underground stowage is not economically practicable under present conditions. (Aberfan Report, 1967.)

Round (1968) estimated that the total cost of underground stowage in the United Kingdom would add between 17.7 and 23.4 pence per ton of face produced coal.

Stowage was considered at Coalcliff, where it has been estimated that the cost of equipment and installation for crushing, conveying, storing and stowing underground would involve an initial outlay of $3.5 million (Turnbull, 1972).

(d) *Tailings Dams* (Lagoons)

The drainage and dewatering of the fine sizes produced in coal washeries poses special difficulties compared with the coarser sizes. This especially applies to the "slimes"—fine clays produced in the flotation process.

Disposal normally involves expenditure without return and it is probably fair to say that increasing public awareness of pollution problems and other hazards, with resulting legislation, were needed to speed up the development and application of more efficient methods of disposal.

Waste fines may be coal plus shale plus clay, etc., from plants where fine sizes are not cleaned or material, predominantly other than coal, from flotation cells.

The publications of the National Coal Board (op. cit.) give special attention to the disposal of fines. In Australia, recent comprehensive reviews are given by Sanders (1972) and Edwards (1972), summarised in part below.

Emplacement may be by the following methods:—

1. Direct disposal, without thickening, may be into watercourses, lagoons or conceivably piped out to sea, as has been done in England. Such practices are no longer acceptable. Disused mine shafts have also been used.
2. Direct disposal, after thickening by appropriate flocculants, may be into tanks or lagoons, or several of these. Clarified, supernatant water is returned to the plant. The fines dry by further drainage and evaporation and are either emptied by mechanical shovels, to be mixed with coarse refuse in a dry state, or are mixed and covered in situ.

   In Queensland, open cuts provide convenient lagoon sites.

   Lagoons are unsightly if in a settled environment. There are records of tailings dams bursting in England and flooding villages.

3. Disposal after dewatering:
   (i) Fine aperture screen decks offer a partial solution only, as the finest particles tend to accumulate in the circuit.
   (ii) Cyclones are commonly used to concentrate waste solids. Again the ultrafines tend to remain in circuit.
   (iii) Complete filtration, either vacuum filtration or pressure filtration, has been successfully employed overseas. It seems probable that this will be more widely used in Australia in the near future.
   (iv) Solid bowl centrifuges with inbuilt flocculation have proved rather costly and only partly efficient, with some fines remaining in the centrate water. There are used in the U.S.A., but only to a very limited degree in Europe.
   (v) Deep cone thickeners have proved successful in England and Germany. Powerful flocculants are employed. Although installation is expensive, operating cost is virtually only the cost of the reagent. Moisture content remains higher than in other methods, but handleability is quite good due to strong physical bonding engendered by the flocculants employed.
   (vi) Various combinations of methods of dewatering may be employed to meet economic demands. More research is needed to find a low cost, efficient method of dewatering.

The following table indicates the range of moisture of dewatered product in (iii) to (v) above.

	Moisture, %
Vacuum filter	25-35
Pressure filter	15-25
Solid bowl centrifuge	40-60
Deep cone thickener	50-60

Water pollution might occur if sulphur compounds which are leached out of the fines penetrate porous dam walls. Sulphur may be acted upon by aerobic bacteria, producing H_2S. Again, if walls are breached, natural streams will be polluted.

An ideal situation for a tailings dam has been employed at Avon Colliery, south of Wollongong, N.S.W. Here the tailings are dewatered, after thickening, by filtering through beds of coarse refuse into a lower dam. The clean water so derived is recirculated into the washery. No water normally escapes into natural watercourses, but in any case, no pollution would ensue.

In the United Kingdom, regulations provide for safety factors, clearing of areas and drainage requirements in tailings dams, as for spoil heaps. All lagoons must be constructed below ground level, except where coarse spoil is used as a base filter. The base topography must have a slope of less than 1 in 12. The top water level must at no point be higher than 18 metres above the natural surface *at the lowest point* of the toe of the bank, unless the cross section of the bank is deliberately greater than safety requires. In certain conditions, the top water level must be much less. There must also be at least 1 metre of freeboard all round.

Full details of slopes, batters and cross sections are spelled out. Terracing, grassing and other stabilising methods are an integral part of lagoon management.

The best solution lies in the hands of the companies concerned to study the processes and do a good job for safety and anti-pollution.

5.3. *Utilisation*

Treatment of colliery waste prior to utilisation is normally restricted to a heat process. A comprehensive summary has been compiled by Gibson (1970). Recent research by Hanquez et al. (1973), Sztaba et al. (1973) and Toubeau (1973) emphasises the growing importance of utilisation studies.

It is anticipated that increasing use will be made of the products of waste material subjected to the fluidised bed combustion process. Within the Division of Mineral Chemistry, C.S.I.R.O., Sydney, research into fluidised bed combustion of colliery waste is actively proceeding. A pilot plant with a one square foot reactor bed is testing a wide range of composition and sizes of local wastes, including tailings, to produce suitable usable material.

Output of the plant is 65 lbs. of inerts for 100 lbs. of input waste. Temperature of operation is in the order of 800°C. After inital gas firing, the combustion is more than self sustaining in the air stream supplied. Cooling is effected by increasing the air supply or by water filled tubes immersed in the bed.

Fortuitous treatment of spoil is provided in burning heaps, referred to below as "burnt spoil".

(a) *Road Materials*

Burnt spoil has been used extensively in Britain for many years for road base and embankments. Tests on unburnt spoil indicate that this will be equally suitable. Adequate compaction is considered to be the main factor, composition being of secondary significance. Cement stabilisation, in which 7% to 10% of cement is added to selected spoils, has given promising results in sub-base. About 1 million tons per annum may be used in Britain for this purpose. Burnt spoil has also been used for skid-resistant surfaces.

(b) *Concrete Aggregates*

Inerts from fluidised bed and crushed burnt spoil appear suitable for use in aggregate. Sulphate is a problem in many unburnt spoils. Lightweight aggregates, from heat-expanded clays and shales, have so far not found very great application in the building industry, but future potential is good.

In Belgium, there is a potential market for lightweight aggregate of approximately 1 million tonnes. Toubeau (1973) has described a new process of lightweight aggregate manufacture. This involves drying, fine grinding, addition of 15% water and the extrusion of 6 mm diameter "spaghetti", which is chopped up into more or less equant shapes. A three-section kiln is employed in the firing of the material, each section with a specific heat range. The pellets so formed are almost spherical, with little moisture absorption and make a satisfactory concrete.

In France, a new four-kiln process has been described by Hanquez et al. (1973). This involves a specialised complex heating cycle for the production of pellets over a large size range between 5 mm and 25 mm without crushing, as in other processes. The vitrified outer layer thus remains intact and consistent mechanical properties are attained.

(c) *Brick Manufacture*

The potential for using colliery spoil in brick manufacture without treatment is limited due to the maximum of 5% carbon acceptable in brick clays. Alternatives are to blend, to pre-roast or to mix with lime, cement and other chemicals. Unfortunately, a grey product usually results. Inerts from fluidised bed combustion may also play a part in this field.

(d) *Sulphur Derivatives*

The use of pyrite from colliery waste has been studied by Sztaba et al. (1973). In Poland, power station coals mostly contain 1.5% to 2.5% total sulphur; 85% to 95% of this occurs as pyritic sulphur (pyrites up to 65%; marcasite up to 35%). Typical discards range from 11.60% yield at 54.4% ash to 17.27% yield at 73.6% ash. Sulphur in these discards ranges from 6.69% to 26.30%. Pyrites is concentrated by jigging, flotation, magnetic and electrical methods. Results to date indicate that pyrites so produced will find application in the smelting of Polish copper ores and in the production of high grade sulphuric acid.

(e) *Heat*

Both hot exhaust gases and heat exchangers immersed in the bed in the fluidised bed combustion process provide a ready source of heat for use in the colliery and the washery. Heat so derived has great potential for application in the drying out of fines, in conjunction with those methods discussed in Section 5.2(d) above. In the United Kingdom, some 5 million tons dry weight of solids is at present flocculated, for which the cost of the flocculants alone exceeds £½ million.

The possibility of using up-cost methane in gassy pits as an additive to the air input in fluidised bed combustion is also being considered.

Commercial units employing fluidised bed combustion as a source of heat are now operating in the U.S.A., and pilot studies in the United Kingdom indicate that the process will soon find commercial activity in that country (N. Y. Kirov, pers. comm.).

(f) *Other Uses*

A wide range of other uses for treated colliery waste, some in conjunction with fly-ash from coal-fired power stations, is under review. These include pozzolans, cheap plastics manufacture, expanded pellets for decoration of grassed areas and pavements, soil conditioners and other uses in agriculture, building boards and insulants (mineral wool). The occurrence of trace elements, e.g., germanium in waste heaps has also received attention.

6. Rehabilitation of Spoil Areas

Various aspects of rehabilitation of spoil disposal areas have been referred to above, particularly in Section 5.2 (a) and (b), where comments on the nutrient status of spoil are included.

It remains to be stressed that research into this aspect should be given substantial priority in spoil emplacement proposals. Of great importance in this field is the careful removal of topsoil and subsoil, the method and period of stockpiling and the later replacement of the soil on the filled area prior to re-vegetation.

7. Conclusions and Recommendations

The most practicable immediate solution to colliery waste disposal lies in the existing trend towards a more sophisticated treatment of large scale landfill projects, with due regard to safety and environmental considerations. A build-up of properly executed examples, appropriately rehabilitated, will attain greater acceptability of such emplacement and so avoid costly public objection.

Underground stowage of waste under Australian conditions of coal seam occurrence and mining practice is uneconomic and impracticable in the light of present technology and no immediate solution is foreseeable.

For the future, research into the further utilisation of spoil, especially as products of fluidised bed combustion in the form of industrial structural material and as a heat source, gives promise of significant inroads into the rate of accumulation of spoil.

8. Acknowledgements

Acknowledgement is made of most helpful discussions with many friends in the mining industry, and associated organisations, in particular Mr. K. G. Mosher and Mr. R. Marshall of The Coalcliff Collieries Pty. Ltd.; Dr. R. A. Durie and Dr. P. Waters of C.S.I.R.O.; Mr. F. J. Gardner and Mr. D. Rolinson of Austen & Butta Limited; Mr. J. Muir, Deputy Chief Inspector of Coal Mines, Department of Mines; Mr. W. E. Foskett, Geological Survey of New South Wales; Mr. C. Stanbury of A.C.I.R.L. Ltd.; Mr. J. Sanders of Cargo Superintendents (A/sia) Pty. Ltd., and Mr. R. Wilson, Australian Iron & Steel Pty. Ltd.

REFERENCES

Aberfan Report, 1967 — Report of the Tribunal appointed to inquire into the Disaster at Aberfan on October 21, 1966. London. H.M.S.O.

Calhoun, F. P., 1968 — Avoiding Pollution from Refuse Disposal. Mining Congress Journal.

Campbell, A. R., 1971 — Fires in Coal Mining Wastes. The 1971 Australian Waste Disposal Conference. Ed. N. Y. Kirov. University of New South Wales.

Davies, W. E., 1968 — Coal Waste Bank Stability. Mining Congress Journal.

Edwards, G. E., 1972 — Current Local Practice in the Treatment and Disposal of Water-Borne Colliery Wastes. Ed. R. T. Wheway. Illawarra Regional Committee of the Water Research Foundation of Australia.

Gibson, J., 1970 — An Appraisal of the Utilisation Potential of Colliery Waste. Mining and Minerals Engineering, November 1970.

Hanquez, E., Boutry, C., and Chauvin, R., 1973 — Colliery Shale as a Raw Material for Building Purposes. Preprints of Sixth International Conference on Coal Preparation. Paris, 1973.

Joint Coal Board, 1972 — 25th Annual Report, 1971-1972. Government Printer, New South Wales.

Longworth, C. R., 1968 — The Aberfan Disaster Findings and their bearing on Tip Conditions at Coalcliff Colliery, Coalcliff, New South Wales. Unpublished Report, Longworth & McKenzie. Consulting Engineers, Sydney.

Longworth, C. R., 1969 — Investigation of Mine Refuse Pile Stability at Coalcliff, New South Wales for the Coalcliff Collieries. Pty. Ltd. Unpublished Report, Longworth & McKenzie, Consulting Engineers, Sydney.

National Coal Board, 1968 — Spoil Tip Management. N.C.B. Production Department.

National Coal Board, 1970 —Spoil Heaps and Lagoons. N.C.B. Technical Handbook.

National Coal Board, 1971, — N.C.B. (Production) Codes and Rules: Tips: 1st Draft.

National Coal Board, 1971 — Management of Tips. Production Department Instruction. P1/1971/1.

Nicholson, S. R., 1970 — Utilisation of Colliery Waste in the Building and Construction Industries and in Agriculture. National Meeting of Minerals Engineering Society. September 1970.

Round, C., 1968 — Dirtless Mining. The Mining Engineer 1968.

Sanders, G. J., 1972 — A Review of Australian and Overseas Practice in the Treatment and Disposal of Water-Borne Colliery Wastes. Ed. R. T. Wheway. Illawarra Regional Committee of the Water Research Foundation of Australia.

Sztaba, K., Blaschke, W., Pudlo, W. and Pilch, W., 1973 — Possible Separation of Pyrites Concentrate from Polish Power Coal Discards. Preprints of Sixth International Conference on Coal Preparation. Paris, 1973.

Toubeau, G., 1973 — Manufacture of Lightweight Aggregates from Washery Screen Discard. Preprints of Sixth International Conference on Coal Preparation. Paris, 1973.

Turnbull, E. S., 1973 — An Environmental Impact Study of Coal Refuse Emplacement at Maddens Plains. Unpublished Report. Wollongong City Council.

PAPER 31

Salvage and Reuse of Textile Waste Material, with Special Reference to the Wool Textile Industry

By H. Pedersen*

Basically, there are two categories of textile wastes: (i) manufacturing waste, which is the waste from the various manufacturing processes, and (ii) recovered and reclaimed waste, which is waste collected after use and then reprocessed.

The wool textile industry can be divided into two sections, the worsted and the woollen industries. Long, fine fibres — the best quality raw material — are used in the worsted industry to produce yarns for weaving good quality clothing and other materials. The woollen industry (producing carpets, blankets, rugs and knitted materials) use lower quality raw material, coarser and shorter fibres and waste from the worsted industry. Because yarn strength is dependent on fibre length, worsted yarns are stronger and finer than woollen yarns.

In both proceses the objectives are removal of dirt, vegetable matter and short fibres and production of an array of parallel fibres which can be twisted into a yarn. Dirt and vegetable matter are mostly removed by wet processes. The processes for producing the parallel array of fibres are carding, combing and drawing, after which the product is spun and twisted into a yarn. To produce worsted yarn one must go through all these processes while to make a woollen yarn only a few are needed. The wet processes are common to both, as is the carding process whereby very short fibres and any remaining dirt and vegetable matter are removed. This waste material is swept up and collected. It is after carding that the two processes diverge.

In the worsted process the fibres are often washed after carding to improve the appearance of the "sliver", as the fibre array is called. The sliver is then combed, drawn, spun and twisted into yarn. During combing short fibres are removed. Their length is controlled by the settings on the comb which in turn is governed by the quality of yarn required. After combing and drawing of the fibre array, a "top" of parallel fibres of uniform mass per unit length is produced. The top is then spun and twisted to the required yarn on the spinning frame.

Woollen yarns do not go through all these steps, only carding and spinning. The sliver from the card goes straight to the spinning frame for spinning and twisting to give the yarn.

Fig. 1 and 2 show these two processes. It is indicated where in the processes waste is produced, and what is done to this waste. These are the manufacturing wastes which also include waste from garment manufacturers.

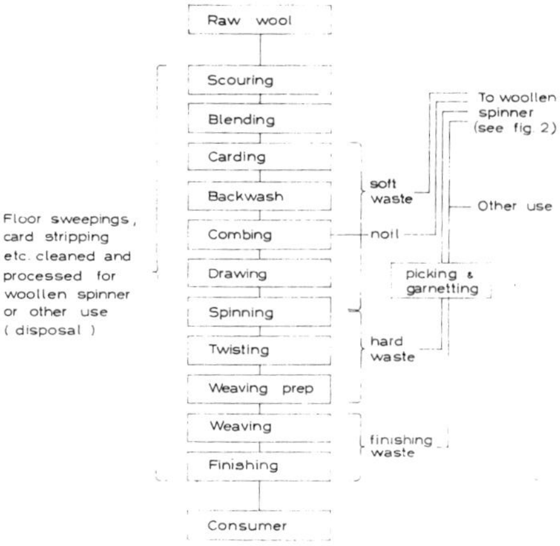

Fig. 1: Worsted Process

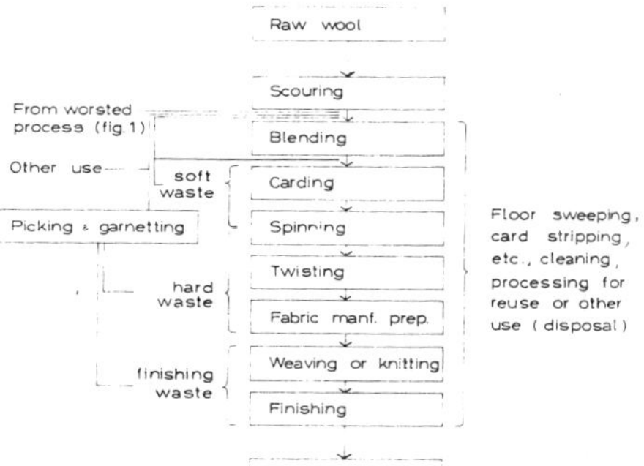

Fig. 2: Woollen Process

* Post-graduate Student, M.App.Sc., Environmental Pollution Control, University of New South Wales.

Manufacturing wastes are divided into four classes: noils, soft waste, hard waste, and finishing wastes. The short fibres removed during combing in the worsted process are called noils. Except for their short length the noils are not reduced in quality compared with virgin wool, and are therefore used as raw material for woollen yarns. Combing produces two products, the top and the noil: the top is further processed to give a worsted yarn, while the noil must be recarded on a woollen card before it is spun to a woollen yarn. Mills producing both worsted and woollen products use the noils themselves, but those producing worsted goods only, sell the noils to woollen mills or to wool traders. The price paid for the noil will depend on previous processing such as dyeing and carbonising.

Soft waste is the fibre waste from carding, combing and drawing and also some of the waste from spinning. It is called soft because it is still in a fibrous state and can go back into the process without any special pretreatment (see Fig. 1 and 2).

While a batch is being processed the soft waste is usually collected and fed straight back into the card hopper until the batch is finished on the card. The soft waste that is collected after this stage is stored in bales and used in the next batch of the same quality and colour.

Hard waste is the waste produced after spinning and twisting, and in the processes of preparing for weaving and knitting. It is called hard waste because twist has been introduced, the fibres no longer being in the fibrous state but in the form of yarns or fabric pieces. This waste cannot be fed straight into the card-hopper as it would ruin or damage the card-clothing. Hard waste is processed by picking and garneting, two processes which tear the yarns or fabric into a fibrous mass. In these processes a great deal of fibre damage and breakage occurs which reduces the length and quality of the fibres, but they are still valuable as a source of raw material for the woollen industry. Some wollen mills use a great deal of such raw material together with other wastes.

Finishing wastes (also called flocks when in the form of very short fibres) are wastes from the finishing process of fabrics. Flocks from cropping and shearing are used for pressed felts, bedding packing and also wallpapers. Wastes from fabric and garment manufacturers are classed as finishing wastes, but they could be also classed as hard waste since they are treated as such to produce the fibrous state. Waste swept up from the floor or collected after cleaning the card is first cleaned to remove dust and dirt and can then be used for upholstery, felting, etc., and even fed back into the woollen process.

The second category of textile waste is recovered or reclaimed waste, mostly in the form of rags and old, discarded clothing (Fig. 3). In this category one can make a distinction between wool wastes, which are usually reprocessed into new textile products, and non-wool wastes, which usually end up as industrial rags. Collection of rags is mainly done through charitable organisations.

Reclaimed Waste

Fig. 3:

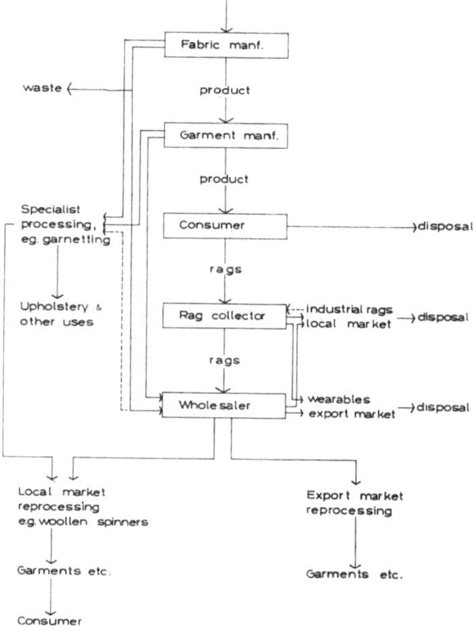

Possible movment of Waste

According to health regulations the rags and clothing have to be washed either before collection or by the rag collectors, who must have washing facilities on their premises. After collection the rags are sorted into categories such as light, medium or heavy cotton, synthetics or wool. At this stage the rags are also cut into small manageable pieces, and buttons and zips are removed. Cotton and synthetic threads are removed from wool rags destined for reprocessing either by chemical means or by pulling them out. This cleaning of threads is usually done by the processor and not the collector.

Manufacturing wastes are the wastes mostly used for reprocessing in the textile industry in Australia. Health regulations in New South Wales prohibit the use of rags and old clothing for reprocessing, only manufacturing wastes can be used. Most of the other waste material is exported. India imports large quantities of used wool clothing and other wool rags from Australia and the U.S. However, the Indian Government stipulates that any rags imported for reprocessing must be cut into small pieces because it does not want a black market in used clothing to develop. Italy, too, imports large quantities of wool rags for reprocessing. One city in Italy, Prato, lives on the reprocessing of wool wastes into fabrics and clothing. The U.S. imports some waste from Australia but is also an exporter of textile wastes.

The majority of firms in this business are only involved in collecting and sorting textile waste. They do no processing except cutting and removal of buttons and zips. A few, however, do specialise to some extent. One deals mainly with woollen hosiery waste bought from rag collectors and garment manufacturers, which it sorts into colours and may garnet before selling it to woollen spinners for reprocessing.

These people do not buy any old clothes and rags, but only manufacturing wastes. Another wholesaler sells industrial rags to the local market and exports most of the wool waste. His best customers are local garages which he supplies with industrial rags. To get the cotton waste for industrial use from the rag collectors he must also take used wool clothing, which he sells to India and Korea. This is the least profitable part of the business, sale of industrial rags and hard waste from manufacturers for export being far more profitable. Spain is one of his best customers for hard cotton waste, which it imports in large quantities.

Specialising in this field is at present not very profitable. One small firm which specialises in picking and garneting just manages to break even now, compared with some years ago when the business was going very well. Now there are only two men employed (apart from the owner) working a normal five-day week; while earlier there were two twelve-hour shifts working seven days a week. There has been a reduction in the amount of waste being used for reprocessing here in Australia, partly due to health regulations and partly to the high costs involved. The garneter gave the following cost figures (cents/lb) on processing hard waste into a fibrous mass ready for carding and spinning:—

	cents/lb
Cost of waste	16
Cost of picking	5
Cost of blending and more tearing	7
Cost of garneting	7
	35

A charge of 40c/lb will be necessary in order to break even. He quoted this price to a customer who was, however, able to get the job done in Melbourne for 38c/lb and then transported the material from there. Labour costs are his highest costs and because of the nature of the work he employs male labour only. Other organisations, such as The Smith Family, employ mostly women since they are a cheaper labour source and the work involved (cutting and sorting) is light and well suited to women. Some men are employed to do the heavy work.

Compared with India, labour costs here in Australia are very high, which is one reason for exporting the waste. Also, countries like India can less afford the high prices of virgin wool and thus use recovered wool waste as a cheaper source of raw material. Under these circumstances export of textile waste becomes a good proposition. So being a specialist in garneting is a drawback. There is no market in export of processed waste. Also, most worsted and woollen mills have facilities for picking and garneting.

The price of 16c/lb quoted above for waste is only an aproximate one. It does vary to some extent although not as much as prices for virgin wool do. High wool prices such as prevail at present tend to force waste wool prices up also. This is caused by the suppliers of waste, such as the garment manufacturers, wanting to recover some of the high cost of their own raw material.

When wool prices were low some years ago the demand for waste wool was also low, because mills bought cheap white wool and stored it for future use. It is better to use white wool, which can be dyed any shade that fashion demands, rather than waste wool, which has already been dyed and the use of which is therefore restricted.

In Australia, buying and storing cheap wool is a more attractive proposition than having to use waste wool. Up to 6 months may pass before the price of waste wool changes due to fluctuations in the price of virgin wool. The price of industrial rags is not so much influenced by changes in the price of new raw material but remains fairly stable. Labour and other costs have probably a greater influence on the price. The types of waste available from garment manufacturers are seasonal in nature, and this may have some influence on the price. During the winter months the garment manufacturers are making clothes for the summer, and hence the waste available for collection will be mostly cotton and other light material. On the other hand, during the summer months mostly wool and other bulky clothing waste is available for collection.

It is not quite true that all rags collected can only be used either for reprocessing or as industrial rags. About 20% of what The Smith Family collect can be used as clothing. It is possible that they do the same as other rag collectors and send this to Papua New Guinea and the South Sea Islands for sale as wearable second-hand clothes. Apparently this is quite a lucrative, although only a very small, part of the business.

In worsted and woollen mills men are employed to collect and sort the waste from the various processing stages. They also sweep the floor and clean the machines, collecting the waste, which is then sorted cleaned and baled. Some of the waste is used in the next appropriate batch if it is suitable for reuse. The rest is sold to people who can make use of it in some way, e.g., for upholstery or felting. No textile mill disposes of fibre waste as garbage — in laboratories and offices they have one bin for textile waste and one bin for other rubbish. In one mill known to the author they sell the hard waste, as they have no waste processing machines except for a cleaning machine for floor sweepings. The soft waste is fed back into the cardhopper or baled up for the next batch. Each shift has three or four men collecting and sorting waste. One cannot afford to let the waste go with the garbage, especially these days when wool is so expensive.

Australia not only exports but also imports textile waste (see Table 1). The imports, however, are in smaller amounts than the exports. The imported waste is mainly cotton and man-made fibre waste but even a small amount of wool is imported. Wastes such as these are blended with new fibres and treated in the normal manner to produce yarns.

In one Australian mill waste nylon was used to a large extent in 80/20 wool/nylon blends. The waste nylon did not perform as well as new nylon during the processing; it was used because of the lower price, and to reduce some of the costs due to the expensive wool.

Table 1
AUSTRALIAN EXPORTS OF TEXTILE WASTES

Type of Waste	**Amount, kg. 1971-72**	**1972-73**
Wool, noils	1,123,093	1,144,284
Wool, sheep and lamb	2,544,707	280,981
Other wool waste	—	1,844,307
Used clothing	593,833	744,630
Used and new rags, scrap twine, ropes, etc.	12,944,000	17,503,000
Man-made fibre	362,083	231,104

Countries of Destination:
Red China, Italy, Japan, U.S., India, Korea, Belgium and Luxemburg, Spain and others.

AUSTRALIAN IMPORTS OF TEXTILE WASTE
1971-1972

Type of Waste	**Amount**
Wool	107,677 lb.
Cotton	6,862,611 lb.
Man-made fibre waste	820,325 lb.

Table 2
U.S. EXPORTS OF TEXTILE WASTES
JANUARY-OCTOBER 1972

Type of Waste	**Amount, lb. x 10^3**
Cotton	142,635
Wool	77,006
Man-made fibre	105,793
Nylon fibre	18,774
Other non-cell. fibre	32,982

Countries of Destination
Canada, Spain, India, Italy and other Common Market Countries.

Table 3
RAW-MATERIAL CONSUMED BY WORSTED AND WOOLLEN INDUSTRIES IN U.S. AND U.K.

Raw Material	**Amount 1939**	**1959**
U.S.	**(lb. x 10^3)**	
Shorn and pulled	296	430
Other wools, noils and waste	26	88
Reprocessed wool	118	58
Reused wools	118	9
Specialty hairs	28	16
Cotton	15	7
Man-made	16	112
U.K.	**1949**	**1959**
Shorn and pulled	502	518
Other wools, noils and waste	64	42
Reclaimed wool	55	67
Specialty hairs	18	30
Cotton	9	4
Man-made	6	69

REFERENCES

1. W. Von Bergen, Ed. *Wool Handbook*. Vol. 1, 1963. L. Wiley and Sons, New York, London.
2. H. Spibey, Ed. *The British Wool Manual*, 2nd ed. 1969. Columbine Press.
3. Commonwealth Bureau of Census and Statistics.

PAPER 32

The Control of Gaseous Wastes and Odours

By N. Y. Kirov*

1. The Problem: The Nature and Sources of Gaseous Wastes

The control of gaseous wastes, organic solvent vapours and fumes, and objectionable odours is one of the most complex of waste disposal problems, causing much concern to environmental pollution control authorities.

Processes which involve the evaporation of organic solvents are largely responsible for the emissions of light volatile hydrocarbons and solvent fumes. These may be produced from operations such as:—

(i) petroleum refining and the petro-chemical industries;
(ii) paint, varnish, enamel and lacquer manufacturing, drying and storing;
(iii) printing and lithographing;
(iv) curing and polymerising;
(v) certain chemical manufacturing industries; and
(vi) the disposal of sewage and refuse.

Objectionable odours arise from certain industries and "the offensive trades" such as the manufacture of soap and detergents, paper making, coffee roasting, fish manure and fertiliser plants, food processing, tanneries, fish markets, abattoirs, fat rendering processes, pigsties, the animal by-products industries, and the like. Such odours are often associated with the presence of sulphides, mercaptans, phenolic compounds, aldehydes, amines, amides, esters and various other organic compounds and acids containing nitrogen or sulphur.

That many of these are objectionable, most people will agree. But often, it is difficult to determine the extent to which an odour is objectionable to the community: response by people is very subjective, there are tremendous variations among individuals, people get accustomed to familiar odours and complain about new ones. Because it is difficult to measure and characterise odours quantitatively (see ASTM D-1391-57)[1], and because odour thresholds in air for certain objectionable compounds are as low as one part in a million parts to one part in 1000 million by volume (see Table 1), detection by the human nose is still more sensitive and more realiable than that by most instruments. It is not therefore surprising that public complaints over recent years have shown a steady increase. For example, the N.S.W. Air Pollution Control Branch reports that more than half of the telephone complaints it receives are concerned with odours.

As the actual amount of contaminant present in the effluent gas is usually very small, much lower than the threshold for toxicity, odours rarely have an adverse affect on health; they are important as pollutants mainly because of their nuisance value.

Reactive organic vapours (solvent emissions) and fumes, on the other hand, may be serious health hazards and may play also a significant part in the formation of secondary atmospheric pollutants and in photo-chemical smog reactions, with all their associated and familiar consequences.

This group of emissions consists principally of compounds of carbon and hydrogen, and their derivatives. These include all classes of hydrocarbons (olefins, paraffins and aromatics) and the compounds formed when some of the hydrogen in the original compounds is replaced by oxygen, halogens, nitro- or other substituent group.

2. Control Methods

The disposal of gaseous wastes usually involves the handling of very large gas volumes. In turn, the gas volume determines the physical size and cost of the treatment equipment, even though the quantity of actual contaminant may be extremely small.

Hence, the first step in any control operation is the containment and capture of the offensive pollutants in order to prevent their further dilution. This is achieved by effective design of hoods, ducts and exhaust systems.

The separation, conversion or destruction of the gaseous wastes may then be carried out by one or more of the following conventional techniques. In so doing, however, one must bear in mind the conclusion of the recent report of the U.K. Working Party (1973)[2] on the sup-

* Head, Department of Fuel Technology, University of N.S.W.

pression of odours, namely "that complete answers to many industrial odour problems do not exist at present".

2.1. *Reduction at Source*: by better housekeeping in plant operation and by process variation and modifications. This depends on the individual processes, but often by improved control or modification of the production process it is possible to eliminate or reduce solvent fumes or odours to an acceptable level; sometimes this may be achieved by investigation and modifying the properties of the raw materials of production, e.g., by elimination of reactive solvents or elimination of combustible solvents (use of water emulsion).

2.2. *Adsorption* of odorous chemicals and organics on a solid material of suitable active properties, i.e., a porous solid which has a very high surface area per unit mass. The adsorbent may be subsequently reactivated by heating or by a solvent, and the odorant recovered. The most useful and most common adsorbent for odour control is active carbon or charcoal in granular form, supported in porous beds. It has the advantage that its effectiveness is not weakened by the presence of moisture. This can be an effective and economical control method for emissions with low concentration of odorous compounds, or for the recovery of hydrocarbon solvents.

2.3. *Absorption* in spray towers or by wet scrubbing in a suitable liquid. The design of absorption equipment is a well-developed chemical engineering technique. Water is a cheap and commonly-used solvent. It may be used, for example, for the removal of hydrochloric or hydrofluoric acid gases, nitrogen oxides, chlorides, and silicon tetrafluoride (SiF_4), which are readily soluble in water. Some contaminants and odorants, however, are insoluble in water or need to be reacted chemically in solution, as with sulphur dioxodie and mercaptans removal in an alkaline solution; hydrogen sulphide (H_2S) or carbon disulphide (CS_2) in an ethanol amine aqueous solution in a bubble column; aldehydes by sodium (or calcium) bisulphite; and amines by hydrochloric (or sulphuric) acid. However, it should be

Table 1

ODOUR THRESHOLDS IN AIR (p.p.m.)*

(Jou. Air Pollution Control Association — Feb. 1969, v. 19, n. 2, p. 94)

Compound	ppm	Compound	ppm
Acetaldehyde	0.21	Ethyl acrylate	0.00047
Acetic acid	1.0	Ethyl mercaptan	0.001
Acetone	100.0	Formaldehyde	1.0
Acrolein	0.21	Hydrochloric acid gas	10.0
Acrylonitrile	21.4	Methanol	100.0
Allyl chloride	0.47	Methyl chloride	(above 10 ppm)
Amine, dimethyl	0.047	Methylene chloride	214.0
Amine, monomethyl	0.021	Methyl ethyl ketone	10.0
Amine, trimethyl	0.00021	Methyl isobutyl ketone	0.47
Ammonia	46.8	Methyl mercaptan	0.0021
Aniline	1.0	Methyl methacrylate	0.21
Benzene	4.68	Monochlorobenzene	0.21
Benzyl chloride	0.047	Nitrobenzene	0.0047
Benzyl sulfide	0.0021	Paracresol	0.001
Bromine	0.047	Paraxylene	0.47
Butyric acid	0.001	Perchloroethylene	4.68
Carbon disulfide	0.21	Phenol	0.047
Carbon tetrachloride (chlorination of CS_2)	21.4	Phosgene	1.0
Carbon tetrachloride (chlorination of CH_4)	100.0	Phosphine	0.021
Chloral	0.047	Pyridine	0.021
Chlorine	0.314	Styrene (inhibited)	0.1
Dimethylacetamide	46.8	Styrene (uninhibited)	0.047
Dimethylformamide	100.0	Sulfur dichloride	0.001
Dimethyl sulfide	0.001	Sulfur dioxide	0.47
Diphenyl ether (perfume grade)	0.1	Toluene (from coke)	4.68
Diphenyl sulfide	0.0047	Toluene (from petroleum)	2.14
Ethanol (synthetic)	10.0	Tolylene diisocyanate	2.14
		Trichloroethylene	21.4

* These recognition odour thresholds for 53 commercially-important chemicals have been determined in the laboratory by an expert panel under standardised conditions. The study was carried out in 1967 by A. D. Little under contract for the Manufacturing Chemists' Association.

noted that disposal of the contaminated waste water from such a process may be a serious problem.

2.4. *Surface and Contact Condensation* techniques operate on the principle of removing heat from the vapour, thus allowing condensation of the offensive vapours, or condensation of the water vapour with the odour—producing substance.

2.5. *Odour Masking or Odour Counteraction* is achieved by mixing or superimposing a stronger and more pleasant odour with the objectionable odour. For example, diesel exhaust odours are reported to be controlled by one transport company by vaporising a compound and reodorising the gases to give them a "fresh woody scent", and at least one local council in Sydney has been experimenting with the spraying of streets with pleasant-smelling odorous compounds.

However, pungent odours are not usually amenable to such treatment and the disadvantage of this technique is that the odorous compounds are not removed and that the additives themselves may become air pollutants.

2.6. *Chemical Oxidation,* by reacting the undesirable compounds with chemical agents such as:—

(i) Ozone—suitable for waste water treatment processes but not yet widely used;
(ii) potassium permanganate—effective, but costly; sometimes used for feedlot odour control;
(iii) chlorine, as:—
  (a) a bleach liquor from chloride of lime,
  (b) sodium hypochlorite solution, or
  (c) chlorine gas.

For example, the use of chlorine in the sewage industry to control odours, as illustrated by the reaction:—

$$H_2S + Cl_2 \rightarrow 2HCl + S$$

Chlorination, as an odour control method, is simple, efficient and comparatively low cost.

2.7. *Dispersal and Dilution* from high chimneys to reduce the odour intensity and its local nuisance. This is relatively cheap as there are no operating costs but must allow for a large factor of safety, taking into account all likely meteorological conditions, i.e., for acceptable odour control, peak rather than average concentrations are important.

2.8. *Incineration.* High temperature oxidation of the organic and toxic compounds present in gaseous wastes is employed to destroy the offensive effluents and odours by converting them into harmless gases, i.e., carbon dioxide and water vapour. This is by far the most effective control method at present available for the disposal of gaseous organic wastes and odours. Its successful application depends on the provision, in the design and operation of the burners, of the conditions needed for satisfying the basic principles of combustion: good contact of the contaminated gas with the burner flame, adequate gas velocities and turbulence to ensure good mixing and adequate time and sufficiently high temperature for combustion to be completed.

Incineration should be considered wherever there is a problem of pollution by combustible matter. It should be noted, however, that with certain compounds, e.g. degreasing solvents which contain trichlorethylene, perchlorethylene or other halogenated compounds, special care is needed as these may, on reaction with a direct flame, produce a poisonous gas, e.g. phosgene. Also, in some cases, if combustion is not complete, the intermediate products produced may be even more undesirable than those which we are trying to eliminate.

3. Basic Types of Incineration Systems

For convenience the basic types of incineration systems suitable for the oxidation of gases and organic vapours may be considered under the following three main groups (in order of decreasing concentrations of combustibles):—

(i) Direct flame or direct-fired incineration systems, at temperatures above 1200°C.
(ii) Thermal incineration, in a temperature range of about 600 to 900°C.
(iii) Catalytic oxidation, at temperatures of 300 to 600°C.

3.1. *Direct-Flame or Direct-Fired Incinerator Systems*

This method is generally used for burning gases of low calorific value when the gases are at or near the lower combustible limit when mixed with air, or when the waste gas is a combustible mixture without air addition. Direct-flame incineration usually takes place at temperatures above 1200°C.

Gases of about 950 kcal/nm^3 calorific value or above may be burnt without auxiliary fuel; if the calorific value is below, as in the case of blast furnace gas, autothermic combustion is still possible provided the gas is preheated. Such wastes may be treated almost as a fuel, and often, standard fuel burning equipment can be used for incineration.

If the combustible component in the mixture is below the lower flammability limit (about 500 kcal/nm^3) then natural gas or other auxiliary fuel must be added to sustain combustion.

Hydrogen sulphide, hydrogen cyanide and carbon monoxide produced as by-products from reactions, or solvent organic vapours mixed with air, may be burned in this way. Generally, direct flame incineration is employed when little auxiliary fuel is required and the combustible waste gas supplies at least 50% of the total fuel value. The key criterion for success in such a system is the satisfactory provision in the design and operation of the basic requirements for combustion, to which a reference has already been made.

This type of combustion system is exemplified by:—

(a) conventional industrial burners fired in the combustion chamber of industrial boilers and furnaces operating on forced and/or induced draught; the mixture is injected into the burner or combustor, mixed with sufficient air to achieve complete combustion, and burned as a fuel; and

(b) the flare-type burner, as used in petroleum refineries, petro-chemical plants and chemical-manufacturing industries: in this case the fuel-air mixture is ignited at the tip of the flare and burns in the surrounding air.

The latter is used for incineration of economically unrecoverable mixtures of refinery waste gases or for disposal of excess gases and vapours, but care is needed when burning highly odorous gaseous impurities (e.g. H_2S), for which an upper concentration limit may have to be imposed, since complete combustion in flares is difficult to achieve. Refinery flares may have a capacity of more than 50 tons of waste gas per hour. A major control problem in such cases is often the wide variation of the gas flow rates. In this system, steam or air jets are usually incorporated to promote rapid mixing and to control smoke emissions, particularly when hydrocarbons of high C/H ratio are burnt. As a result, the noise sometimes associated with such flares and the light from the flame may give rise to a nuisance problem. In a typical flare design the steam jets surround the flame and introduce air together with the steam to assist rapid combustion.

Velocities in the pipe carrying the gases should be kept above the flame propagation of the mixture, to prevent flashbacks; flame arrestors may be placed in the line but they are not a complete protection. A pilot flame is also provided. Some of the demands which must be met in order to ensure that the operation of a flare does not constitute an air pollution or a safety hazard have been discussed by Trobisch[3].

An interesting new development by the British Petroleum Co.[4] is the Coanda flare tip. It makes use of the wall adhesion effect (discovered by H. Coanda) to achieve air aspiration and hence good combustion characteristics. In the B.P. Coanda flare tip, gas at high pressure (around 2 atm.) is ejected from an annular slot over a curved surface to which the gas stream clings. The resulting change in momentum causes air to be induced into the gas stream and the gas/air mixture burns at the end of the Coanda surface. Low pressure, smoke-producing gas (propane, butane, etc.) is injected into the aerated flame produced by the Coanda injector, and is consumed smokelessly (limit ratio for low pressure to high pressure gas should not exceed 1:3). This burner is small in size, simple and cheap to construct and easy to install.

3.2. *Thermal Incineration*

Thermal afterburners are suitable for the treatment of non-flammable weak mixtures of organic vapours, solvent fumes or other odorous emissions. The calorific value of such gases may be in the range of 10 to 200 kcal/nm^3 (say, concentrations up to about 25% of the lower explosive limit) and hence if incinerated directly they would require too much additional thermal energy.

The gases are destroyed by combustion in an independently-heated enclosed refractory-lined combustion chamber or afterburner, at temperatures usually in the range of 700°C to 900°C, but sometimes as low as 550°C, in the presence of flame. Natural gas or other fuel may be used in an industrial burner to preheat the chamber, preferably to a temperature of over 760°C. The contaminated gas is injected into this chamber (at velocities of 4.5 to 9.0 m/sec.) and allowed a retention time of from 0.3 to 1.0 sec. at this maximum temperature; for effective control a minimum of 0.5 sec. is often specified. Such systems are capable of operating continuously at efficiencies of 90 to 99+%.

An effluent gas of high oxygen content (say, over 16%), can be used as either primary or secondary air in the combustion process. The hot combustion gases are usually passed through a heat exchanger, where much of their sensible heat (50% to 80%) may be recovered by direct recuperation, with substantial savings in fuel costs. The amount of heat recovered is largely a function of the heat exchanger costs and its design.

A major advantage of thermal afterburners is their flexibility and the ease with which they can be adapted to meet changing emission standards, or emission composition, by simply adjusting their operating temperatures. However, sufficient care must be exercised to ensure that such adjustments are made within the design temperature range and capabilities of the furnace refractories. They can also be readily adapted for automatic temperature control.

If a boiler is available as part of the plant, it may be more expedient to discharge the waste gases directly into the boiler furnace as part of the combustion air supply, provided sufficient dwell-time and adequate temperatures for the destruction of the unwanted gaseous pollutants could be ensured at all times.

The desired afterburner temperature varies with the design conditions, the nature of the contaminants and the efficiency of cleaning required. High afterburner efficiencies generally demand higher discharge temperatures and longer residence times.

Recommended (after Danielson[5]) minimum afterburner operating temperatures for various types of industrial operations are indicated in Table 2.

Table 2

Operation	Recommended Temperature °C
Carpet laminating	650-760
Core oven	760
Cloth carbonisation	980
Deep fat fryers	650
General opacity problems	650-760
Odour control	700-820
Oil and grease smoke	650-760
Paint bake ovens	650-820
Pipe wrapping	760
Rendering operations	650
Smokehouse	650
Solvent control	700-820
Varnish cookers	650
Vinyl plastisol curing	650-760

For the destruction of odours from sewage sludge, slightly higher temperatures may be required. Sawyer and Kahn[6] tested 29 sludge samples for 174 odour evaluations and reported that only a few sludges were adequately deodorised at 680°C. Most sludges required temperatures of 700 to 735°C, while a few required 760°C, and over. The intensity of odour was appreciable at 30°C below the threshold level, and severe—at 60°C below. Temperatures required for the destruction of odours from volatile, low-molecular weight fatty acids commonly present in sewage sludge were as given in Table 3.

Table 3

Organic Acid	Temperature for Destruction of Odour °C
Acetic	745
Propionic	745
Butyric	775
Caproic	775

3.3. *Catalytic Oxidation*

Catalytic afterburners, or catalytic vapour incinerators as they are sometimes called, are particularly useful for the destruction of gaseous wastes and odours containing low combustible concentrations of organic effluents (say, below 10 kcal/nm^3) which would not directly support combustion and would otherwise require large quantities of additional fuel to complete their combustion. The principle of catalytic burning may also be applied to higher calorific value gases (e.g. 280 kcal/nm^3) with energy production. While this is feasible, the adiabatic temperature rise of 880°C at complete combustion gives rise to a number of engineering problems (Vollheim[7], also Trobisch[8]).

Here we are concerned with a heterogeneous chemical reaction, which is surface limited, and the function of the catalytic agent is to provide a large surface area so as to help accelerate the surface reactions. As a result the combustion process is completed at substantially lower temperatures (say 300 to 600°C) and with less excess air; hence auxiliary fuel requirements are much lower.

In this system, the odorous gases or organic fumes are passed through a heat exchanger where they are preheated by the outgoing hot deodorised gases, before entering the combustion chamber. In the combustion chamber, they are further heated to the temperature required to achieve catalytic oxidation (gas, oil or even electric heating may be used) and then are passed through the catalyst bed. The "clean" gases are discharged through the heat exchanger. Generally, catalytic oxidation units are smaller and weigh less than comparable thermal incinerators, since the process is carried out at lower temperatures and at much shorter residence times (say 1/30 sec.) (see for example Searles[9]) the combustion chamber is usually smaller and constructed of lightweight material. This gives a low heat capacity with a fast warm up to full operating conditions after shutdown.

The catalysts usually employed are noble metals such as Pt, Pd or Pt Rh, or the oxides of various base metals (Cu, Cr, Mn, Ni or Co), finely divided on various supports, such as a heat resistant steel or a ceramic body, e.g. gamma alumina. The efficiency of various basic oxide catalysts in the oxidation of a number of sulphurous air pollutants was studied by Harkness et al.[8]

The various catalysts require different operating conditions with regard to temperature and residence time. A good catalyst support needs to be a material of high intrinsic surface area that is shaped to give a high geometric surface area and a configuration that gives a low pressure drop.

Commercial catalysts may be supplied in many different forms in a manner designed to ensure a low pressure drop and a high surface to volume ratio. The following are four basic configurations commonly used (see Ross[10]):—

(a) nichrome wire screen support, with a coating of the precious metal;

(b) a number of air foil-shaped alumina rods on which a noble metal catalyst has been deposited;

(c) a bed of spheres of pellets coated with the catalytic agent; and

(d) an alumina honeycomb on which the catalyst has been deposited.

These latter have a high resistance to thermal shock, high mechanical strength and freedom from wear by attrition.

Because of the cost of the catalysts and of the additional equipment required, catalytic afterburners require a higher initial capital outlay than thermal afterburners but the savings in operational cost in terms of fuel requirements may be substantial, particularly if a long catalyst-life period can be achieved.

It is important to ensure that temperatures are kept below the safe limit of the catalyst, say below 800°C maximum, and that the effluent gases do not contain impurities (e.g. arsenic, lead, zinc, mercury or phosphates, sulphur and chlorine) which will "poison" and deactivate the catalyst.

This system appears to have been most successful for application to gases containing low concentrations of organic contaminants or particulates such as those from:— food preparation processes, oxidation of paint solvents, wire enamelling ovens, lithographing ovens, chemical manufacturing processes and the like (see e.g. Ross [11]).

However, wider use of this method must await the development of more economical, robust, and poison-resistant catalysts.

4. Case Histories of Practical Control

Cleary, Pottinger and Ferrari[12] reported the following successful applications of afterburners to destroy industrial odours.

(i) Exhaust from coffee and chicory roasters, containing complex odorous organic compounds (mainly aldehydes and organic acids), was effectively deodorised in a propane-heated catalytic afterburner at temperatures of 380°C (for the coffee) and 480°C (for the chicory).

(ii) Exhaust from an oven, drying phenol formaldehyde-bonded plastic sheeting, gave completely satisfactory performance (over 98% reduction in phenol concentration) after passing through a catalytic afterburner at an outlet temperature of 590°C.

(iii) Exhaust gases containing aldehydes (particularly acrolein) and aerosols from a foundry core oven was passed through a thermal afterburner in which at temperatures above 800°C the aerosols were effectively removed, but the removal of the aldehydes was less efficient.

Strauss and Cock[13] reported on operating experience and the relative economics of three different types of direct flame afterburners, treating gases:—

(i) from tinplate lacquer baking ovens, in an oil-fired (operating temperature 760°C) unit with an elaborate heat exchanger;

(ii) from a P.V.C. coating plant, using an oil-fired unit with a simple heat exchanger; and

(iii) from another P.V.C. coating plant, but with a gas-fired unit, and also with a simple heat exchanger.

It was shown that the major operating cost factor was the fuel consumption.

Catalytic combustion was not practicable for the latter two cases because of the presence of an organic plasticizer in the gases. However, data were given for two catalytic combustion afterburners in Victoria, which were used for:—

(i) incineration of formaldehyde and associated fumes from the manufacture of laminated resin-impregnated-paper surfacing material at about 360°C, and

(ii) control of odours and heat recovery from wire-enamelling ovens.

These authors concluded that thermal afterburners are quite high in operating costs, and that catalytic combustion is cheaper, particularly when effective heat exchangers are used.

Thermal incineration is said to be the most effective method of controlling varnish kettle emissions, which may contain several thousand compounds (see Di Giacomo[14]). The afterburners were operated at 680°C with 0.67 sec. residence time. The development of certain improvements involving the use of an integral-blower burner or a non-powered raw gas burner are claimed to reduce installation, operating and maintenance costs.

The following data are given by Arnold[15] for the performance of an improved type-burner in the fume stream:—

(a) *Wire enamelling oven* (0.3 sec. residence time)

Outlet Temperature °C (Combustion Chamber)	Hydrocarbons, ppm in	Hydrocarbons, ppm out	Conversion %
490	1400	480	66
590	1410	300	78
680	1420	130	91
700	1400	110	92

(b) *Metal coating oven* (0.3 sec. residence time)

Outlet Temperature °C (Combustion Chamber)	Hydrocarbons, ppm in	Hydrocarbons, ppm out	Conversion %
425	3300	1400	58
540	3200	480	85
650	3300	250	92
730	3300	90	97

(c) *Plastic coating machine*

Outlet Temperature °C (Combustion Chamber)	Efficiency, % Conversion Hydrocarbons only
760	99.9
720	99.3
695	97.7

Sawatani (quoted by Bethea[16]) is reported to have considered various odour control methods as applied to fisheries, pigsties and poultry houses in Japan and to have found thermal incineration effective at 700°C. Mills et al. (Bethea[16]) used a 650°C thermal incinerator with 0.3 sec. contact time for controlling odours from a rendering cooker. Catalytic combustion at a temperature of 350°C, using a platinum wire net, was suggested as a possible control technique for these applications, provided that the inlet gas could be free of dust. Other workers have found that catalytic oxidation of the odorous emissions from animal rendering with platinum or palladium catalyst at 350°C to 400°C can achieve 90 to 95% efficiency.

5. Summing-Up

Complete answers to many objectionable industrial odour, solvent and fume problems do not exist. Design criteria are not easy to detail, as knowledge is limited and much of the data is proprietary.

With regard to odour control, there is a need for standarisation of measurement techniques, instruments and odour panel procedures in order to be able to evaluate reliably the effectiveness of odour removal by various control devices.

Catalytic oxidation is not universally accepted at present.

The simplest equipment for direct incineration is a combustion chamber in which the process gases mix turbulently with a flame produced by an oil- or a gas-fired burner.

Thermal incineration is capable of giving a sustained performance level if the flows, temperatures and operating conditions are held constant, and appears to be becoming the main control technique for solvent and other combustible gaseous wastes. However, fuel costs are critical in many applications, and it is important to determine accurately the conditions needed for optimum combustion.

REFERENCES

1. American Society for Testing Materials Method D-1391-57 (Standard Method for Measurement of Odour in Atmospheres — Dilution Method).
2. U.K. Working Party on the Suppression of Odours from Offensive and Selected other Trades; Department of the Environment, Part 1: Assessment of the Problem in Great Britain, 166 pages. 1973.
3. Trobisch, K. "Oxidative Procedures for Cleaning Waste Gases in the Chemical Industry" Twelfth Symposium (International) on Combustion, p. 645, 1969.
4. The British Petroleum Co. Limited, Sunbury Research Centre, Chemistry and Industry, Sep. 25th/Oct. 2nd, pp. 1082-1118, 1971.
5. Danielson, J. A. (ed.) "Air Pollution Engineering Manual" U.S. Environmental Protection Agency, Publication AP-40, Cincinnati, Ohio; 1973.
6. Sawyer, C. N. and Kahn, P. A. "Temperature Requirements for Odour Destruction in Sludge Incineration" Jou. W.P.C.F., Dec. v. 32, n. 12, pp. 1274 (1960).
7. Vollheim, G. "Catalytic After-Burning — A Method of Purifying Flue Gases", Twelfth Symposium (International) on Combustion, p. 651 (1969).
8. Harkness, A. C., Murray, F. E. and Girard, L. "Catalytic Oxidation of Sulphurous Air Pollutants", Atmospheric Environment (Pergamon Press) 1968, v. 2, pp. 303-320.
9. Searles, R. A. "Catalysts Cut Costs by Reducing Temperature for Incinerating Plant Odours", Process Engineering, Feb. 1973, pp. 100-102.
10. Ross, R. D. Paper 15, pp. 4. The 1971 Australian Waste Disposal Conference, Feb., Department of Fuel Technology, University of New South Wales, Sydney, Australia (1971).

11. Ross, R. D. "Incineration of Solvent-Air Mixtures", Chemical Engineering Process, August, pp. 59-64, v. 68, no. 8 (1972).

12. Cleary, G. J., Pottinger, J. F. and Ferrari, L. M. Paper 12, pp. 20. First Australian Refuse Disposal Conference, Aug., Department of Fuel Technology, University of New South Wales, Sydney, Australia, 1967.

13. Strauss, W. and Cock, W. H. "Practical Odour Control in Industry", 1965 Clean Air Conference, held at the University of New South Wales, 19th Aug., Sydney, Australia (1965).

14. DiGiacomo, J. D. "New Approaches to the Design of Afterburners for Varnish Cookers" Jou. Air Pollution Control Association, v. 23, n. 4, pp. 287-290. 1973.

15. Arnold, G. D. "The Incineration of Waste Liquids, Gases and Fumes". Product Finishing, pp. 21-27, April 1973.

16. Bethea, R. M. "Solutions for Feedlot Odour Control Problems" Jou. Air Pollution Control Association, v. 22, n. 10, Oct., pp. 765-773, 1972.

Acknowledgement

The author is indebted to Mr. J. D. Court, Air Pollution Control Engineer, Health Commission of N.S.W., for helpful information and useful comments.

PAPER 33

Resource Recovery in Solid Waste Management

By Ronald S. Rosich*

1. Introduction

As a result of the continuing publicity being given to the problems of pollution and the exhaustion of natural resources, renewed attention is being paid to the recycling of a greater proportion of industrial materials.

In many industries, particularly the metals industry and the paper industry, a substantial proportion of requirements has always been met by the re-use of waste material. This is often so because the waste is reasonably uniform in composition and a reasonably steady supply is assured.

By contrast, municipal waste is a complex mixture and one whose amount and composition varies significantly with respect to locality, seasonal changes, social and economic factors and changes in these social and economic factors with time, etc. (see Ref. 1). It is clear, nevertheless, that the amounts of municipal waste being produced are increasing while the currently used methods of disposal are becoming less acceptable (on environmental grounds) and more expensive (e.g., as suitable landfill sites become less accessible). Furthermore, the *energy crisis* has focused attention not only on our dependence on oil and gas as fuels but also as a source of reduced carbon for the petrochemical industry, an industry which plays a major role in our modern lifestyle.

2. Sources of Waste

2.1. *Process Waste*

Process waste consists of materials which occur in the normal operation of industrial processes. Typical examples are scrap metal from casting and machining and plastic and paper offcuts. In many cases little further work is needed to produce material fit for re-use and a large proportion of the non-ferrous metals recycled is recovered from this source.

2.2. *Process Effluent*

Increasingly stringent regulations regarding water pollution will lead to the production of increasing amounts of sludges containing a wide range of metal salts which result from their precipitation from liquid effluents. These liquid effluents may arise from industrial processes or from gas-cleaning plant installed to meet air pollution control requirements. Presently most such sludges are dumped but the recovery of metals from these sludges is technically feasible. Some of the sludges, e.g., *red mud* and related effluents such as furnace ashes, may find more direct uses, especially in the construction industry (see also Table 3).

2.3. *Capital Waste*

Capital waste arises from consumer and industrial goods which have ended their useful lives, e.g., scrapped cars and ocean liners. A considerable proportion of ferrous metals is recycled in this way.

2.4. *Municipal Waste*

Based on a figure of 0.8 Kg/person/day (see Ref. 1) about 3.5m tonnes per year of domestic waste is collected by local authorities in Australia. This consists mainly of paper, putrescibles, metals, glass, rags, ashes and dirt and plastics.

2.5. *Agricultural and Animal Wastes*

Combined agricultural (500 million tonnes) and animal wastes (1500 million tonnes) have been estimated to make up more than half the weight of the total yearly American solid waste.[8] As much as 50% of the animal waste is generated in concentrated husbandry operations close to urban areas. For example, one feed lot with 10,000 head of beef cattle can produce 260 tonnes of manure per day which results in the same sewage disposal problem as a city of 164,000 people.[8] Similar problems are now being encountered in Australia.

2.6. *Some Relevant Data*

The rate of extraction of some non-ferrous metals from primary sources has increased rapidly in the last two decades (Table 1) and concern is now being expressed about the life of these primary sources.

Table 1

ANNUAL WORLD TONNAGES OF METALS MINED, 1951 TO 1968
(in thousand short tons)[2]

Metal	1951	1968	% increase
Copper	3100	5893	90
Lead	1890	3309	75
Zinc	2600	5471	110
Tin	169	227	34

* School of Applied Science, Canberra College of Advanced Education, A.C.T.

A significant proportion of the consumption of these metals has been provided by the recycling of waste (Table 2).

Table 2

RECYCLING OF NON-FERROUS METALS IN THE U.K. (1969)
(in thousand short tons)[3]

Metal	Consumption	Recycled	% Recycled
Copper	680	287	42
Lead	360	229	64
Zinc	365	88	24
Aluminium	610	238	39
Tin	19	1.5	8

In Australia, the amounts (in round figures) of these metals recycled have been estimated to be: copper and copper base 50,000 tonnes; lead 30,000 tonnes; zinc 10,000 tonnes; and aluminium 30,000 tonnes.[4, 7]

Table 3 indicates the main solid wastes in the U.K. and the uses to which they can be put.

A similar but more detailed compilation for Australian industrial wastes is contained in a recent paper by Beretka.[6]

3. Resource Recovery Processes for Industrial Waste

3.1. *Hand Sorting*

Hand sorting is still a widely used technique, e.g., especially in sorting of metal and metal

Table 3

THE MAIN SOLID WASTES IN THE U.K.[5]

NAME	ORIGIN	Approx. quantity produced Mtonne/yr.	METHOD OF DISPOSAL	Stock in spoil heaps or lagoons Mtonne	MODE OF USE
Blastfurnace slag	Manufacture of iron	9	Complete utilisation		Aggregate, cements, roadstone, abrasive, slag wool.
Steel slag	Manufacture of steel	4	Major utilisation		Roadstone, 2Mtonne/year returned to blastfurnace.
Pulverised fuel ash (PFA)	Power station boilers	9-10 includes 2 of furnace bottom ash	Major utilisation. Remainder in lagoons and spoil heaps. Some dumped in sea.		Fill, lightweight aggregate, bricks, cement, soil stabilisation.
Clinker	Power station boilers	2	Complete utilisation		Aggregate, fill.
Colliery shale	Coal mining	50	Minor utilisation. Remainder in spoil heaps.	3000	Road fill, lightweight aggregate, cements, bricks.
China clay waste. Quartz sand Rock and overburden	Extraction of China clay	10 10	Minor utilisation. Remainder in spoil heaps and lagoons	>125	Aggregate, fill, bricks.
Micaceous waste		1.5			
Slate waste	Slate quarrying	1.4	Minor utilisation. Remainder in spoil heaps	>300	Lightweight aggregate, filler for paper, fill.
Waste glass	Waste glass within domestic refuse	2	Minor utilisation. Remainder within refuse dumped on land		Soil conditioner, source of reactive silica, tiles, sand, paper.
Incinerator ash	Incineration of domestic refuse	1.25	Utilisation, if any, not known, dumped on land.		
Phospho-gypsum	Phosphoric acid manufacture		Dumped in sea and on land		
Fluorspar waste	Mining fluorspar	0.35	Spoil heaps and lagoons		Some back fill.
Mine tailings (Tin)	Mining tin ore	0.75	No utilisation. Lagoons		
Red mud	Extraction of aluminium	0.15	Very minor utilisation. Remainder in lagoons		Pigment for roadstone, paints, and plastics.
Zinc, lead, copper and phosphorus slags	Residues from blastfurnaces	0.1-0.35	Minor utilisation. Remainder dumped on land		Aggregate, grit, blasting.
Plastics waste	Containers, packaging and so on	0.25	No utilisation. Stock. within refuse, dumped on land.		
Domestic wastes		18			

alloy scrap, but with the increasing cost of labour its role is bound to diminish.

3.2. *Magnetic Separation*

The use of magnets, following crushing or shredding where necessary, is widespread in the separation of ferrous metal from non-ferrous materials. By using much more powerful magnets it is possible to separate certain non-ferrous materials, but work in this area is still largely at the laboratory stage.

3.3. *Air Classification*

Air classification of suitably ground or shredded material is receiving considerable attention, especially in municipal waste (see Section 4). Air classification separates particles in a moving airstream according to their density and shape and several systems utilising this principle have been developed.

3.4. *Pyrometallurgical Processes*

Pyrometallurgical processes are widely used in the recovery of non-ferrous metals. For example, most of the copper which is recovered is completely re-refined by pyrometallurgy, i.e., by essentially the same processes used in extracting and refining the metal from its natural ores.

3.5. *Hydrometallurgical Processes*

In hydrometallurgy the materials are dissolved in a suitable aqueous solution and then separated by a variety of chemically based operations. Examples of these include the processing of copper and nickel ores and scrap copper using ammonia in the separation step. A recent development is in the use of an aliphatic tertiary amine in the separation of cobalt from nickel in a solution obtained from an industrial sludge.[3] Another recent development is in the use of water and certain organic solvents (e.g., acetonitrile) for the separation and purification of copper and silver.[9]

4. Resource Recovery Processes for Municipal Waste

Only two methods are currently fully developed and practised for the recovery of resources from mixed municipal waste—heat recovery from incinerators and composting. Heat recovery from incinerators has been practised in Europe and Japan for some time and recently has been introduced into the U.S.A. and Canada. Nevertheless, there are still some significant technical problems with these systems, e.g., erosion and corrosion of the boilers. The technology of composting is well established but only in special circumstances has there been a sufficient demand for the finished product to make the process economically viable.

There has been a marked increase in the development of new techniques for resource recovery from municipal waste during the last few years and these include:

(1) direct materials recovery processes,
(2) direct energy recovery processes,
(3) pyrolysis processes,
(4) chemical processes.

4.1. *Collection Systems*

Although most discussions on waste disposal concentrate on the *disposal* aspect, we still have a long way to go in upgrading the collection side of the problem. Even the great technological achievements of the moon landings saw the astronauts dumping their refuse on the moon by the same ancient methods used on earth. We no longer carry our sewage out in buckets, yet our technology of solid waste collection is still at much the same stage.

Nevertheless, pipeline systems (analogous to sewerage systems) are being used (Table 4) and considered (Table 5). From the point of view of recovery of resources and also the maintenance of water resources System 4 would seem to offer the most promise.

Table 4

SUCTION PIPELINE SYSTEM AS BEING USED AT THE LISSON GREEN PROJECT, U.K.[5]

Basic Data:	
Area of site	Approx. 24 acres
Number of flats	1,904
Number of persons	4,532
Population density	192 persons/acre
Daily refuse output	40-50 cu. yd.
Indicative Costs:	
Installation	108 Pounds Sterling/flat
Running cost	5 Pounds Sterling/annum
Mechanical Data:	
Pipe	5/16″ gauge, top quality drawn steel, 20″ diameter
Speed of refuse in pipe	90 ft./sec.
Suction in pipework	1/4 atmosphere
Maximum power consumption	110 Kw/turbine (2 turbines + 1 reserve)
Active period	Approx. 1 hr./day

4.2. *Direct Materials Recovery Processes*

4.2.1. *As primary process.* Cellulose comprises from 40 to 50% of typical municipal refuse, mostly as paper. A water based pulping process has been in operation at Franklin, Ohio, since 1971 in a 150 tonne/day demonstration plant producing fibre pulp. Equipment to recover aluminium and glass (by air classification) has recently been installed. An optically based automatic sorter is being used to separate clear glass from coloured glass.

4.2.2. *Incinerator residue recovery.* The U.S. Bureau of Mines has developed a method for processing incinerator residues on a continuous basis to reclaim metallic iron concentrates, non-ferrous metal composites, glass fractions, and fine carbonaceous ash. The process utilises conventional and proven mineral separation equipment and techniques including froth flotation (glass fraction) and magnetic separation. A demonstration plant is being built.

4.3. *Direct Energy Recovery Processes*

4.3.1. *Heat Recovery Incinerators.* Combustion of the raw refuse in suitably designed incinerator-boiler plant. Steam is produced for heating and/or for the generation of electrical power. The Paris plant was completed in 1965 and there are now plants in other parts of Europe, Japan, and North America.

In addition to steam or power generation, other uses for the heat recovered are possible. It may be used to dry sewage sludge and thus allow the sludge to be used as fuel for the furnace. Again, the heat may be used to desalinate water or to produce steam to power equipment such as shredders and fans.

4.3.2. *Fuel Recovery.* The feasibility of using raw or separated refuse as a partial substitute for conventional fuels in power plants and industrial furnaces has been under study for several years.

In one demonstration plant (300 tonne/day) the refuse is milled and magnetic material separated out before being fed to the furnace along with pulverised coal. The refuse contributes 10 to 20% of the total fuel. An air classification system has recently been added to allow the removal of heavy and largely non-combustible material. This will reduce wear in the fuel feed components and improve the quality of the furnace ash (which is sold for road construction).

In another proposed system a fibre pulp

Table 5

PIPELINE COLLECTION SYSTEMS CONSIDERED BY BUILDING RESEARCH ESTABLISHMENT, U.K.[10]

(Cost in pounds sterling)

SYSTEM	Collection Freqency	Access Point	Condition of Refuse	Frequency Ancillary[1] Collection	Domestic Equipment	Technical Problems & Development Needs	Present Cost
0. Storage in paper or plastic sacks and manual collection (included as a reference base)	Weekly	Outside kitchen	Dry whole	None	None	None	170 240
1. Disposable containers floating in a modified sewerage system	Immediate	Inside kitchen	Dry compressed	Quarterly	Crusher	Container design. Crusher inlet design. Transport. Effect on sewers.	
2. Pulverised refuse in water pumped as a slurry	Immediate	Inside kitchen	Wet pulverised	Quarterly	Grinder pump	Slurry concentration. Grinder design. Pipe wear.	320
3. Crude refuse sucked through pipes	Daily	Probably outside kitchen	Dry whole	None	None	Commercially available system. Pipe wear unknown.	550[2]
4. Pulverised refuse sucked through small pipes	Daily	Inside kitchen	Dry pulverised	Quarterly	Grinder fan	Power requirements. Grinder design.	490[3]
5. Organic refuse pulverised and discharged to sewers. Metals and glass crushed, collected manually	Immediate	Inside kitchen	Wet pulverised	Monthly	Grinder Crusher sealer	Transport characteristics Grinder and sealer design.	270

NOTES: (1) The reference here is to an ancillary collection for small items (e.g., dry batteries) which could not be removed by a mechanical system. A collection for bulky discards such as furniture would be needed in each case.

(2) The inlet to this system is shared between two households and so the cost basis here is different.

(3) If the grinder were shared between two households (of System 3), it is estimated that this cost could be reduced to 310 pounds sterling. System 4 is the one being tested by BRE.

(similar to that in 4.2.1.) would be dewatered to give a combustible cake (50% solids content).

4.3.3. *Generation of Electricity.* In this proposed system the refuse would be shredded and separated, burned, and the hot exhaust gases used to drive a gas turbine/generator to produce electricity.

4.3.4. *High Temperature Incineration.* Combustion of raw refuse, with the assistance of an auxiliary fuel such as oil or natural gas, at a sufficiently high temperature (1700°C) so that the non-combustibles leave the furnace in a molten state. The products recovered are steam, ferrous metals, and a slag. A demonstration plant is in operation.

4.4. *Pyrolysis Processes* (Indirect Energy and Materials Recovery)

Heating of the organic portion of municipal, industrial, agricultural, or animal wastes to a suitably high temperature (400-900°C) in the absence of oxygen, leads to the production of lower molecular weight compounds which may be used as fuels and/or base chemicals and feedstocks for the petrochemical industry.

The process leads to three component streams:

(1) a gas component which includes hydrogen, methane, carbon monoxide, and carbon dioxide,

(2) an *oil* which includes organic chemicals such as acetic acid, acetone, methanol, and furfural, and

(3) a char consisting of carbon plus any inerts (glass, metals, rock, etc.) that are not removed in the primary separation steps. The proportion of the three components can be controlled to a considerable extent by varying the pyrolysis conditions. Both the gas and the oil are sulphur-free and have calorific values comparable to other fuel gases and fuel oil.

The process is being actively investigated by a number of groups from around the world and one of these groups estimates that a fuel gas equivalent to 100,000 tonnes of coal could be obtained from every million tonnes of domestic refuse (see also pages 35 and 36 this conference).

Particularly in the present climate of an *energy crisis* the fuel potential of the pyrolysis process is receiving most attention. Nevertheless, the reserves of reduced carbon (petroleum and coal) are finite and in the long term the major importance of pyrolysis may lie in its use to produce carbon from renewable forest reserves.

At least three demonstration plants are being built which incorporate pyrolysis processes.

4.5. *Composting* (Indirect Materials Recovery)

There are no major technological barriers to making compost, modern digesters being able to compost refuse with a paper content of up to 90%. The major problem is (usually) the lack of a viable market for the compost.

4.6. *Chemical Processes* (Indirect Energy and Materials Recovery)

Various chemical methods, such as hydrolysis, hydrogenation, wet oxidation, photodegradation, and aerobic and anerobic digestion have been suggested for converting municipal refuse and other types of organic matter into usable products such as glucose, hydrocarbons, protein, etc.

Few of these processes are yet at the pilot plant stage and all are in an early stage of development.

4.7. *Economics of Basic Resource Recovery Processes*

The following is a brief summary of the results of an examination of the economic aspects of eight basic resource recovery processes, along with two primary non-recovery solid waste disposal methods, carried out in the U.S.A.[11] The conventional disposal techniques —sanitary landfill and incineration—were used as bases for comparison.

Under the conditions used in the generalised economic analysis the net operational costs per input tonne ranged from about U.S.$3.00 to U.S.$9.00 compared to U.S.$2.50-6.00 for sanitary landfill and about $U.S.8.00 for incineration only (see Table 6). Nevertheless, it must be emphasised that these costs can be significantly affected by local variations in a variety of factors, including the value of the resources recovered.

Furthermore, at least one reclamation plant has been set up as a profit making venture, recovering glass, metals, and paper. The city will pay the company about U.S.$5.85 per tonne to take the refuse whereas it currently pays about U.S.$12.00 per tonne to incinerate its wastes.

The eight basic processes examined were:

(1) Incineration with recovery of materials from the incinerator residue.

(2) Incineration with heat recovery, used to produce steam for industrial applications.

(3) Incineration with heat recovery, used to

produce steam for industrial applications, and recovery of materials from the incinerator residue.

(4) Incineration with heat recovery, used to generate electric power.

(5) Pyrolysis, with recovery of oil, char, and inorganic materials.

(6) Composting of the biodegradable organic portion of the mixed refuse and recovery of inorganic materials from the nonbiodegradable portion.

(7) Materials recovery: paper pulp, aluminium and ferrous metals, and glass.

(8) Recovery of organics for use in public boiler furnaces as a supplementary fuel, along with ferrous metal recovery.

Table 6

SUMMARY OF RESOURCE RECOVERY PROCESS ECONOMICS

Process Concept	Investment ($000)	Total Annual Cost ($000)	Resource Value ($000)	Net Annual Cost ($000)	Net Cost per input tonne ($)
Incineration only	9,299	2,303	0	2,303	7.68
1. Incineration and residue recovery	10,676	2,689	535	2,154	7.18
2. Incineration and steam recovery	11,607	3,116	1,000	2,116	7.05
3. Incineration plus steam and residue recovery	12,784	3,508	1,535	1,973	6.57
4. Incineration and electrical energy recovery	17,717	3,982	1,200	2,692	8.97
5. Pyrolysis	12,334	3,287	1,661	1,626	5.42
6. Composting (mechanical)	17,100	2,987	1,103	1,884	6.28
7. Materials recovery	11,568	2,759	1,328	1,431	4.77
8. Fuel Recovery	7,577	1,731	920	811	2.70
Sanitary Landfill (close-in)	2,472	770	0	770	2.57
Sanitary Landfill (remote)[1]	2,817	1,781	0	1,781	5.94

Source: Midwest Research Institute (Ref. 11) — Costs are quoted in US$.

1. Located 100 miles from the municipality.

2. For comparison purposes landfill costs in Australia have been estimated at about A$1.90 tonne and incineration costs at about A$5/tonne.

The economic evaluation was based on a municipally-owned and operated facility with an economic life of 20 years, receiving 1000 tonne/day of mixed domestic and commercial refuse, and operated on a 24-hour/day, 300-day/year schedule.

5. Conclusion

Considerable effort has been devoted in recent years to the development of new techniques for dealing with solid wastes, using existing technology (or developments from it), or developing new technology. Nevertheless, much of the work is still at an early stage and further rapid changes can be expected.

There are at least two areas which seem to merit special attention:

(1) The development of systems for the production of reduced carbon as alternative sources to petroleum and coal.

(2) The development of new approaches to allow complete separation of the constituents of the inorganic fraction of wastes.

6. References

1. Broek, E. van den, and Kirov, N. Y., *The Characterisation of Municipal Solid Wastes,* Paper 6, 1971 Australian Waste Disposal Conference, Sydney, 1971.

2. Bureau of Mines, *Minerals Yearbooks,* 1951 to 1968, U.S. Government Printing Office, Washington.

3. Pearson, D. and Webb, M., *The Salvage and Recycling of Useful Materials,* The Chemical Engineer, p. 89, February, 1973.

4. Lyons, L. A., *Salvaging and Utilisation of Metals,* Paper 31, 1971 Australian Waste Disposal Conference, Sydney, 1971.

5. Kinnersly, P., New Civil Engineer Special Feature, August, 1973.

6. Beretka, J., *Industrial Wastes and By-Products Generated in Australia: the Results of a Survey,* Proceedings of the Royal Australian Chemical Institute, Vol. 40, p. 357, December, 1973.

7. Lyons, L. A. and Greenup, R. S., *Salvaging and Utilisation of Metals,* Environment 73 Conference, Sydney, 1973.

8. Lyons, C. J. and Morrison, D. L., *Solid Waste 1: Where Does It All Come From?,* Battelle Re-Search Outlook, Vol. 3, No. 3, p. 2, 1971.

9. Parker, A. J., *Hydrometallurgy of Copper and Silver in Solvent Mixtures,* Search, Vol. 4, No. 10, p. 426, 1973.

10. Courtney and Sexton, *Refuse Collection from Houses and Flats by Pipeline,* Building Research Establishment, U.K., current paper, CP 4/73.

11. *Resource Recovery,* Midwest Research Institute, February, 1973.

PAPER 34

Collection and Sale of Recyclable Goods from Households

By O. R. Taylor*

A report on the trial scheme conducted over 20 months by The Smith Family in selected suburbs of Sydney, describing methods of operation, community response, containers, disposal of goods and reasons for cancellation of scheme.

The Smith Family has over many years built up a substantial collection of clothing, furniture and household goods in the Sydney Metropolitan Area, Wollongong and Newcastle. This collection is by far the largest in Australia and is organised solely on a basis of the donor ringing and asking us to call. Our service is reliable and the call is made on the day arranged.

From this collection we retain all goods that are suitable to be given away to people in need, and the residue is sold. The Smith Family is the largest material welfare agency in Australia and in the year ended 30th June, 1973, it distributed free, goods worth more than $4,000,000. All profits from the businesses operated by the Family, which dispose of the goods unsuitable for welfare in various ways, provide 24% of income, and adequately cover total operating costs of 16% of income.

With this background of an experienced, large and successful collection of goods surplus to people's needs, and a proven ability to use these goods for the benefit of people in need, the administration has always been on the lookout for ways of extending these activities to earn more money and so help more people.

These thoughts led to an examination of the feasibility of a collection of recyclable goods on a door-to-door basis and eventually a decision was made to introduce a trial scheme. It is emphasised that the operation was a trial scheme only, for since we closed it down many people have erroneously assumed that such a collection is uneconomic. Ours was not profitable but we did many things that would be discontinued if a full scale operation was set up and we did not negotiate long-term prices for many of the goods.

We set out to ascertain in detail what the public response would be to such a collection and we obtained that information, together with first-hand practical information on the problems of picking up and disposing of the goods. This information was not available anywhere else here or overseas.

The Trial Scheme began in March 1971 and the following goods were collected:—

Clothing and Rag, Bottles and Jars, Paper and Cardboard, Steel and Aluminium Cans, Household Goods and Books, and Scrap Metal.

13,000 households were called on by two trucks each four weeks in the following suburbs:—

Mosman, Chatswood, Naremburn, Cammeray, Lindfield, Strathfield, Burwood, Stanmore, Miranda, Kogarah, Sylvania, Woollahra, Paddington, Bondi, Bellevue Hill.

Five months later the scheme was extended to include the whole of the Municipality of Ku-ring-gai, involving a further 28,768 homes and five more trucks.

Methods of Operation:

(a) The runs to be covered in both the first Pilot Scheme and the Ku-ring-gai development were allocated on the estimated capability of 300 households to each truck per day.

(b) Following the marking of runs on suitable maps, female canvassers were employed and about one week before the first call of the collectors, they delivered leaflets to the householder explaining the Scheme — personally if possible. In Ku-ring-gai, the introductory leaflet was accompanied by a letter from the Mayor.

(c) The first call was then made by the collectors on the date which the canvassers had advised the householder. A copy of a calendar with the dates of subsequent collections marked for the following six months was left on the first call to every householder on the runs. On successive calls, a Newsletter on which the date of the next call was prominently displayed was left at every home on the run.

(d) A strict four-weekly cycle was maintained and calls were made on public holidays (including Easter and over the Christmas/New Year period).

* Commercial Manager, The Smith Family, Sydney.

(e) The trucks used were 7-ton Bedfords with a tray of 30 ft. x 8 ft., the maximum size of a non-articulated vehicle. On each truck were placed nine containerised pallets. Of these, three were used to receive bottles, four — paper, one — clothing and rag, and one — salvage and household goods. The remainder of the tray area was used to receive the overflow of any of these items, together with a wire basket to contain steel and aluminium cans. There was a "hungry board" over the cabin which was occupied by a further containerised pallet used to receive larger items, particularly scrap metal and smaller pieces of furniture.

(f) The trucks were manned by a driver and two collectors. Once in the area the collectors called at every home and the work was equally divided by the driver, who acted as team leader and assisted in loading the goods. As previously stated, the collectors were instructed to call on every householder whether they had previously responded or not.

(g) Householders were encouraged wherever possible to leave the goods just inside their front gate. However, goods which were stored elsewhere on the property were also collected, as many people did not leave them out as requested.

(h) We employed a roadside supervisor whose duties included the co-ordination of the activities of all vehicles on the road. As these were usually operating in a confined area, he was able to arrange for trucks with light collections to assist other teams unable to complete the work of the day.

(i) The driver and the collectors were paid under the section of the Transport Workers' Award which applies to garbage collectors and sanitary workers.

(j) On return to the yard, the drivers and offsiders had completed their duties for the day. The vehicles were unloaded by a nightshift foreman and offsider using a forklift.

Collection Response and Community Attitude to Scheme

Of the total calls on all routes the average percentage response was:—

Bottles	41.4%
Paper	39.2%
Clothing	25.4%
Cans	19.0%
Non-Donors	28.6%

While The Smith Family has dropped the trial Scheme it has not abandoned the idea, and because detailed collection data would be invaluable to a possible competitor, it is not proposed to provide herein the actual volume and value of the goods collected.

From the inception of the project, the response from the community was most favourable. We had widespread support not only from the householders who were visited by our trucks, but also from such organisations as the Metropolitan Waste Disposal Authority, Ku-ring-gai Council and companies from within the Packaging Industry. Requests for a similar service to be established in other areas were widespread, both from individuals and local councils.

The attitudes of the people included in our collection runs were also examined in a survey carried out by The Mitchell College of the University of New South Wales.

Summarised, this revealed almost universal support for the service, particularly when it was provided by a charity. Many people had reservations about the support they would give to a similar service by their council.

Containers and Segregation of Goods

The biggest problem in operating a collection of this kind is the cost of sorting the goods either at the point of collection or depot. We decided on a combination of the two, by segregating the main items when they were loaded on the truck.

We did not attempt to organise the householders to do the sorting before leaving the goods out mainly because we had doubts about being able to rely on them to do so and, in addition, we did not provide containers for them. We consider, from our experience over many years of collecting from the public, that detailed segregation of goods for collection by the householder will not be successful probably because of lack of interest. Of course legislation could change this. The question of containers was considered in detail but there are many problems, e.g., size, shape, compartments, etc. Exchange containers may be feasible. We provided a cardboard container in one trial area and it was an instant success in that it increased the volume per house and also increased the response rate by 20%.

We would have liked to have trial plastic containers in each household but they are expensive and we could not afford to purchase the number required.

Disposal of Goods Collected:

The prices received for the various items were the subject of individual negotiation between The Smith Family and the companies concerned and we feel they should remain confidential.

The various items were disposed of as follows:

Paper

Sold to Australian Paper Manufacturers Ltd., the only major outlet in Australia. There is a strong demand from them and they state they will buy as much paper and cardboard as they can get. It is recycled by them. Countries like Indonesia also import newspaper for wrapping purposes but it has to be sorted and packed in bundles. The cost of doing this plus the freight makes the operation uneconomic in the current cost climate.

Bottles

We used all available means to dispose of bottles at different times and for varying periods so that we gained a full knowledge of the market.

Methods used were:

Segregation of beer bottles and their sale back to the N.S.W. Bottle Co. There is no limitation to the number they will accept.

Segregation and return of all soft drink bottles on which a deposit has been paid.

Segregation into brands and sale of wine and spirit bottles and flagons. This was subsequently discontinued because wine and spirit bottles were made "one trip" bottles by Australian Glass Manufacturers Ltd. and their value now is only as cullet.

All other bottles and jars were broken and sold to A.G.M. as cullet after being sorted into the base shades of brown, white and green. A.G.M. state they can recycle as much cullet as they can get.

Sale of bottles totally unsorted on a fixed price per container basis to J. McCarthy & Co. Pty. Ltd.

This system was tried after wine and spirit bottles were made "one trippers". This decision substantially changed the used bottle business and, when allied with the current high wage rates, the profitability of a bottle sorting operation is questionable.

Cans

Aluminium cans, which during our Scheme had a very small percentage of the market, were sold as scrap metal for recycling.

Steel cans were sold back to the "Steel Can People", a consortium of can manufacturers and B.H.P. Because the steel can has little economic value we were also paid a subsidy to assist with our collection costs.

Clothing, Rag, Books, Household Goods

These were retained by The Smith Family, and items suitable for our welfare purposes were given away to people in need. The residue was sold through our business outlets.

Scrap Metal

Proved to be almost non-existent and is not worth taking into account.

Frequency of Service

As mentioned earlier we called each four weeks, irrespective of holidays, but we did receive many requests for a more frequent service. This was tried on a fortnightly basis in some areas but without any real increase in the volume of goods collected.

We think the ideal service would be weekly because many householders do not have the space to accumulate goods for longer periods. Unfortunately, our resources did not allow us to run a weekly trial scheme.

Cancellation of the Scheme

We operated the collection for 15 months until November 1972, when it was cancelled for the following reasons:

The trucks proved to be unsuitable. The constant stopping placed great strain on clutches and differentials, and caused overheating. We were not in a position to provide a replacement fleet at the time.

The whole operation was labour intensive and wages were rising.

Our premises were inadequate for that type of operation on a long term basis.

The question of higher prices for most of the products sold was one that would take time to resolve, and we felt it preferable to close down the scheme while all matters were investigated. We had, after all, operated it only as a trial and believed we had accumulated enough information from it.

The Smith Family still has hopes of implementing a door-to-door collection of goods, and currently is having a professional invesigation made.

PAPER 35

Pyrolysis of Organic Wastes

By D. Barrett*

Introduction

The world produces enormous amounts of organic wastes, both agricultural and synthetic, and they provide a major problem of disposal. Disposal is met in many different ways, ranging from crude dumping procedures to sophisticated incineration processes. In general, these methods represent the local cheapest solution rather than an attempt to utilise any potential energy which may exist in these wastes. Pyrolysis, or the destructive distillation in which organic wastes are split into solid, liquid, and gaseous by-products, offers the opportunity of recovering some of this energy.

The destructive distillation of wood and coal has been practised for centuries and a great deal of information is available, but outside this area the information is much more meagre. In this paper, the pyrolysis of several cellulostic waste materials, e.g., sawdust, peanut shell and rice husks, is considered and also the pyrolysis of rubber as an example of a polymeric or synthetic type waste. The work has been carried out under conditions which simulate fluidised bed conditions since it seems probable that this type of process is more likely to have industrial significance, rather than static bed pyrolysis. Under fluidised bed conditions, the products of decomposition are rapidly removed from the hot zone and the product distribution is different from that obtained in a static system.

Organic wastes vary widely in type, but in considering their pyrolysis it is important to see if common ground can be found to make processing simpler. The first broad division considered has been into cellulostic and polymer type wastes, since it seems possible that the cellulostic materials will show strong similarities during pyrolysis, and similarly for polymer materials. This investigation has gone some of the way to examine the validity of such a division.

Pyrolysis of Cellulostic Materials

Wood, peanut shells and rice husks are all cellulostic materials. Wood has been considered to consist of 3 components, cellulose (50%) hemicellulose (25%) and lignin (25%). Other cellulostic materials are comprised basically of these components but in different proportions.

Brown[1] has found that the yield of products when wood is completely pyrolysed is about that obtained by pyrolysing the proportional amounts of the major wood constituents. The hemicellulose is said to decompose first, mainly between 200 and 260°C, followed by cellulose at 240-300°C, and finally by lignin at 280-500°C.

This belief in the additive nature of the products is not supported by Wenzl,[2] who considers that more interaction takes place between the products of pyrolysis from the wood components. It seems probable that the conditions of pyrolysis such as surrounding atmosphere, residence time of products, etc., may be expected to affect the interaction between components. No decisive answer is available at present, and no systematic work on this problem has been done on other cellulostic materials.

Pyrolysis of Polymers

Considerable investigation on the pyrolysis of polymers has been undertaken extending back for over a century, work being reported on rubber as early as 1860[3]. It seems likely that in the majority of cases depolymerisation is a primary reaction and that the size of the fragments obtained is related to the residence time in the reaction zone. In many cases therefore it could be expected that a high yield of monomer and dimer would be produced, and only small amounts of char and gas. Recent work in this area is reported by Wall in a survey of the literature on polymer pyrolysis[4].

Pyrolysis of Organic Mixtures in Practice

Practical pyrolysis of synthetic waste materials and urban refuse has received attention only in the last few years. The investigations have been carried out usually in batches in non-agitated retorts. The work has mostly been non-selective with respect to composition but has examined the effect of pyrolysis tem-

* Senior Lecturer, Department of Fuel Technology, University of New South Wales.

perature. The most detailed work has been reported by the United States Bureau of Mines using the BM-AGA pilot plant[5]. Union Carbide[6] have been experimenting with a process of compressing, melting and pumping waste plastic into a heated pyrolysis tube. The pyrolysis yielded hard and soft waxes and greases as well as a carbonaceous residue and gases. On pilot plant scale, destructive distillation of tyres has been shown to be a feasible method of obtaining valuable products from natural and synthetic rubber sources[7]. The product distribution was shown to be strongly temperature dependent.

Experimental Work

The apparatus used in our work is a batch-type reactor stirred by a paddle stirrer; the material to be pyrolysed is introduced from a hopper, after the reactor has reached operating temperature. The recovery train is quite simple, a cylindrical knock-out pot which catches the bulk tar, liquor and any fines carryover, followed by a tar mist trap and a gas collection system. The details are supplied in Figure 1.

Experimental runs on cellulostic wastes were carried out at constant temperature (600°C). The runs on rubber considered the effect of varying the temperature. The results are examined separately for the cellulostic and polymeric materials even though the products formed have some points of resemblance.

Results on Cellulostic Materials

The analyses of the waste materials are shown in Table 1. Also included is an approximation of the proportions of cellulose, hemicellulose

FIG. 1 LEGEND

1. Thermocouple Inside Stirrer.
2. Hopper Stoker.
3. Blanket Gas Feed Line.
4. Feed Hopper.
5. Reactor.
6. Gas Burner.
7. Paddle Stirrer.
8. Insulating Brick Furnace.
9. Ice-Water Bath.
10. Copper Tar Trap.
11. Volatile Products Offtake.
12. Stirrer Shaft Bearing.
13. Variable Speed Motor.
14. Calibrated Galvanometer.
15. Cotton Wool Collector.
16. $CaCl_2$ Bottle.
17. Copper Orthophosphate Absorber.
18. KOH Bottle.
19. Three-Way Valve.
20. Aspirator.
21. Outlet Water Valve.

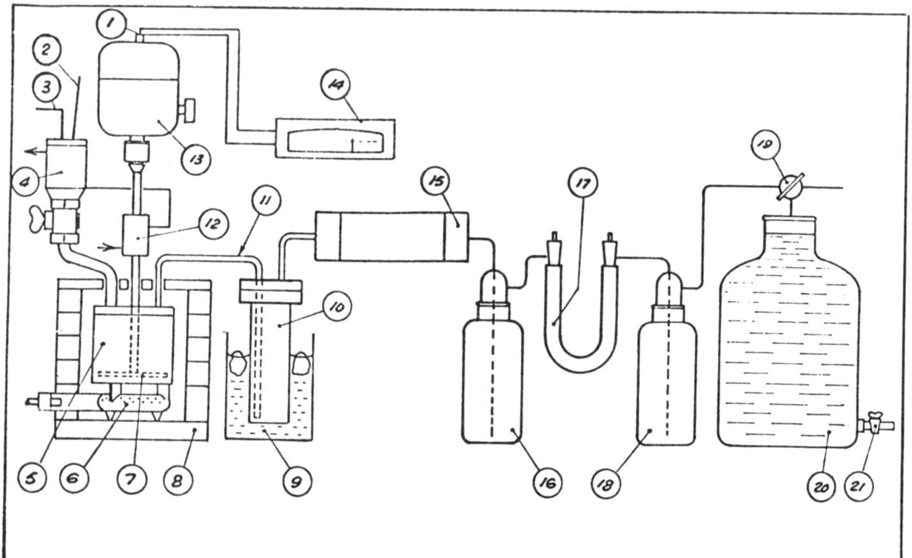

FIG. 1 SCHEMATIC PYROLYSIS APPARATUS.

and lignin. In Table 2 are shown the yields of pyrolysis products.

In yield of products, sawdust and peanut shells seem similar when compared on a moisture- and ash-free basis; rice husks appear quite different. This is illustrated more clearly by comparing the distribution of the elements in the products, shown in Table 3.

It seems probable that the process of decomposition of the rice husks is quite different; however, the sample of rice husks had a high ash content due to the inclusion of fine soil. Reported typical values for the ash content of rice husks range from 19% to 24%. The results obtained here may therefore be unusual. In any event more information is required on other cellulostic materials to see if some useful generalisation may be made.

Table 1

ANALYSES OF CELLULOSTIC MATERIALS

	Sawdust a.d.	Sawdust d.a.f.	Peanut Shells a.d.	Peanut Shells d.a.f.	Rice Husks a.d.	Rice Husks d.a.f.
Proximate Analysis, wt. %						
Moisture	11.4	—	7.2	—	6.2	—
Volatile matter	77.3	88.0	72.0	79.6	45.4	88.5
Fixed Carbon	10.6	12.0	18.5	20.4	6.0	11.5
Ash	0.7	—	2.3	—	42.4	—
Ultimate Analysis, wt. %						
Moisture	11.4	—	7.2	—	6.2	—
Ash	0.7	—	2.3	—	42.4	—
Carbon	40.3	45.9	49.2	54.3	25.3	49.3
Hydrogen	5.8	6.6	7.2	7.9	4.9	9.5
Nitrogen, Oxygen, Sulphur	41.8	47.5	34.1	37.8	21.2	41.2
Cellulose	—	50	—	45	—	50
Hemicellulose	—	25	—	20	—	25
Lignin	—	25	—	35	—	25

Table 2

YIELDS OF PRODUCTS OF PYROLYSIS

Pyrolysis Products, Weight %	Sawdust a.d.	Peanut Shells a.d.	Rice Husks a.d.	Sawdust d.a.f.	Peanut Shells d.a.f.	Rice Husks d.a.f.
Char	12.5	18.5	55.0	13.4	17.7	23.6
Tar	12.6	13.2	6.2	14.3	14.4	12.2
Liquor	17.5	13.0	20.0	7.0	7.4	27.2
Gas	56.2	53.0	17.5	64.0	58.0	34.5
Losses	1.2	2.3	1.3	1.3	2.5	2.5

Table 3

DISTRIBUTION OF ELEMENTS IN PRODUCTS OF PYROLYSIS

	Carbon wt. %	Hydrogen wt. %	$O_2 + N_2$ wt. %	% of Total Heating Value
Char: Sawdust	27.4	6.2	0.7	29.3
Peanut Shells	29.6	5.5	3.0	25.4
Rice Husks	51.1	17.7	2.6	48.7
Liquor: Sawdust	—	26.8	30.5	—
Peanut Shells	—	17.5	28.5	—
Rice Husks	—	40.0	66.5	—
Gas: Sawdust	46.2	47.7	66.0	55.8
Peanut Shells	45.1	50.0	61.5	48.2
Rice Husks	32.3	19.1	30.5	19.7
Tar: Sawdust	26.4	19.3	2.8	14.9
Peanut Shells	25.3	27.0	7.0	26.4
Rice Husks	16.6	23.2	0.4	31.6

Results on Rubber Material

The ultimate analysis of the rubber material is shown in Table 4; an assessment is also made of the proportions of natural rubber material and other substances such as carbon black, sulphur and inorganic fillers. This assessment permits some insight into the decomposition which takes places during pyrolysis.

Table 4

ANALYSIS OF RUBBER MATERIAL

Ultimate Analysis (air dried), wt. %	
Carbon	60.4
Hydrogen	6.4
Sulphur	2.8
Ash	23.4
Nitrogen and Oxygen	7.0 by difference
Natural Rubber	54.6
Carbon Black	12.2
Vulcanisation Sulphur	2.8
Inorganic Filler	23.4
Nitrogen and Oxygen	7.0 by difference

The overall material balances for the pyrolysis runs are given in Table 5 and the results suggest that the yield of char declines as the temperature increases; the converse occurs with the gas. It seems likely, although more data is required, that the tar yield goes through a maximum. This is not unexpected since the pyrolitic reactions may be divided into primary and secondary stages. At relatively low temperatures only primary cracking occurs and at 400°C even this may be incomplete whereas at 500°C primary cracking should be rapid and comprehensive and produce a lot of material in the liquid range. At higher temperatures, secondary reactions take place and convert much of the liquid material into products of lower molecular weight.

Table 5

YIELDS OF PRODUCTS FROM PYROLYSIS OF RUBBER

Temperature °C	400	500	600
Char wt. %	41.6	31.4	25.9
Liquor and Tar wt. %	49.6	57.5	49.7
Gas wt. %	4.2	7.2	16.9
Unaccounted	4.6	3.9	7.5

The formation of char as a result of secondary cracking is probably not significant under the operating conditions. The residual solid material may be expected to comprise essentially all the inorganic filler plus the carbon black, about 35%, and the weight % of char plus losses for the runs at 500°C and 650°C are 35.3 and 33.4 respectively. Within experimental tolerances it seems probable therefore that the natural rubber substance is converted to liquid and gaseous products only, at these temperatures.

The structure of the rubber material was completely changed during pyrolysis and the residue was a very fine lightweight black char. Much of this char was of dust size and carry-over was a significant problem and would presumably be so in a commercial fluidised bed operation. In static bed operation the residue could be anticipated to be a flaky bulk[7] but in fluidised bed operation the size of the residue would probably be independent of the feed size.

Conclusion

To be able to predict the extent of the overall reaction of pyrolysis on an engineering scale is a formidable task. For polymer-type materials, similar to rubber, it seems possible that a consideration of primary and secondary pyrolysis reactions related to such factors as temperature, residence time, particle size and atmosphere could prove adequate. For the cellulostic materials the picture is less clear and no discernible pattern has emerged. The fluidised bed type reactor, with its high heat transfer rates and short residence times for the volatile products, offers a simpler system for study of the pyrolysis process than is possible by static bed experiments.

REFERENCES

1. Brown, F. L. Work cited by Roberts, A. F. A Review of Kinectic Data for the Pyrolysis of Wood and Related Substances. Comb. and Flame, *14*, pp. 261-72 (1970).
2. Wenzl, H. F. J. The Chemical Technology of Wood. Academic Press, N.Y. pp. 253-300 (1970).
3. Williams, C. G. Phil. Trans. 150, 241 (1860).
4. Wall, L. A. Pyrolysis of Polymers. Nat. Bureau Standards, Special Publication 357, June 1972.
5. Sanner, W. S., Ortiglio, C., Walters, J. G. and Wolfson, D. E. Conversion of Municipal and Industrial Refuse into Useful Materials by Pyrolysis. U.S. Bureau of Mines. Report of Inv. 7358 (1970).
6. Showyin, L. The Problems Associated with the Disposal and/or Reuse of Plastic Containers. The 1971 Australian Waste Disposal Conference, Paper 33, University of N.S.W., Sydney.
7. Wolfson, D. E., Beckman, J. A., Walters, J. G. and Bennett, D. J. Destructive Distillation of Scrap Tyres. U.S. Bureau of Mines, Report of Inv. 7302 (1969).
8. Anon. The Utilisation of Rice Husks. The Agricultural Gazette, Vol. 51, p. 552 (1940).

Acknowledgement

This work is part of a general and continuing Fuel Department project, and thanks are offered to Messrs. Z. Nagy and J. H. Gray for their part in the investigations.

PAPER 36

Pyrolysis — Resource Recovery from Solid Waste

By P. R. Bell* and J. J. Varjavandi*

The ever-increasing production of solid waste, shortage of metropolitan land space, fossil fuels and raw materials has forced scientists all over the world to review seriously the much thrown around word — recycling. Not only here are we talking about recycling of the materials themselves but also the recycling of the energy content of the materials. For example plastics which are difficult materials to recycle directly are produced from non-renewable fossil fuels and have a high heat producing potential which should be recovered. The methods of solid waste disposal in common use, i.e., open dumping, sanitary landfill and incineration are being questioned as to whether they are adequate solutions to this growing solid waste problem.

The availability of dumping and landfill areas within the boundaries of our cities is becoming quite a problem now and looking to the future, with predictions [1] of solid waste production being six times what it is now by the year 2000, it is obvious that this method of disposal is not the ultimate answer. Incineration reduces the waste to a sterile residue with an approximate volume of one-tenth of that of the original refuse, thus extending the life of existing landfill sites. The need for sophisticated air pollution control equipment makes such incinerators quite expensive initially and the corrosive nature of the gases leads to high maintenance costs. Heat recovery from such disposal plants has not yet been fully developed on commercial basis. [2] Incineration merely destroys the waste materials with no positive contribution to the world's dwindling resources.

Pyrolysis is a process which has the advantages of incineration, i.e., volume reduction and sterilisation of the waste, but with the added advantages:

(1) It has been shown [3] to be a nett producer of energy which is in a convenient usable form (gas, oil and char).

(2) The gaseous pyrolysis products are produced in an oxygen deficient atmosphere and hence the formation of NO_x and SO_x are negligible. Any noxious gases such as HCl (from PVC degradation) and any particulate matter can be removed simply by scrubbing before combustion of the gaseous products. Hence this process poses none of the air pollution problems associated with incineration.

(3) The process can be designed so that any valuable chemicals formed can be recovered.

(4) Reclamation of the metallic constituents after this disposal process is possible since minimal oxidation takes place.

The pyrolysis process consists of heating the waste material in an oxygen deficient atmosphere so that thermal decomposition occurs rather than combustion. The major reacting component in municipal waste is cellulose and a possible pyrolytic reaction is:

$$3(C_6H_{10}O_5) \xrightarrow{\Delta} 8H_2O + \text{“}C_6H_8O\text{”} + 2CO + 2CO_2 + CH_4 + H_2 + 7C$$

where "C_6H_8O" represents a family of liquid products. It is a well documented fact that the products of pyrolysis are strongly temperature-dependent and hence this equation would be representative of a uniformly heated homogeneous sample at a particular temperature. In a large scale municipal pyrolysis unit where the temperature distribution would be anything but uniform and the waste would be anything but homogeneous there would be a great many such reactions occurring in parallel. Thus the design of such units must be based upon empirical data.

Progress to date in pyrolysis has been the result of two lines of research:

(1) Laboratory batch studies from which detailed scientific data has been obtained and published, but little of which is directly applicable to continuous processes.

(2) Large scale continuous process studies, carried out by private firms from which little data is available.

Experiments [4] carried out by the U.S. Bureau of Mines show that the pyrolysis of

* A.C.I. Environics, a Unit of A.C.I. Technical Centre Pty. Ltd.

one ton of municipal garbage at temperatures of 500°900°C produces:—

Solid Residue	154-424 lb	— porous carbon structure.
Tar	0.5-6 gall	— high molecular weight hydrocarbons.
Light Oil	1-4 gall	— mainly benzene.
Liquor	97-133 gall	— mainly water with dissolved organic compounds.
Gas	7,380-18,058 scf	— similar in composition to town gas.

Two basic designs have been proposed for the large scale pyrolysis of municipal refuse:

(a) *Indirect Heating*

The waste is heated externally in gas tight retorts, the vapours and gases are drawn off; the combustibles are either recycled to heat the retort, used in waste heat boilers or simply flared.

(b) *Direct Heating*

Pre-heated air or oxygen, in stoichiometrically deficient quantities, is passed through the bed of refuse, reacting exothermally with the char — the heat from which drives off all of the volatiles and pyrolyses the bulk of the waste, producing combustible gases which can be used to heat the incoming oxidising atmosphere.

A.C.I.'s interest in the pyrolysis of solid wastes dates back to 1970 when a small laboratory scale, batch pyrolysis unit was constructed and operated at the A.C.I. Technical Centre.

A variety of solid industrial waste materials was chosen for the initial runs, namely:

(a) low density polyethylene
(b) polyvinyl chloride
(c) phenolic resin
(d) nylon 66
(e) acrylonitrile butadine styrene polymer
(f) paper
(g) automobile tyres

In the initial trials, the gas produced was flared and the calorific value of the oils produced was measured. In all cases, except for paper and phenol formaldehyde where a large quantity of water was produced, the calorific value of the condensate was similar to that of fuel oil.

Based on the results of the initial trials, a second laboratory scale pyrolyser was constructed. This second model was designed to run continuously with the product gas being used as the fuel source for the unit. The successful operation of this small continuously operated plant with a variety of waste feedstocks led to the conclusions that pyrolysis:

(i) is a feasible process for volume reduction of many bulky wastes including urban and industrial waste into a sterile residue;
(ii) can produce some useful by-products in the form of combustible gases, oils, waxes and activated carbon;
(iii) is an inherently clean process which should produce minimal environmental pollution especially when compared with the incineration techniques;
(iv) has potentially lower construction and operation costs than incineration.

The encouraging results from these investigations prompted the decision to go ahead and construct a large scale continuously operated plant with a capacity of a half ton per eight hour day.

Description and Operation of the A.C.I. Pilot Plant

This plant (see Fig. 1) is designed to convert, on a continuous basis, solid waste into usable by-products by thermal degradation in an oxygen-deficient atmosphere using the indirect heating method.

Shredded solid waste is fed into a drag link conveyor which elevates the waste and deposits it into the top entry section of the reaction tube. The waste falls under gravity through the top of the reaction tube to form a bed 6 ft. deep in the reaction tube. The height of the bed is regulated by a set of "roll-crushers" which serve (a) to break up the residue; (b) to remove it from the reaction tube and (c) to regulate the rate of removal of the residue.

In the reaction tube, the solid waste is heated to high temperatures (500°-900°C) in an oxygen deficient atmosphere, resulting in the degradation of the waste material into a range of solid, liquid and gaseous products, depending upon the original starting material and the temperature in the reaction tube. The heat is provided by encasing the tube in a refractory-lined gas-fired furnace. Any other method of heating could be used but gas heating is considered to be best because the gaseous product of the process can be used as the heat source once the process has been initiated.

After passing through the roll-crusher the solid product falls under gravity past the gas off-take into the quench tank.

The water in the quench tank serves:—

(a) to quench the residue and cool it as it leaves the system;
(b) to act as a water seal to prevent air entering the reaction tube from below; and

(c) to provide sufficient head to compensate for the slight negative pressures in the reaction tube due to the suction build up of the vacuum pump.

At the gas off-take, product gas vapours are removed from the reaction tube via a manifold arrangement by suction and are piped away for vapour separation and cleaning. This is done by passing the product gas and vapour through a chilled water-cooled condenser by which most condensables are removed. The remainder of the gas stream is passed through a scrubber where any residual solids, vapours and water soluble gases are removed.

The cleansed gas is then passed back to the furnace, via a gas pump which provides the suction required to remove the gas from the reaction tube and is used to heat the reaction tube to sustain the process. The plant is equipped with a sophisticated automatic gas flow control system which regulates the gas flow rates both in the main and the pyrolysis burners. When enough gas is being produced for self sustained operation of the plant, the main burner switches off automatically. If too much gas is being produced the flare system is activated and the excess gas is flared. In some cases the gas production may fall below that required to maintain the temperature in the reaction tube; if this happens the main burner cuts in to provide the additional heat. A variety of shredded solid waste feedstocks was used:

(a) *Municipal Garbage*

The products from the pyrolysis of municipal garbage consist of:

Solid Residue — A reduction in volume of about 85%, depending on the actual composition of the feed. The residue has a porous carbon structure and can be disposed of as a safe, sterile landfill or possibly used as a low grade activated carbon in water purification. This is collected in the quench tank at the bottom of the reaction tube.

Tar (or Grease)	— a green coloured high molecular weight hydrocarbon
Light Oil	— similar to fuel oil
Liquor	— mainly water with dissolved alcohols and organic salts.

The tar, light oil and liquor are collected at the outlet of the condenser.

Gas — a fast burning gas indicating a high percentage of H_2.

(b) *Rubber Motor Tyres*

Shredded rubber motor tyres were successfully pyrolysed under a variety of operating conditions producing large volumes of gas and oil:

1 ton of rubber (= 100 tyres) yields 100 gall. of oil, or 22,000 s.c.f. of gas + ½ ton char.

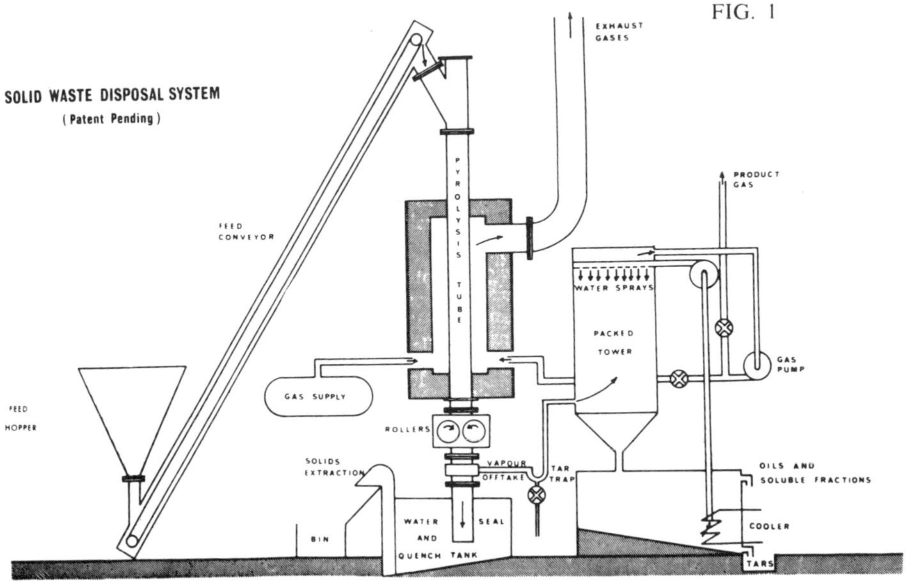

FIG. 1

The gas analysis for two runs at different pyrolysing temperatures is given below:

Material Pyrolysed	Rubber	Rubber
Ave. Temp. in Tube	300°C	410°C
Approx. gas analysis (not incl. N_2 and O_2)	%	%
H_2	4	7
CO_2	20	15
CO	1-5	1-5
Cl Hydrocarbons	16	33
C2 ,,	30	26
C3 ,,	12	8
C4 ,,	13	7

Note that for the higher temperature run, higher percentages of H_2 and CH_4 were produced — this being due to the further cracking of the carbon chains C_2 — Cn. Enough gas was produced at 400-500°C for the self sustained heating of the retort. Oil was collected from the tar trap and scrubber, and its physical properties were:

Oil Collected	Tar Trap	Scrubber
C.V. BTU/lb.	17,570	17,690
S %	1.75	0.96
Viscosity at 100°F Cs	8.5	7.3

The solid carbon residue retained the original shape of the shredded rubber although it was only 50% of the original weight. This residue formed a very porous bed and it was noted that it exhibited an affinity for oil.

(c) *Polyethylene*

With a feedstock such as plastic which forms virtually no solid residue it is necessary to have a porous inert bed (e.g., coke) in the retort so that the liquified plastic waste does not run straight out into the quench tank. By varying the height of this bed it was found that the degree to which the feedstock is pyrolysed can be controlled. With a 2-3 ft. bed into the hot zone large volumes of gas were produced. Enough gas was being produced at 400°-500°C for self-sustained heating of the system. With a 1 ft. bed into the hot zone mostly oil was produced.

Products

1 ton of polythylene yields 200 gall. of oil, or 44,000 s.c.f. of gas.

Future Developments

The immediate program involves the development of waste disposal units for specific types of wastes, e.g., rubber, plastic, wood chips. The throughput capacity of these units will be of the order of ½ ton/hr. It is anticipated that such units would be available commercially in the near future. Work will then continue on the development of much larger units (of the order of 5 tons/hr.) for the disposal of municipal refuse.

REFERENCES

1. "The First National Survey of Community Solid Waste Practices, Australia, 1972-1973", ACI Ltd., July 1973.
2. Watson, R. H., Burnett, J. M., "Refuse as a Fuel", Proc. Instn. Mech. Engrs., Vol. 183, Pt. 1, No. 24, 1968-69, pp. 519-526.
3. Hoffman, D. A., "Processing of Solid Municipal Waste by Pyrolysis", ASME Winter Annual Meeting, Nov. 16-20, 1969, preprint, 69-WA/PID-10.
4. Sanner, W. S., Ortuglio, C., Walters, J. G., and Wolfson, D. E., "Conversion of Municipal and Industrial Refuse into Useful Materials by Pyrolysis", U.S. Dept. of the Interior, Bureau of Mines, Report No. 7428, 1970.

ACKNOWLEDGEMENT

The authors are grateful to ACI Management for permission to publish this paper.

PAPER 37

Studies on the Production and Utilisation of Gas from Coconut Wastes in the Philippines

By Ibarra E. Cruz*

1. Introduction

The Philippines is the world's top coconut producer. In 1971, the world output was 30,000 million nuts of which Asia's share was 24,600 million. The Philippines contributed 7814 million of these or 32 per cent of the Asian and 26 per cent of the world total.

Coconut products are exported from the Philippines in the form of copra, coconut oil, dessicated coconut, and coconut husk fibre known as coir. In 1972, copra from the Philippines accounted for 53 per cent of Asia's total and 44 per cent of the overall world total.

In the processing of coconut into copra, roughly equal weights of copra, coconut shells, and husks are produced. The coconut husks are usually further processed into coir fibre (mattress and bristle fibres) if a coir factory is located in the vicinity of the coconut plantations. Roughly again, half of the husk goes into fibre and half ends up as waste coir dust during the defibring process.

The principal sources of wastes from coconut therefore are coconut shells and coconut husks. Present methods of utilising these wastes are rather primitive. In most parts of the Philippines, the coconut husk is used as fuel in native copra drying kilns locally known as "tapahan". The "tapahan" is simply a big hole dug in the ground inside which the husks are burned. Over the hole is a slatted platform on which the coconuts split into halves, are laid to dry in the hot gases and smoke that emanate from the rather inefficient combustion of coconut husks. After 36 to 38 hours the coconut is dried into copra in lots of 1000 to 2000 nuts.

The coconut shells recovered from copra-making are commonly burned in primitive carbonisation retorts to produce charcoal. The gas by-products of carbonisation are discarded, a practice which is wasteful, considering that 66% of the coconut shell is volatile matter and only slightly over 20% is recovered as charcoal.

In certain regions of the Philippines where coconut plantations supply the raw material for dessicated coconut factories, the coconuts are dehusked before being sold to the factory. The husks are in turn sold to coir defibring plants. The defibring process utilises the retting method whereby the husks are soaked in water for some 12 hours before being fed to the defibring machines. The resulting fibres therefore are wet and require drying by natural means (under the sun) and/or by mechanical means. At present, mechanical dryers for coir utilise diesel fuel.

Production output in coir factories is limited by three factors: (a) the supply of coconut husks; (b) the rate of disposal of the waste coir dust; and very recently, (c) the availability of diesel fuel for mechanical drying. The problem of waste coir dust disposal is particularly acute in that this voluminous waste now occupies large portions of factory sites. In many instances, the availability of other dumping sites dictates whether a factory can continue production once its current dumping site is filled up. Thus, the largest coir factory in the Philippines is able to produce only about 120 metric tons of coir per month (during the rainy months production is much less). Furthermore, the entire Philippine coir producers are able to supply barely 1% of the foreign demand for coir (which in 1970 was at 300,000 metric tons annually), although the Philippines have the potential of supplying the total world demand.

Therefore, it seems that many of the problems that beset the coconut industry in the Philippines may be solved by a more efficient utilisation of its waste products. This thinking spurred the College of Engineering of the University of the Philippines to initiate, as early as 1969, studies on the efficient utilisation of coconut wastes to help, in particular, the copra and coir fibre industries in the Philippines.

2. Producer Gas from Coconut Shells

Experiments were carried out in the U.P. College of Engineering in 1969 to gasify coco-

* Dr. I.Cruz is director of Graduate Studies, College of Engineering, University of the Philippines.

nut shells with air and steam. An experimental, updraft gas producer with a one-square-foot grate area was used. The effects on performance of steam content in the blast, the fuel bed depth, and the rate of gasification were studied. The parameters used to evaluate gasification performance were cold gas thermal efficiency and percentage decomposition of combined water.

Coconut shell contains about 50% combined water by weight, as can be calculated from the ultimate analysis of coconut shell shown in Table 1. Because of the considerable amount of combined water, the combustion zone in the reactor during gasification did not become too hot as to cause clinkering of the ash, and therefore addition of steam in the blast was not necessary for clinker control. However, addition of steam increased the hydrogen content of the gas, hence improved its calorific value and the efficiency of gasification. From stoichiometric calculations it was found that there was more water coming out with the gas products in the form of vapour than water entering the reactor in the form of steam. This was necessarily so because of the incomplete decomposition of the combined water of coconut shell. In effect, steam in the blast went through the reactor undecomposed and after reaction, was increased by an amount equal to the combined water that remained undecomposed.

Table 1

ULTIMATE ANALYSIS AND HEATING VALUE OF COCONUT SHELL AND COIR DUST COMPARED WITH WOOD

Ultimate Analysis, % (Dry, Ash-Free Basis)	Coconut Shell	Coir Dust	Wood
C	49	46	50
H	6	7	6
O	44	46	43
N	1	1	1
Gross Calorific Value, Btu/lb.			
(a) From Dulong's Equation	8020	7200	
(b) From Bomb Calorimeter Test	8090	7260	8010
Net Calorific Value, Btu/lb.	7400	6580	7430

Figure 1 shows the experimental results of test runs to determine the effects on gasification of steam content in the blast. It can be seen that the percentage decomposition of combined water increased with percentage steam in the blast up to a certain point, from 45% decomposition when there was no steam to an optimum 80% decomposition at 9% steam in the blast.

Similarly, the cold gas thermal efficiency increased with blast steam content from 67% at no steam to an optimum 82% at 9% steam. Cold gas thermal efficiency was taken to be the ratio of the net calorific value in Btu/cu. ft. of the producer gas saturated at 60°F, to the heating value of coconut shell plus the heat content of steam in the blast to produce one cubic foot of gas.

Fig. 1

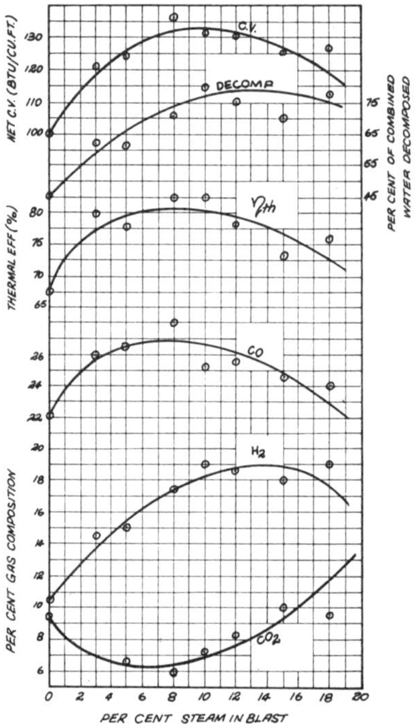

Fig. 1. EFFECTS OF STEAM CONTENT IN BLAST (GASIFICATION RATE = 40 lbs./hr./sq.ft.)

Figure 2 shows the influence of gasification rate on thermal efficiency and percentage decomposition of combined water. The gasification rate was directly measured by weighing the fuel fed and consumed in the gas producer during a given period of time. The actual rate varied from 20 to 120 lb./hr./sq. ft. The optimum condition was at a gasification rate of 40 lb./hr./sq. ft. at 9% steam in the blast. The cold gas efficiency was 82% and the percentage decomposition of the combined water was 76%.

Fig. 2

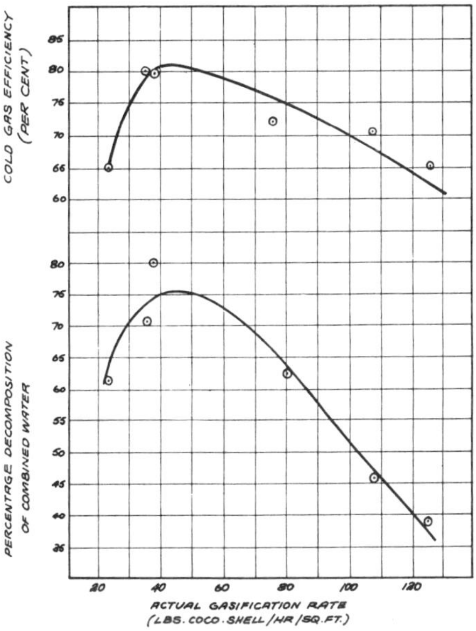

Fig. 2. EFFECT OF GASIFICATION RATE ON PERFORMANCE OF GAS PRODUCER (8 TO 10% STEAM IN BLAST)

The effect of fuel bed depth was also studied by obtaining data in test runs at different bed depths, varying the depth from 8 inches to 48 inches in increments of 4 inches. It was found that no signficant correlation existed between fuel bed depth and gasification performance. In 11 test runs at a gasification rate of 80 lb./hr./sq. ft. and 9% steam in the blast, the mean value of thermal efficiency was 72% with a standard deviation of 2.5, the average combined water decomposition was 62% with a standard deviation of 10, and the average net calorific value of the gas was 117 Btu/cu. ft. with a standard deviation of 6.

3. Producer Gas from Coir Dust

The same experimental gas producer as used for coconut shell was also used to gasify coir dust. The initial moisture content of the waste coir dust was 85%, wet basis, and it was necessary to expel some of the water by mechanically pressing the coir dust to form compressed briquettes of about 1 x 2 x 2.5 inches on the edges with moisture content reduced to about 50%.

It was also possible to gasify coir dust in its loose form in an updraft gas producer. The short fibres mixed with the dust prevented most of the dust from being blown away from the fuel bed.

Both compressed briquettes and loose coir dust were used in separate experimental runs lasting from 4 to 6 hours each. Experimental results are shown in Table 2. Gasification rate was obtained by calculation knowing the air flow rate and the gas analysis.

It was observed that the thermal efficiency, the calorific value, and the percentage decomposition of combined water were higher when gasifying loose coir dust than when gasifying briquetted coir dust.

It was observed further that when the fuel contained moisture above 50% (wet basis), gasification performance was unreliable. Every

Table 2

EXPERIMENTAL RESULTS — GASIFICATION OF COIR DUST

	Briquetted Coir Dust			Loose Coir Dust		
Moisture Content, %	50	40	30	50	40	30
Gas Analysis, %—						
CO_2	11.5	7.2	7.3	13.7	11.2	11.4
CO	15.8	23.1	23.0	14.0	18.2	18.7
H_2	7.5	8.4	7.8	12.0	10.7	15.2
CH_4		0.3	0.5	0.9	1.3	0.6
N_2	65.2	61.0	61.4	59.4	58.6	54.1
CV, Btu/cu. ft.—						
Gross CV	74.1	103.0	103.2	91.4	105.2	113.6
Net CV	70.3	98.6	98.7	84.5	98.6	105.5
Thermal Efficiency, %	55.5	69.5	69.0	62.2	68.8	74.0
Combined Water Decomposition, %	18.7	20.5	19.7	43.6	37.8	50.0
Gasification Rate, lb./hr./sq. ft. (as fired)	72.9	67.9	52.4	70.2	62.2	56.7

Fig. 3

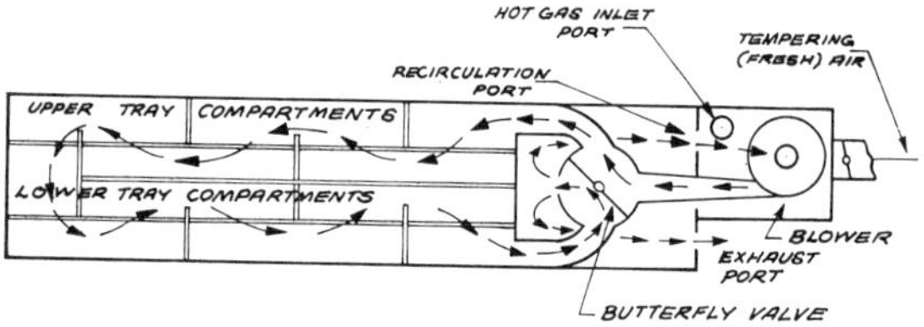

Fig. 3: Copra Dryer Showing Direction of Flow of Air.

(Direction of flow is reversed when the butterfly valve is rotated through an angle of 90 degrees from position shown.)

time the producer was charged with fresh fuel there was a tendency for the gas to become incombustible, and about 10 minutes' wait after charging was necessary for the gas to burn again. At moisture contents of 40% or less, the gas produced burned continuously even right after charging. Frequent poking of the fuel bed, however, was necessary to bring down the fuel to the grate, as the coir dust bed had a tendency to stick to the side of the reactor walls even after the bottom part of the bed had been consumed.

4. Producer Gas as Fuel for Copra Drying

A mechanical dryer for copra (Figure 3) was designed to utilise raw producer gas from coconut shells as fuel, dimensions of the drying chamber being 16 ft. x 4ft. x 5 ft. Direct heating was employed, i.e., producer gas together with the tar was burned in a furnace producing a smokeless flame, and the combustion products came into direct contact with the material being dried. To predict its performance, a mathematical model of the dryer system was developed.

The results of theoretical calculations predicting the total drying time and efficiency at various loads of nuts and blower rates, to remove 48% moisture from an initial 51%, wet basis are shown in Table 3.

Table 3

THEORETICAL RESULTS OF DRYER PERFORMANCE

Dryer Capacity No. of Nuts	Blower Rate, W_8 lb./min.	Inlet Temperature T_3, °F	Recirculation RC, %	Exhaust Air Relative Humidity RH, %	Drying Time Hours	Efficiency %
1000	90	180	64	40	12	66
1000	90	180	92	60	16	79
2000	100	180	63	50	18	68
2000	150	180	85	48	16	73
2000	150	200	87	54	12	79

4.1. *Experimental Results of Mechanical Dryer Performance*

Test runs on the experimental dryer were begun in January 1973 and lasted until January 1974. After having dried some 6000 nuts in batches of 500 to 1500, the predicted performances of the dryer were found to be in close agreement with the actual performances. As an example, the predicted performance for a test run to dry 1587 nuts is shown in Table 4 and the actual performance data are given in Table 5. At an average dryer temperature of 149°F and an ambient temperature of 85°F, theoretical calculations predict that 47% moisture can be removed from an initial 51% moisture content in 18 hours at a blower air rate of 150 lb./min. and an average recirculation of 61%. Actual results show that 48% moisture was removed from an initial 52% in a total drying time of 18 hours. The average dryer temperature was 141°F, ambient temperature was 80°F, and average recirculation was 56%. Coconut shells used as fuel was 76% of the shells recovered from the copra produced.

Table 4

THEORETICAL RESULTS OF COPRA DRYER PERFORMANCE BASED ON OPERATING CONDITIONS AS FOLLOWS:

Average Dryer Temperature = 149°F
Average Ambient Temperature = 85°F db, 78° wb
Initial Coconut Moisture Content = 51%
Number of Nuts = 1587
Air Circulated = 150 lbs/min.
Heat Loss, H = 7%

Drying Time	Inlet Temperature	Exit Temperature & Relative Humidity		Recirculation	Drying Rate, Dr, Grains per Min-Half Nut	Moisture Removed	Dryer Efficiency
Hours	°F	°F	% RH	% RC		% MR	%
0						0	
1	130	113	45	0	1.22	2	36
2	130	113	43	0	1.18	4	35
3	130	114	42	0	1.14	6	33
4	130	114	50	60	1.10	8	53
5	130	115	48	60	1.06	10	52
6	130	115	47	60	1.03	12	51
7	160	133	50	80	1.94	15	69
8	160	134	49	80	1.81	18	67
9	160	136	45	80	1.59	21	65
10	160	138	41	80	1.57	24	63
11	160	132	39	60	1.95	27	55
12	160	135	32	60	1.76	30	52
13	190	153	38	80	2.64	35	68
14	190	159	29	80	2.18	39	63
15	190	165	23	80	1.75	42	57
16	190	171	17	80	1.05	46	42
17	190	175	15	80	0.78	47	34
18	190	179	12	80	1.37	44	49

Table 5

EXPERIMENTAL RESULTS OF COPRA DRYER PERFORMANCE UNDER ACTUAL OPERATING CONDITIONS AS FOLLOWS:

Average Dryer Temperature: 141°F; Average Ambient Temperature: 80°F; 71°F wb;
Initial Coconut Moisture Content: 52%; Number of Nuts: 1587(1); Air Circulated: 150 lbs/min.

Drying Time	Inlet Temperature	Exit Temperature & Relative Humidity		Recirculation	Moisture Removed; Heat Loss		Dryer Efficiency	Notes
Hours	°F	°F	% RH	% RC	% MR	% H	%	
0					0			
1	125	97	60	0		17	52	Test continuous for 8 hours, Dec. 20, 1973
2	133	98	63	0		11	59	
3	130	110	35	0		9	37	
4	145	108	60	60		7	71	
5	148	126	29	60		6	51	
6	160	120	60	80		6	79	
7	166	137	30	80		5	68	
8	147	116	55	80	24	5	77	
9	135	102	50	0		17	50	
10	161	127	43	80		11	70	Dec. 21 1973
11	136	96	50	0		17	59	
12	145	113	42	80		11	74	
13	162	132	33	80		9	68	Test continuous for 8 hours, Dec. 26, 1973
14	178	142	26	80		7	69	
15	185	148	20	80		6	69	
16	190	156	16	80		6	65	
17	191	164	12	80		5	59	
18	190	170	10	80	48(2)	5	50	

1. Weight of fresh coconut (without shells) = 707 kilograms.
2. Weight of copra produced (4% moisture content based on initial weight of fresh coconut) = 368 kilograms.

5. Coco Gas and Its Uses

Coco gas is the term used for the gas by-products of carbonisation of coconut shells and coir dust when heated in closed retorts in the absence of air. As previously mentioned, the present method of charcoal production in the Philippines is quite wasteful since only 20% to 25% of the coconut shell is recovered as charcoal and the gas by-products are discarded. Research projects on improved designs of carbonisation retorts and the proper utilisation of the gas by-products are presently being undertaken at the College of Engineering, University of the Philippines. For instance, in a study on an integrated system of copra drying and charcoal production, a carbonisation retort is being designed to burn dirty or waste lubrication oil. (The waste oil burners have been designed and successfully tested to burn waste oil at rates of 2 to 20 gallons per hour.) The charge of coconut shells are to be heated indirectly by hot combustion gases that pass through flues that vertically traverse the inside of the vertical retort. The coco gas by-products will be used as fuel for copra dryers.

Studies on a pilot plant scale of the performance of a horizontal retort for carbonising coir dust have been in progress in the University of the Philippines for several years now, the coir dust char output being either in the form of briquetted charcoal or in powder form which can readily be processed into activated carbon. The coco gas by-products are partially used as fuel for heating the retort. However, since coco gas is a clean fuel (there is no sulphur and the tar is burned with the gas when sufficiently hot) it can be more useful as fuel in processes where clean combustion is required, for instance in directly heated dryers for copra and dessicated coconut, and recently, as fuel for internal combustion engines.

The recent scarcity of petroleum products has focused interest in the Philippines on coco gas and producer gas as alternative sources of energy. Research projects are now under way to (a) utilise coco gas in large (60 h.p.) stationary internal combustion engines, (b) convert oil-fired steam boilers to burn coco gas, and (c) synthesise motor fuel from coco gas. The last mentioned project is of considerable interest since alcohol as a motor fuel has been proved feasible by extensive Philippines studies. The main drawback is that alcohol is more expensive than gasoline. However, production of alcohol from coco gas synthesis may prove more economical than the traditional fermentation methods using sugarcane and starchy root crops.

Typical analyses of yields from the coir dust carbonisation retort are shown in Table 6.

Table 6

YIELDS OF PRODUCTS FROM CARBONISATION OF COIR DUST AND WOOD

Product	Coir Dust %	Wood %
Charcoal	40.9	41.3
Pyroligneous Liquor	28.0	31.7
Tar	3.5	10.0
Non-condensable Gas	27.6	17.0
Composition of Non-condensable Gas, %		
CO_2	18.7	54.0
CO	22.5	28.0
H_2	47.6	1.0
CH_4	10.4	15.0
C_nH_m	—	2.0
Air	0.8	—

6. To Sum Up

Coconut shells can be gasified with air (with or without steam) into producer gas with a net calorific value of about 130 Btu per cubic foot. The raw producer gas together with the tar can be burned efficiently and the absence of sulphur makes it most suitable as fuel for copra dryers and other mechanical dryers for foodstuffs. The coconut shells recovered from copra making are more than enough to meet the fuel requirements of the copra dryer. The surplus shells can be converted into charcoal, for which there is a growing market abroad. To improve charcoal recovery from coconut shells, integrated systems of copra drying and charcoal making are recommended, whereby the coco gas by-products from the carbonisation retorts can supply the fuel requirements of the copra dryers.

Use of coco gas and producer gas from coconut shells (instead of coconut husks) for copra drying, would make available more raw material for the coir fibre industry. This industry, which is at present supplying less than one per cent of the world demand for coir fibre, has the potential to secure the lion's share of the world market. A solution to the problem of disposal of its waste coir dust (the by-product of the defibring process) is gasification of coir dust and the utilisation of producer gas to meet fuel requirements for coir fibre drying.

7. Acknowledgement

The author acknowledges the support given by the National Science Development Board of the government of the Republic of the Philippines in the form of research grants for various projects on gasification, carbonisation and mechanical drying.

8. References

1. Cruz, I. E., Dimagiba, E. P., and Garcia, E. B.: "Producer Gas From Coconut Shells and From Coir Dust." U.P. Engineering Research Journal, Vol. 1 (pp. 1-10), July, 1972.

2. Cruz, I. E.: "Studies on the Performance of a Mechanical Dryer for Copra." Paper presented at a Symposium on Copra Drying and Charcoal Production sponsored by the National Science Development Board, Manila, Philippines, August 23, 1973.

3. Casillan, M. D.: "Thermochemistry of Coir Dust Carbonisation" — Master's Thesis, College of Engineering, University of the Philippines (1968).

4. Festin, T. F., Professor of Chemical Engineering, University of the Philippines. Private communications regarding pilot plant studies on the performance of a horizontal retort for coir dust carbonisation.

5. Galvez, J. Jr., President, Luzon Coir Industries, Inc., and past President, Federation of Coir Producers and Exporters of the Philippines. Private communications.

6. Anonymous: "The Philippines: World's Top Coconut Producer" — Philippine Farmer's Journal, pp. 6-7, September, 1973.

PAPER 38

The Disposal of Tyres

By R. L. PINCOTT*

Imagine going back to a world without motor vehicles. This is probably not very difficult for our grandparents but for today's "generation" motor vehicles are an accepted part of life, and an integral part of motor vehicles are rubber tyres.

From the child's scooter to Jumbo jets, from forklift trucks to the largest earthmoving equipment, you will find rubber tyres being used.

The question troubling us all is the best use to put these tyres to when they are no longer able to perform the specific job for which they were designed.

In this paper it is not my intention to present to you the answer, but to relate to you information which I have gathered over the years about tyres that are presented for disposal, etc., their weights, volumes, uses, reclaiming, etc.

How Many Tyres Are There Annually?

The following figures of tyres consumed in the Australian market were supplied by the Commonwealth Statistician's Department for the year 1972-73:

	Passenger	Truck	Earth-mover	Air-craft
Locally Manufactured	7,444,000	575,000	203,000	9,890
Imported	1,764,219	277,642	20,084	5,822
	9,208,219	852,642	223,084	15,712
Exported	249,466	49,476	11,510	—
Total	8,958,753	803,166	211,574	15,712

This quantity is increasing at approximately 7¾% annually. However, these figures do not indicate the number of tyres that are imported into Australia already fitted to motor vehicles. Remember, too, that there are five, not four, tyres to every motor car.

How Many Tyres Have to be Disposed Of?

All of the tyres that are manufactured and imported into Australia (less the exported tyres) will end up as scrap and have to be disposed of. The tyres may, in their useful life, be worn out, retreaded, worn out and retreaded again, but there comes a time when the tyre can no longer be retreaded and therefore will be presented for disposal.

In 1972-73 New South Wales consumed approximately two million passenger tyres, and of these about one and a half million were in the Sydney metropolitan area, giving 29,000 passenger tyres per week to be disposed of and approximately 3000 truck tyres.

Of the original new passenger tyre weight, 80% is presented for disposal as a scrap tyre, the other 20% representing the original tread surface that has been worn away.

How Much Room Do They Occupy?

The following figures, based on personal experience, have been found reasonably accurate over the year.

There are approximately 20 passenger tyres per cubic yard and three truck tyres per cubic yard, which means that annually we have to dispose of in Australia of 447,937 cubic yards of passenger tyres and 267,722 cubic yards of truck tyres. I am unable to present the equivalent figure for aircraft tyres and earthmoving tyres. If we averaged it out at three cubic yards per earthmover tyre we have 634,722 cubic yards, and took an average of one aircraft tyre per cubic yard, we have 15,712 cubic yards. Totalling the volume of the passenger, truck, earthmover and aircraft tyres we get 1,366,093 cubic yards of tyres annually for disposal.

The above figures are not based on the volume of air space the tyres occupy but the space they would occupy when stacked in a truck in a particular way, which achieves the maximum tyres per cubic yard. If they were just thrown into a heap on the back of a truck, they would occupy approximately 50% more space. On the other hand, if the tyres were compressed in bundles, there would be a saving of 60% in volume. If the tyres were shredded, they would only occupy approximately 15% of the original volume.

How Much Do They Weigh?

Again, the following are figures that I use in my day-to-day business of handling scrap tyres.

* R. L. and A. T. Pincott, Scrap Tyre Merchants.

There are approximately 130 passenger tyres to the ton and 25 truck tyres to the ton. This gives us 68,913 tons of passenger tyres per annum and 32,086 tons of truck tyres per annum, a total of truck and passenger tyres of 100,999 tons per annum, plus my estimate of half a ton average for the total of earth-mover and aircraft tyres of 113,643 tons, giving a grand total of 214,642 tons of scrap tyres for disposal annually.

How Many Tyres Presented for Disposal are Actually Scrap?

In this context scrap tyres are defined as tyres which cannot be returned to a condition to be used for the specific job that they were designed to perform, i.e., be retreaded or safelv repaired.

Of the tyres that are presented for disposal our estimates are that only approximately 60% come under this definition of scrap tyres. The other 40% are tyres suitable for retreading.

Why Are Retreadable Tyres Presented for Disposal as Scrap?

The demand for the particular size of retreaded tyre determines whether the worn-out case is worth keeping by the tyre service for retreading. For instance, there is no demand for retreaded tyres of Rolls Royce cars.

The demand for retreads of a particular size of tyre can vary from tyre service to tyre service, suburb to suburb, city to city, metropolitan to country, state to state.

Storage space is another major factor. When a new size of tyre comes onto the market and there is not the immediate demand for retreads, the cases need to be stored for long periods of time until retread moulds are produced and the demand for the retread size of tyre comes about. Storage space is expensive in retail areas, where the tyre dealers are located, so tyre dealers elect to throw the retreadable cases away.

The Scrap Tyre Dealer

He is a dealer who collects the tyres that are presented for disposal and is able to recognise the 40% of tyres that are retreadable and disposes of the "scrap" tyres. He knows where particular sizes of retreadable cases are in demand and channels the retreadable cases back into the system. He also makes arrangements for the storage of retreadable cases, that are not in demand at that particular time, for future use.

He is a valuable asset to the community, saving a great deal of waste; in this regard he has counterparts in the scrap metal, waste paper and rag merchants.

Disposal by garbage operators of good retreadable tyres is wasteful. It would be desirable to channel all tyres for disposal through a scrap tyre dealer first before they are finally discarded.

Reclaiming Rubber from Scrap Tyres

In Australia there are two plants which reclaim rubber from scrap tyres. These two companies reclaim the rubber from approximately one and a half million passenger tyres per annum.

However, it is understood that these companies are having great difficulty in competing price-wise with reclaimed rubber that is being imported from England and other overseas countries for use by the rubber product manufacturing industry in Australia. The reclaimed rubber imported into Australia is sold at approximately 10 cents per lb., whereas the local manufacturers need approximately 14 cents per lb. to continue to operate.

Figures supplied by the Commonwealth Statistician's Department show that in the year 1972-73, 46,586,599 lb. of reclaimed rubber was imported into Australia, which is the equivalent of 4,159,517 passenger tyres, or approximately half of the scrapped passenger tyres in Australia per year. If the two rubber reclaiming plants in Australia cease to operate, there will be another 1½ million passenger tyres per year to dispose of. We are, in fact, importing 4,159,517 scrap tyres into Australia in the form of reclaimed rubber.

I feel there is good reason for a government body to have a close look at this industry to see whether the reclaimed rubber that is being imported is being dumped here, and whether an import duty or some other incentive can be introduced to enable the rubber reclaiming plants in Australia to continue to operate and expand in order to use all of the scrap tyres available. This will not only result in a reasonable saving of money flowing out of Australia, but will also help solve a major disposal problem.

Disposal of Tyres by Burning

Technically, it is quite feasible to destroy scrap tyres by incineration. Equipment is available to achieve this without infringement of the requirements of the Clean Air Acts. The problem is, however, cost. It costs 3-4 times more to incinerate than to dump tyres in land-fill sites.

The heating value of tyres varies between 12,000 to 15,000 Btu/lb. In comparison, normal garbage generates 3000 to 4000 Btu/lb. Scrap tyres are already a good source of fuel but the handling, storing, collecting of the tyres, the low demand for steam and high costs are against this method of disposal.

Fishing Reefs

There is an increasing interest in the use of scrap tyres to construct artificial fishing reefs. The tyres are put together in small bundles of approximately 10 passenger tyres each, then five of these bundles are strapped together and taken out on barges where divers place the tyres on a sandy bottom and thus build up an artificial reef.

Fish like to inhabit an area where they can feed and be close to an area where they can dart to safety if a larger fish comes along. The artificial reef built of tyres provides the perfect safety as the larger fish cannot fit into the nooks and crannies that bundles of tyres provide.

The tyres do not rust away as car bodies did when they were first used to build artificial reefs. After a period of time, useful organisms attach themselves to the tyres and produce food for the fish. Artificial reefs built of tyres can be placed in a barren area, and fishermen are able to know the exact spot where the fish are biting.

It should be understood that while using scrap tyres for artificial reefs is putting scrap tyres to a use, it is not the complete answer. One such reef, currently under construction in Sydney, will use approximately 600,000 passenger tyres, which is equal approximately to the number of scrap tyres generated in Sydney over a period of twenty weeks.

This is not a cheap method of disposal: the tyres have to be collected, each tyre has to have a hole cut in it, and the tyres have to be bundled together and carried out on barges and handled by divers to be placed in position — a very high labour cost project — which is neither an economic nor a long term answer to the problem.

Swamp Reclamation

The reclamation of swampy ground lends itself as a method of scrap tyre disposal. The general complaint of tip operators, who use landfill with general garbage and tyres, is that the tyres will not consolidate in the tip because the inside cavity of the scrap tyre never fills with hard fill, hence the ground remains spongy and the tyres work their way to the top of the tip when traffic travels over it.

Swampy, useless ground that is on river level and the water level of which fluctuates with the rising and falling of the river, can be successfully reclaimed by using the tyres to raise the level of ground, so that the fluctuations of the normal river level would not affect the reclaimed land. Tyres will not pollute the water or land in any way and will remain there indefinitely.

Trenches are dug in the swamp and the earth removed is placed alongside the trench. A pump is used to keep the trench dry and then scrap tyres are stacked into the trench in a particular way. The earth previously removed is placed back on top of the tyres and another trench is dug parallel to the previous one. The type of earth being river silt oozes down through the tyres and into their cavities and consolidates into a solid mass. The rising and falling water level helps to move the silt into the cavities.

The site, when completed, is raised up above the high water level of the nearby river; it is solid and is able to be put to use.

Shredding

One of the great problems of scrap tyres is the volume they occupy and the difficulty of handling.

If tyres could be shredded easily into a form where they could be handled by conveyor belts, hoppers and the general equipment available to reduce labour, they would be in a form that should be acceptable into normal landfill sites.

Tyre engineers strive to produce a tyre that will not wear out, be abrasive and absorb impact. They use high tensile steel in the bead case and now there is a general swing to use steel cords in the radial belts in both passenger and truck radial ply tyres. All these qualities make for a better tyre but play havoc with any machinery used to try and shred tyres.

The machines that I have seen so far for shredding tyres will not cut the beads up into sections and each tyre has to be handled individually into the machines and makes the use of such machines uneconomical.

The Jacques shredder is reputed to be able to shred passenger and truck tyres in quantity and I hope to see one of these shredders in operation in the very near future. There is no reason why shredded tyres would not be acceptable into any landfill sites.

Novel Uses of Scrap Tyres

Scrap tyres have been used for various purposes. One that springs to mind is that associated with the mini bike boom. Bike clubs have been using scrap tyres for edges of mini bike tracks. With a small trench digger, such as a Ditch Witch as used by the P.M.G. to lay telephone cable, a trench is dug along each side of the bike track to approximately half the depth of a normal passenger tyre. Passenger tyres are then placed standing up in the trench and sand is used to fill in the rest of the trench. This forms a definite barrier around the edges of the track but will not cause any damage if a bike runs into it.

Tyres have been used to form a barrier around plants like a normal pot, they have been stuck together and used as table legs in trendy furniture, placed on sides of wharves as buffers for ships, but this latter use is decreasing as rubber companies manufacture now specially-produced buffers.

PAPER 39

Recycling of Metals from Scrapped Automobiles in Australia

By L. A. Lyons* and C. S. Tonkin*

1. Recycling Problems

Some of the technical and commercial problems involved with recycling do not always appear as obvious to some people as they should be:

(i) Costs associated with recycling may exceed those incurred in production from a primary raw material. In recycling metals from old motor vehicles, engineering and metallurgical technologies of a high order are required. Especially in the case of a stripped car, handling and transport costs may be prohibitive and encourage abandonment even in city streets by unscrupulous owners.

(ii) If there is no inherent profit in the operation, someone — generally the community — must pay if recycling at a financial loss is desirable. Recycling at a loss may be warranted in order to husband primary resources or to improve environmental conditions.

We understand that interment of car bodies costs some Australian municipal authorities more than $25 each. We do not attempt to weigh the relative merits of resource husbandry, an uncluttered environment and investment for profit — they are all important.

(iii) Recycling operations may reduce litter or pollution in one area only to introduce similar problems elsewhere, e.g., scrap metals recycled in an urban environment may cause a nuisance to more people than production from ores in a remote part of the country.

2. Metals from the Automobile

We are attempting to survey a wide and variable field which includes cars small and large; old and new but unusable due to accident; imported and locally made; trucks and buses of steel and aluminium. We can probably only guess at the annual tonnages of metals available.

According to investigations in the U.S.A., where much more data of a statistical nature are available than in Australia, the percentage make-up of new cars is as follows:

Table 1

COMPOSITION OF A NEW MOTOR CAR[1]

			%
Ferrous Metals	Steel		71.6
	Cast Iron		12.6
Non-Ferrous Metals*	Zinc		2.1
	Aluminium		1.9
	Copper		1.0
	Lead		0.7
	Nickel	mostly plated	0.1
	Chromium		0.1
	Tyres		3.5
	Other organic and inorganic materials		6.4
			100.0

* A trend which is not obvious from a table such as the above is the increasing usage of zinc and aluminium in the motor vehicle industry.

While Table 1 shows ferrous and non-ferrous metals as comprising 84.2 per cent and 5.8 per cent, respectively, of new car composition, the following Table 2 illustrates the motor vehicle industry as having a greater impact on the domestic consumption of non-ferrous metals than ferrous metals. A comparison is drawn with consumption in the U.S.A.

Table 2[5, 6]

METAL CONSUMED IN THE MOTOR VEHICLE INDUSTRY AS A PERCENTAGE OF CONSUMPTION (1969)

		Australia % of consumption*	U.S.A. % of consumption
Cu		9.0	8.3
Al		13.2	9.9
Zn		14.4	32.5
Pb		38.2	50.0
Ferrous	Steel	7.5	20.0
	Iron	1.5	

* Including recycled metal.

The U.S.A. (production 11,000,000 motor vehicles annually) produces four times as many cars as trucks and buses. In Australia (production 500,000 motor vehicles annually) the ratio is 20 to one.

* Sims Consolidated Ltd.

3. When a Car Becomes Scrap

Generally, unwanted vehicles pass through the hands of wreckers who remove reusable parts (which are eventually scrapped in any case) and sell radiators, batteries and other components as "scrap metal".

Engine blocks, generally cast iron, are also likely to be removed for separate sale as scrap metal. Iron blocks have satisfactory weight to volume ratio, are relatively easy to break and the product is of value to foundries.

Finally, there remains the really objectionable residue: a steel frame and body of considerable volume and low weight, contaminated with upholstery and other organic materials as well as inorganic matter such as glass.

In the U.S.A. hulks range from 273 kg (if everything is stripped) to 1455 kg (if nothing is stripped). The average hulk weighs about 693 kg. On the same basis the average Australian hulk is estimated at 590 kg.

The installation and successful operation of two scrap steel fragmentising plants (often referred to as motor car shredders) in Australia has attracted more than usual interest. In this paper, we do not intend to refer solely to fragmentisers as they are only one aspect of automobile recycling and, also, because they are used for other scrap too.

Conservative estimates of the weight of the average Australian car and the average Australian truck/bus are 1182 kg and 5455 kg respectively.

Scrappage rate (as a percentage of vehicles on register) is 5.0-6.0 per cent[4] annually. For the purposes of Table 3, 3 per cent truck/bus scrappage rate is assumed. However, this makes little difference to the total.

Table 3

ESTIMATED METAL AVAILABLE IN ONE YEAR FROM SCRAPPED MOTOR VEHICLES
(Excluding motor cycles and assuming 6% scrappage rate for cars)

	Thousand Tonnes
Steel	280.2
Cast Iron	46.0
Zn	5.8
Al	5.5
Cu	3.4
Pb*	2.1
	343.1

* This is in fact a small proportion of the lead recycled from motor vehicles. Secondary lead is available mostly as battery scrap and the greater proportion (about 16 thousand tonnes annually) comes from batteries replaced in vehicles still on the road (refer Table 4).

As a comparison with the above estimates we offer figures showing estimated annual consumption of metals in the automotive industry.

Table 4

ANNUAL METALS USAGE IN AUSTRALIAN AUTOMOBILES[5, 6]

		Thousand Tonnes
Ferrous	Steel	487.5
	Iron	84.0
Pb		26.2
Zn		15.9
Al		14.4
Cu		8.4
		636.4

4. Non-Ferrous Metals

Just as fortunate concentrations of metallic minerals (ores) in parts of the earth's crust permit economic extraction of metals, so concentrations of some metals in old motor vehicles enable us to recycle them economically, i.e., aluminium, copper, lead and zinc. These four non-ferrous metals are recyclable if not 100 per cent recoverable. It is of interest to compare their abundance in the earth's crust with the proportions in motor cars, Table 5.

Table 5

METALS — RELATIVE ABUNDANCE IN THE EARTH'S CRUST AND IN THE MOTOR VEHICLE

	Metals in the Earth's Crust Percent	Relative Abundance	Metals in the Motor Vehicle	Relative Abundance
Al	8.130	61.80	Steel	81.6
Fe	5.010	38.08	Cast Iron	13.4
Cu	0.010	0.08	Zn	1.7
Zn	0.004	0.04	Al	1.6
Pb	0.002	0.02	Cu	1.0
			Pb	0.6
	13.156	100.00		100.0

These are "everyday" metals. For comparison, the unfamiliar metal, silicon (used in various alloys), is said to be 27.72 per cent of earth's crust, sodium (which is not often seen as a metal but which is widely used as common salt) is 2.85 per cent, and magnesium 2.09 per cent (which, amongst other things, is in, Epsom's salt).

It is impossible to quote stable prices for metallic scraps. Non-ferrous scrap values tend to follow refined metal prices in a given area, but scrap steel is more affected by sudden changes in demand by steelmakers, who may have a constant intake of ore for their blast furnaces at fairly stable prices, but use steel scrap as a means of quickly increasing or decreasing output when markets for their end products rise or decline.

(i) *Aluminium*

Aluminium alloys are used in transmissions, pistons and sometimes in the engine block. Alloys of this metal are also extensively used in transport units of various kinds.

The aluminium recycler does not only receive metal from old automobiles. A wide variety of physical shapes of scrap and a multiplicity of alloys create many difficulties which are overcome by classification, baling and smelting with additives in rotary or reverberatory furnaces. Close control of alloy compositions is achieved by spectrographic and other methods of analysis.

In Australia, there are seven important secondary aluminium alloying plants which produce specification ingots for the diecasting industry. This industry is very heavily dependent upon automobile manufacture for prosperity. Fumes from secondary aluminium melting and refining are extremely difficult and costly to control, but the industry in Australia has spent considerable sums of money in this area (see Section 7).

Prices for aluminium and its alloys are governed by world supply and demand factors. Different scrap categories have ranged in value between 36 cents a kilogram for extruded aluminium to 11 cents a kilogram for aluminium gauze or venetian blinds (which give high melt losses on smelting). Until recently, there has been a world surplus with intense competition for sales and uneconomic operation for many primary and secondary producers. As with most other commodities, shortages have recently developed and prices have risen.

(ii) *Copper*

Copper from automobiles is only a small part of the copper recycling industry. It is the most expensive of our four non-ferrous metals and the most widely-fluctuating in price, being affected by supply-demand factors and often by politics as well. Much of the world's copper is produced in relatively unstable countries, e.g., Chile, Zambia and Zaire.* Australia is an exporter of copper and generally relates copper prices to quotations of the London Metal Exchange.

Whatever the copper price, there is some value for, and trade in, copper-bearing scrap. Radiators, in recent years, have been worth from 70 cents to 85 cents per kilogram in Australia (depending on ruling A.C.P.**). Since the radiator has adhering solder (tin and lead) and maybe brass (copper and zinc), it is an important raw material for gunmetal (red brass) manufacturers.

Radiators, with or without prior "sweating" to remove solders, are treated with other types of scrap in reverberatory, rotary or electric furnaces. Molten metal is sampled and analysed before casting into ingots which are used for valve manufacture, plumbing supplies and many other applications.

There are ten secondary copper ingotting plants capable of producing a full range of specified copper alloys. Surplus scrap is converted to refined copper by a large refining company.

(iii) *Lead*

This metal is very heavily dependent upon the automobile industry, as illustrated by the data in Table 6.

Table 6 [5, 6]

AUSTRALIAN LEAD CONSUMPTION IN THE MOTOR VEHICLE INDUSTRY

	Percent of Annual Lead Consumption
In new cars	7.93
In tetraethyl lead†	5.93
In automotive batteries for other than new cars	24.31
Total	38.17

†Consumed, but not produced in Australia.

Overseas a large part of total usage is in motor fuel and does not concern us here. The remainder is applied principally as antimonial lead in the storage battery, a small heart in a comparatively large monster — car, truck or other vehicle. There has been more than usual publicity lately regarding the possible use of storage batteries as the total energy source of automobiles, but this is probably far in the future. Undoubtedly there will be increasing use of storage batteries in other areas such as railways, forklift trucks, tractors, auxiliary power in buildings and telephone systems.

Sometimes, batteries are recycled simply by rebuilding. There is a place in our economy for this and, in some countries, a great proportion of recycling is achieved in this way. What we have described as the "small heart" is one of

*Some of these countries may question the term "unstable": hence "relatively"!

**Australian Copper Price.

the most difficult management problems for the processor who converts scrap batteries to reusable alloy in developed countries conscious of "environment".

The smelter-refiner receives batteries in cases of hard rubber or plastic or simply the contents of batteries broken by someone else. There are in the world today a number of plants which mechanically break scrap batteries and battery scrap, separating non-metallic and metallic fractions by a sink-float procedure. Such plants require a high tonnage of batteries (20,000 to 40,000 t.p.y.) for economic viability, as their capital cost is likely to reach $A1 million. Whatever means of breaking may be used, there are then required furnaces of various kinds for smelting, followed by a complex series of refining operations before metals and alloys suitable for converting to new battery components can be produced.

Australia has three smelters dependent upon battery scrap for the production of lead alloys.*

Lead prices are more stable than in the case of copper and follow London Metal Exchange quotations. Battery scrap prices are influenced by the price of lead and substantial costs incurred in recycling batteries, including special pollution controls. Old batteries, depending on size, are worth from under a dollar to several dollars each.

(iv) *Zinc*

There is an increasing use of alloyed zinc in automobiles for locks, door latches, trims, carburettors, fuel pumps, electrical features, window frames and grilles.

Considerable investment and technical control are necessary to convert zinc-bearing scrap back to diecasting alloys, but it is less difficult to produce a grade of zinc suitable for galvanising or for the production of zinc oxide. Zinc oxide, interestingly enough, is utilised in rubber tyres as a tread compound (about 3% by weight in an average passenger vehicle tyre). Perhaps, one day, zinc will be recovered from the ash of incinerated tyres.

5. Ferrous Metals

(i) *Cast Iron*

Cast iron contains 3 to 4½% of carbon, has a relatively low melting point (compared with steel) and is easily formed into shapes in moulds (casting).

Where possible, cast iron is separately recovered from automobiles. Most engine blocks are made from cast iron and, after crushing to remove aluminium pistons and other materials, can be readily melted in cupola furnaces. Thus, there is no need for cast iron to be processed in a full-scale steelworks.

(ii) *Steel*

The most familiar form of "iron" is steel, an alloy with up to 1.2% carbon plus other elements. There are many forms from mild steel (low carbon) to tool steel (high carbon) and stainless, containing nickel, chromium and other expensive elements. Steel is generally rolled or forged into shape.

Steel scrap is classified into many types which do not concern us here. Operations involve shearing, baling and other procedures such as fragmentising.

Fragmentisers, or shredders, can produce a very clean product of greater value than most other forms of scrap. However, unless a sufficient tonnage of scrap is available for an economically viable operation, it is unlikely that a shredder will be built in any given location. We are not in a position to disclose Australian feasibility studies but some excellent papers are available from overseas. Note that conditions in Australia differ greatly from those in the U.S.A. and Germany. Vital factors are population density, cars scrapped per head of population and distances between shredder and steel scrap consumer.

Shredders are complicated and expensive assemblages of electrical, mechanical and pollution control equipment.[3] The prime unit is a swing-hammer mill with a rotor weighing about 50 tons. A "small" hammer mill requires an electric motor of 1000 h.p.; a medium-sized machine a motor of 2-3000 h.p., and there are even shredders using 6000 h.p. motors in the U.S.A. Motor size determines capacity (tons per hour throughput). The intermediate machines referred to shred 30 to 40 tons of motor car bodies per hour.

Following the hammer mill are belt conveyors, dust collection equipment such as cyclones, magnetic separators and other units designed to produce metal of over 99% steel. Cost of a fragmentiser and ancillary equipment, with purchase of and preparation of land, may be anywhere from $1 million to $5 million or more.

It is apparent that with Australian populations and distances between centres, there cannot be many shredders added to the two now in operation in Sydney and in Melbourne. Therefore, there will be some unpalatable

*Primary smelters are excluded. One such treats some battery scrap.

choices for civic bodies and transportation authorities. In some areas, cars will need to be burnt to remove organic materials, a cloud of smoke when the wind is "right" being less objectionable than an indefinite unsightly and unhealthy accumulation of old car bodies. Transportation costs may need reduction by subsidy.

The quality of scrap from baled burned cars is inferior to that from shredders, but it can still be sold although at a lower price than that for shredded steel.

Whether car bodies are burned or not, there is the problem of transportation. If we estimate the quantity of steel in a body at 0.5 tonnes and value at terminal market $50 per tonne, there is only $25 worth of content in the hulk, wherever it may be. Transport costs, cost of depreciating and operating a shredder or baling equipment plus cost of despatching the product to an Australian or overseas market, thus exceed the original value. Certainly this is a simplified exercise and the price of scrap fluctuates — but it may be less than $50 per tonne just as easily as above that figure.

On the subject of shredders, a brief mention should be made to cryogenic fragmentising, i.e., embrittlement of ferrous and non-ferrous metals by application of sub-zero temperatures. There is an operating plant in Belgium where liquid nitrogen is used on baled car scraps to facilitate shredding. It is said to be economically viable under conditions there and to produce a very clean and dense scrap.

6. A Footnote on Non-Metals

There are many problems, such as the disposal of tyres and waste from upholstery, glassware and other non-metallics.

Motor vehicles contain roughly 10% non-metallics, which in turn contain up to 50% rubber (or rubber compounds) and 20% glass. Plastics usage is increasing, for example in grilles, fittings, insulation and batteries.

Certain metals (in compound form) may one day be recoverable, e.g., the zinc used as a tread compound in rubber tyres.

7. Recycling, Energy Conservation and the Problem of Shrinking Resources

A timely survey in the U.S.A.* has shown that only a small percentage of the energy required to produce a metal from its ore is required to recycle that metal or its alloys from scrap. This has particular importance in a world where energy conservation has become of prime importance.

For example, the energy required to make a ton of aluminium from recycled metal is only 4% of that needed when using bauxite. Similarly, copper alloy and iron/steel scrap recycling require 13% and 39% respectively of the energy consumed when using ores.

The day has come when we must face the problem of declining economic raw materials resources. We must rely on increasingly lower grade materials. This means greater energy requirements, higher operating costs and higher prices.

In terms of energy conservation and as a source of raw materials, recycling now plays a vital role which will be even more important in the future.

8. Pollution Control in Metals Recycling.

The recycling of each of the metals referred to in this paper has its own special problems of pollution control in addition to the technical and commercial problems involved in raw materials preparation, smelting, refining, alloying and marketing.

In general, the capital cost of a metallurgical plant will be increased by 36% by appropriate pollution control equipment. Operating, i.e., production costs (excluding amortisation of the capital sum) has been increased by 5 to 10% in plants with which the authors are personally familiar.

Pollution control standards and costs are increasing rapidly. What is now "acceptable" waste may tomorrow contain undesirable pollutants. Regulations concerning solid, liquid and gaseous effluents can and are being changed from time to time. Regulations also cover noise "pollution".

Consider a typical secondary smelter, located in an urban environment, operating on lead, copper or aluminium alloy scrap. Smoke emission must be controlled within certain limits and calls for the installation of equipment such as baghouses or scrubbers. In many cases there is no real guarantee that this equipment will operate effectively (or at all) until it is installed. Conditions have been found to vary between "similar" smelters. Imagine the problems facing a smelter which has installed a system which doesn't work (at a possible cost of $100,000), or a smelter which has a system which complies with 1974 regulations but not 1975 regulations. There is no alternative than to try another system. There is no simple solution.

---

*U.S. Environmental Protection Agency.

In some cases noxious gases are emitted with smoke, and pass through the particulate collection system. This requires a method for gas removal such as a scrubbing system.

Liquid effluent must also abide by strict limits. Settling tanks, mixers, thickeners, centrifuges, filters and neutralisation are some of the methods used to combat water pollution.

Secondary smelters sooner or later need systems to control solid, liquid and gaseous wastes. As most plants were not originally designed for this purpose, the costs are understandably great and in some cases crippling. Future plants will need to be designed with pollution requirements as the number one priority, but what happens to a plant or industry which finds it impossible (economically or otherwise) to bring a certain contaminant below the required limit?

Pollution control represents vast problems, in terms of technology and economics, for the metals recycling industry. In the end someone, almost certainly the community, will have to pay for a clean environment.

REFERENCES

1. "Automobile Disposal: A National Problem." U.S. Government Printing Office, 1969, 569 pp.
2. Publications of the National Association of Recycling Industries, Inc. (N.A.R.I.), 330 Madison Avenue, New York.
3. Newell Manufacturing Company, San Antonio, Texas: "Technology and Economics of Large Shredding Machines."
4. Webb, W. R. G. "Automobile Disposal". A Thesis for Masters Degree, University of N.S.W., 1972.
5. The Australian Mineral Industry Annual Reviews (e.g., see 1969), Bureau of Mineral Resources. Canberra, A.C.T.
6. Publications of the Bureau of Census and Statistics, Canberra, A.C.T.
7. A publication of Fairchild Publications, Inc.. "Metal Statistics 1972 — The purchasing guide of the metal industries."
8. Reclamations Industries International Number 3, September/October 1973. P. 45.